Distance Between Receivers and Transmitters

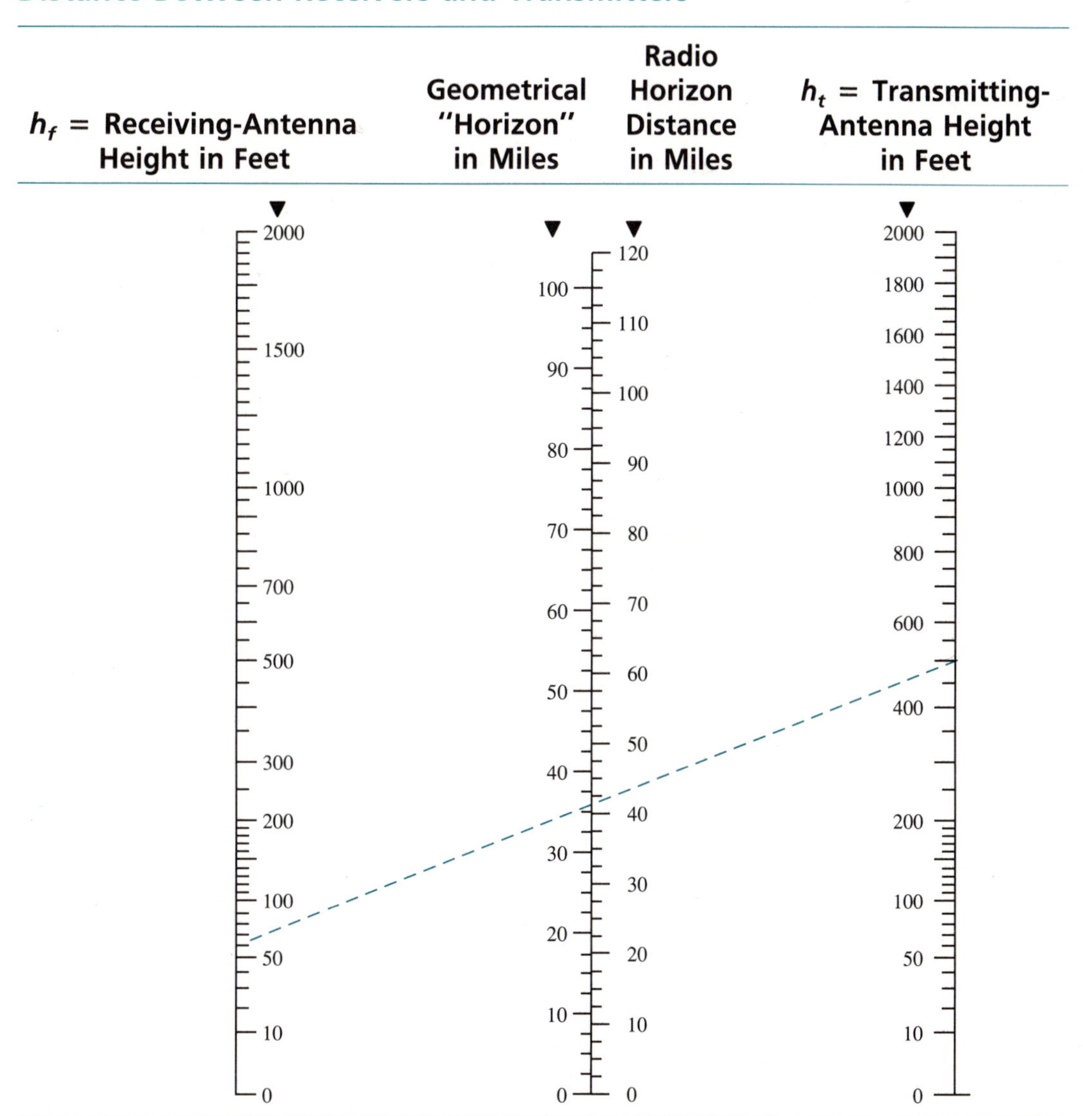

The Saunders College Publishing Series in Electronics Technology

Garrod and Borns DIGITAL LOGIC: ANALYSIS, APPLICATION AND DESIGN

Harrison TRANSFORM METHODS IN CIRCUIT ANALYSIS

Hazen EXPERIENCING ELECTRICITY AND ELECTRONICS
Conventional Current Version
Electron Flow Version

Hazen EXPLORING ELECTRONIC DEVICES

Hazen FUNDAMENTALS OF DC AND AC CIRCUITS

Ismail/Rooney DIGITAL CONCEPTS AND APPLICATIONS

Kale INDUSTRIAL CIRCUITS AND AUTOMATED MANUFACTURING

Laverghetta ANALOG COMMUNICATIONS FOR TECHNOLOGY

Ludeman INTRODUCTION TO ELECTRONIC DEVICES AND CIRCUITS

Oppenheimer SURVEY OF ELECTRONICS

Prestopnik DIGITAL ELECTRONICS: CONCEPTS AND APPLICATIONS FOR DIGITAL DESIGN

Spiteri ROBOTICS TECHNOLOGY

Analog Communications

FOR TECHNOLOGY

Analog Communications

FOR TECHNOLOGY

Thomas S. Laverghetta
Indiana University/Purdue University at Fort Wayne

Saunders College Publishing
A Division of Holt, Rinehart and Winston, Inc.
Philadelphia Chicago Fort Worth San Francisco
Montreal Toronto London Sydney Tokyo

Text Typeface: Times Roman
Compositor: Waldman Graphics
Acquisitions Editor: Barbara Gingery
Project Editor: Marc Sherman
Copyeditor: Linda Davoli
Managing Editor: Carol Field
Art Director: Christine Schueler
Manager of Art and Design: Carol Bleistine
Art and Design Coordinator: Doris Bruey
Text Designer: Tracy Baldwin
Cover Designer: Lawrence R. Didona
Text Artwork: Rolin Graphics
Director of EDP: Tim Frelick
Production Manager: Charlene Squibb
Marketing Manager: Denise Watrobsky

Cover Credit: Courtesy of Tony Stone Worldwide/Chicago LTD., Phil Jason
Photo Researcher: Teri Stratford

Printed in the United States of America

ANALOG COMMUNICATIONS FOR TECHNOLOGY

0-03-029403-7

Library of Congress Catalog Card Number: 90-053272

0123 016 987654321

THIS BOOK IS PRINTED ON **ACID-FREE, RECYCLED** PAPER

Preface

The topic of electronic communications is so vast in scope that there should never be an attempt to cover the entire topic in a single text. The field of electronic communications deserves more attention to detail than most books provide. Most books have a wealth of information, but only present the material in an introductory manner with no depth to the coverage. This book came about as a result of the frustrations that arise when an instructor attempts to find a text that will describe to students the concepts of analog communications and *only* analog communications. I encountered these frustrations while trying to find a text for my own communications course. I would find a text that, basically, had the material, but presented it in such an erratic order that we hopscotched through the chapters attempting to make some sense out of it. It is much easier to follow, and creates a better learning environment, if the chapters are followed in succession. Also, many topics in other chosen texts would begin very well but, just as the student would become interested, the discussions would stop and the author would move on to another topic. I decided to rectify these problems by putting my notes and experiences into an analog communications book.

This text is unique in two ways. First, it covers only analog communications: It can take topics which apply directly to analog circuits and provide in-depth comprehensive coverage of this material. Second, the material is organized so that each chapter builds on the preceding chapters, and concepts learned in early sections are used throughout the book. This will be very evident to students as they see frequent references to previous material. Also, the theoretical material presented in the early chapters helps the student to understand how the final analog systems—such as amplitude modulation, frequency modulation, single sideband, and television—perform the tasks they are designed to do.

I have found through 23 years of practical experience in industry in which I designed and worked with a wide variety of communications systems, that the basic concepts such as propagation, transmission lines, signal spectra, noise, and an understanding of individual components, will give a student an excellent base for understanding communications systems. For this reason, these basic topics are presented first in the text with specific analog systems examined in the second half of the book. The student will also see my industrial experience evident in many practical examples presented throughout the book.

Prerequisite/Suggested Course Syllabus

The book is designed for a one-semester course for students at a sophomore or junior level in an Electrical Engineering Technology program. Prerequisites should be AC and DC theory, at least one electronics course, and college algebra and trigonometry. It is not necessary for the student to have taken calculus. Recommended presentation during a semester is as follows (based on a 45 hour semester):

Chapter	Title	Approximate Number of Hours
1	Introduction to Communications	1
2	Wave Propagation	2
3	Transmission Lines	3
4	Antennas	2
5	Signal Spectra	3
6	Noise	3
7	Communications Components	6
8	Phase-Locked Loops and Synthesizers	4
9	Amplitude Modulation	5
10	Frequency Modulation	5
11	Sideband Systems	5
12	TV, Satellite Communications, and Fiber Optics	3
	Exams	3
	Total	45

Learning Aids

The twelve chapters of this text all follow a basic, easy learning technique. Each provides an introduction to the topic; detailed discussions of all related areas are presented with appropriate examples to show the student how to interpret the theory presented; and abundant questions and problems are available at the end of each chapter. Also, each chapter has a list of **key terms,** and there is a glossary of nearly 200 terms used in analog communications.

Content Overview

Chapter 1 is, appropriately, an introduction to communications. It begins with some very important terms defined that are used throughout the text to describe a variety of systems. Students will facilitate their learning by familiarizing themselves with these terms. Following the definition of terms, the basics and requirements of every communications system—not just analog—are presented, which gives the student an overview of electronic communications. In this section, methods of **transmission** are shown in tabular form and frequency band allocations are presented.

Chapter 2 transfers the student from the world of current flowing through a wire to a wave propagating down a transmission line. The **electromagnetic wave** is discussed in detail with direct references to the TEM mode, wavefronts of the electromagnetic wave, power density (as opposed to a single power figure for low frequency circuits), and characteristic impedance.

After the electromagnetic waves are defined and described, the student will realize that they behave much differently than a current of voltage. Thus, there are concepts of **refraction, reflection, diffraction,** and **interference** that have marked effects on the path and behavior of the waves. These terms are defined and explained so that the student can understand how many external and internal conditions will affect the propagation of electromagnetic waves.

Finally, the chapter presents the concepts of **ground-wave, space-wave,** and **sky-wave propagation.** Explanations of each type of propagation are presented, with advantages and disadvantages discussed. Also, whenever propagation occurs there is a certain amount of loss between the transmitter and the receiver. Losses such as **attenuation** and **absorption** are addressed in this chapter.

In Chapter 3 the student is introduced to the devices which tie together the communications components to make up a communications system: the **transmission lines.** Five widely used types of transmission lines are presented: open-wire, twin-lead, twisted-pair, shielded cable, and coaxial cable. Although the open-wire transmission line is used infrequently, it is presented as an excellent example of the theory and composition of transmission lines.

Following the introduction of specific transmission lines, a variety of their characteristics are presented. Students should become very familiar with these characteristics: characteristic impedance, standing waves, reflection coefficient, and return loss.

The chapter then presents one of the most useful aids for use in transmission line operation, the **Smith chart.** The chart is explained and specific examples are presented to illustrate the value and versatility of the aid.

In Chapter 4 the topic of **antennas** is covered. From the most basic type (an open-ended transmission line) to the complex arrays, the theory and applications of various communications antennas are covered. The chapter begins with discussions and definitions of antenna terms such as radiation pattern, directive gain, power gain, antenna efficiency, polarization, and many more. These terms are used to describe the operation of such antennas as the half-wave dipole, the log-periodic antenna, various antenna arrays, and the helical antenna. After completing this chapter, students will have a much greater understanding of what occurs when the electromagnetic energy of a modulated carrier is applied to such a device and radiated out into the air.

Chapter 5 begins with the distinction between **time domain** and **frequency domain.** With such a distinction understood, one can characterize and analyze signals that are used in complex communications systems. The **Fourier series** is presented and explained so that the student can utilize this powerful tool to break down complex signals into basic components. Finally, the idea of **distortion** in the communications signal is examined. By using the ideas of the Fourier series and having the signals evaluated in the frequency domain, we can obtain a clear picture of what signals should be present and which ones are causing distortion.

In Chapter 6 the topic of **noise** is examined. This topic is difficult for many students because it is not the periodic type of signal that the student is familiar with, but rather a random signal with very different characteristics. The chapter covers both **correlated** and **uncorrelated** noise, with the uncorrelated noise divided into *external* (solar and atmospheric) and *internal* (thermal noise, for example) sources. Such areas as the **noise bandwidth** of a system and the parameters used to describe noise (noise figure, noise factor, and noise temperature) are presented and defined. Examples of how these parameters are used in particular situations are also presented.

Chapter 7 presents the ''chess pieces'' of the communications system. In the game of chess each piece can perform only certain functions. If a player knows these functions, he can execute a winning game. Similarly, the components used in communications systems are designed to perform certain functions. If the student knows the functions of these components, he can put together a working (winning) system. Areas covered are **oscillators** (R-C, L-C, and crystal), **rf amplifiers, mixers,** and **filters.** Each component's theory, associated parameters, and applications are explained.

In Chapter 8, the **phase-locked loop** (PLL) is presented. This component has revolutionized many systems as the concept of integrated circuits has been refined. (The idea of the PLL was known for many years but was impractical to use in many applications.) The individual parts of the PLL are examined separately and then assembled to show the entire loop and its operation.

A result of the PLL being used more extensively is the **frequency synthesizer.** This circuit is presented in this chapter and explanations of both direct and indirect synthesis are explained along with applications of each.

Chapter 9 is the first chapter to apply all of the material covered in the first eight chapters. The theory of **amplitude modulation** is presented as the most basic type of modulation available. An amplitude modulated transmitter is then explained block by block with a great deal of emphasis on the modulator portion of the system. Following the transmission of an AM signal, the chapter then describes the process of demodulation and the reception of this signal. The main component in this scheme is the **peak detector.**

Chapter 10, although called frequency modulation, is actually a discussion of angle modulation. **Angle modulation** is a process which may be either frequency or phase modulation. The theory of both modulations is presented and the subtle differences are elaborated. The transmission, or modulation, process is then described, which again centers on the modulator for the angle modulated system. The reception of an angle modulated signal is the next portion of the chapter with emphasis placed on discriminators and ratio detector circuits. The final portions of both the transmission and the reception portions of this chapter also include FM stereo discussions.

Chapter 11 deals with **sideband systems.** In this scheme there is usually only one sideband and either a **full carrier, reduced carrier,** or **suppressed carrier** transmitted with intelligence on it. The **balanced modulator** is spotlighted because of its importance to the operation of sideband systems. Transmission systems such as the **filter method** and the **phase shift method** are presented and discussed in detail.

Reception methods for sideband systems are discussed in the final portion of the chapter and applications of such systems are presented.

Chapter 12 begins with a detailed explanation of the television signal and how a TV picture is produced. Following this explanation monochrome television is explained, followed by color TV. Details of the various circuits used to perform specific functions within a TV transmitter and receiver are explained with references made to high definition TV (HDTV).

The last two sections of Chapter 12 are intended to be introductory sections for the analog communications student. The concepts and terminology of satellite communications are presented, with applications and link budgets to be used to calculate the required parameters for a satellite system. Finally, fiber optics is introduced and its association with the communications system is explored.

The appendices of the text are dedicated to the SPICE program. This is a very powerful program that finds many applications for communications systems. Appendix A is the program itself with explanations of all the terms used along with sample files used to evaluate specific circuits. Appendices B and C are for semiconductors. Appendix B displays semiconductor models that are used in the files for evaluation of active circuits. Appendix C identifies semiconductor parameters.

Acknowledgments

Many people have contributed to this book and its completion. I would first like to thank my many students over the years who have asked many questions and posed a variety of problems so that I could put together a text that will be useful to other students. Their questions are always welcomed, encouraged and appreciated. Next, I would like to thank my department chairman, Ron Emery, for all of his encouragement throughout this project.

Thank you also to the reviewers: Samuel Derman, City College of New York; James Chapel, Lincoln Land Community College; Gaby El-Khoury, Parks College; John Dyckman, Pennsylvania Institute of Technology; and Joseph Tlana, DeVry-Chicago. Your comments and suggestions were greatly appreciated.

To the professional staff of Saunders College Publishing I extend a special note of appreciation. To Barbara Gingery, Laura Shur, and Marc Sherman, thank you for your expert guidance and suggestions along the way.

Finally, as any of you who have written know, the family is a very important part of completing a project such as this. I want to thank my wife Pat, to whom I have dedicated this entire book, for her patience and understanding, and my son Andy, who understood that there were times when I could not go out and play with him because a deadline had to be met.

Thomas S. Laverghetta
Auburn, Indiana
October 9, 1990

Contents

1

Outline

Objectives

- To provide the basic concepts of an electronic communications system
- To present and define terminology used in electronic communications
- To present a basic block diagram of a communications system and explain each section
- To present communications requirements with regard to frequency allocation, emission characteristics, bandwidth, and information capacity

Key terms

analog communications
modulating signal
demodulation
time domain
propagation
informational bandwidth
oscillator
modulator
angle modulation
FCC
bandwidth
carrier
modulation
sidebands
frequency domain
medium
intelligence
rf source
amplitude modulation
information capacity
emission
transmission medium

Introduction to Communications

1.1 Introduction

The complete electronic communications system at its most basic consists of a form of **intelligence** (data, audio, video, or message) being placed on a radio-frequency (rf) signal by means of a modulator, transmitted through a particular medium, received by the appropriate receiver, and finally removed from the carrier by a demodulator and processed. Figure 1.1 shows a block diagram of such a system.

1.2 Terminology

Our description of electronic communications is very elementary indeed; there are a great many details that we need to consider in order to produce or even understand the basics of working with a communications system. To accomplish this we must

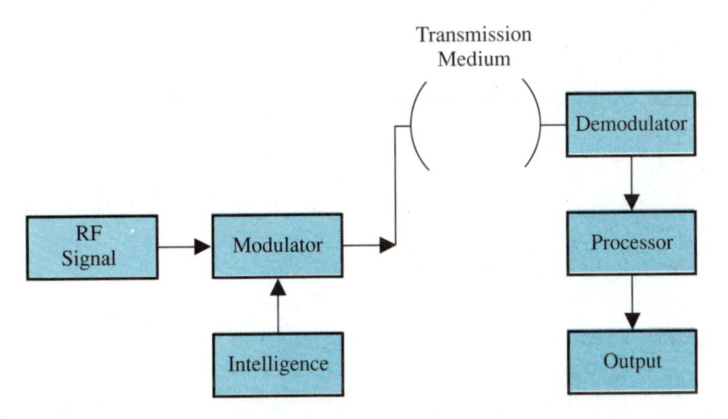

Figure 1.1 Basic communications system.

understand both the language used to describe communications systems and the equipment and technology that constitute such systems. We will now define some of the terminology (buzz words) that are needed to provide the student with the fundamentals for understanding communications.

The first term we need to define is the title of this text. **Analog communications**, until a few years ago, was the only type of communications system available and, therefore the only type studied. The key word here is *analog*. If you relate this to the wrist watch that you wear it may be more understandable. If you have a watch that shows distinct numbers for every minute (and sometimes tenths of a second) of the day, then you have a **digital** watch. That is, everything displayed is in *discrete* steps, there is no in-between level. If, on the other hand, you have one of the "old fashioned" type of watches that has 12 numbers (or markings) and a second hand on it, you have an **analog** watch. With this watch you can see a **continuous** change in time as opposed to the distinct steps shown in the digital type.

This distinction also holds true for **analog communications**, in which the modulating signal placed on a carrier varies continually, rather than in discrete steps. The amplitude, frequency, or phase may change but the change is continuous when it occurs. Thus:

> Analog communications is a form of electronic communications where the modulation used is a continuously varying signal.

The next term to be defined is the carrier. The **carrier** in an electronic communications system is an rf signal used to transport the intelligence from one point to another. An rf signal is used because it travels at the velocity of light (or near to it, depending on the medium it is traveling through), as opposed to the velocity of sound or a video signal, which travels at a considerably slower speed. This slower speed decreases the range that can be used for transmission. With an rf signal, the range of the transmission is increased considerably. Thus, the carrier is a very important part of the overall communications system.

With the carrier term defined, we must now define what is placed on that carrier. We have been calling it, to this point, the intelligence, and that is what is contained in this signal. The actual signal is the **modulating signal**. It is the portion of the signal that contains the intelligence to be transmitted. This signal **modulates**, that is, varies the amplitude, frequency, or phase of the wave on the rf carrier to make the completely modulated signal. This modulated signal may be a single frequency, an audio voice signal, a video signal, or any other form of lower frequency signal that contains some sort of intelligence.

The terms **modulation** and **demodulation** have also been used in our presentation. **Modulation** applies to the process of placing the modulating signal onto the rf carrier, and **demodulation** refers to the process of removing the modulating signal from the rf carrier. These two terms describe operations of initial and final importance in the process of electronic communications.

Another term that comes into play when talking about electronic communications is sidebands. **Sidebands** are those signals, occurring either above or below the carrier frequency, that develop when the process of modulation is performed. We will see

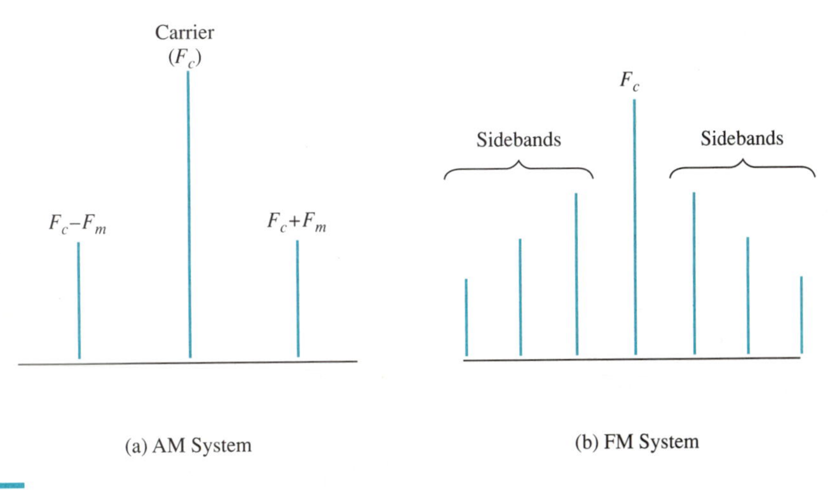

Figure 1.2 Communications sidebands.

later in this text that the sidebands in an amplitude-modulated (AM) system are the signal that includes the carrier frequency plus the modulating signal, and the signal that is the carrier frequency minus the modulating signal. Similarly, when a carrier is frequency-modulated (FM) signals are generated by the combination process and are determined by Bessel functions. These are also called sidebands. Bessel functions will be discussed in detail in Chapter 10 of this text. Figure 1.2 shows sidebands for both AM systems (a) and FM systems (b).

The next two terms to be defined are time domain and frequency domain, both of which are used extensively in communications systems. **Time domain** refers to the oscilloscope presentation that everyone is familiar with. This presentation is one that shows the signal as a function of time. The horizontal axis of the oscilloscope is calibrated in milliseconds, microseconds, or seconds. The vertical axis is calibrated in volts. When a signal is applied to the vertical input of the oscilloscope, you will see a presentaton that may be similar to that of Figure 1.3(a). The only items you can determine from this presentation are the peak (or peak-to-peak) amplitude, the time of one cycle, and the frequency ($1/t$). You can tell nothing about harmonic content unless it has some very severe distortion associated with it.

Frequency domain is the graphic representation of a signal as a function of frequency. If the same signal discussed above was placed at the input to a spectrum analyzer, the display shown in Figure 1.3(b) would be seen. You can see that much more information is shown on a frequency-domain presentation than on a time-domain display. First, we can tell the signal's amplitude, since the spectrum analyzer displays an absolute power. Absolute power is a specific value of power in watts, milliwatts, or microwatts. A relative power is some value compared to another reference level. Second, we can tell the frequency of the fundamental frequency (shown as point A in the figure). These were the only items we could tell from the time domain display. In addition, the frequency-domain display tells us the amplitude of the second harmonic (point B) and its frequency, as well as the amplitude and frequency of the third

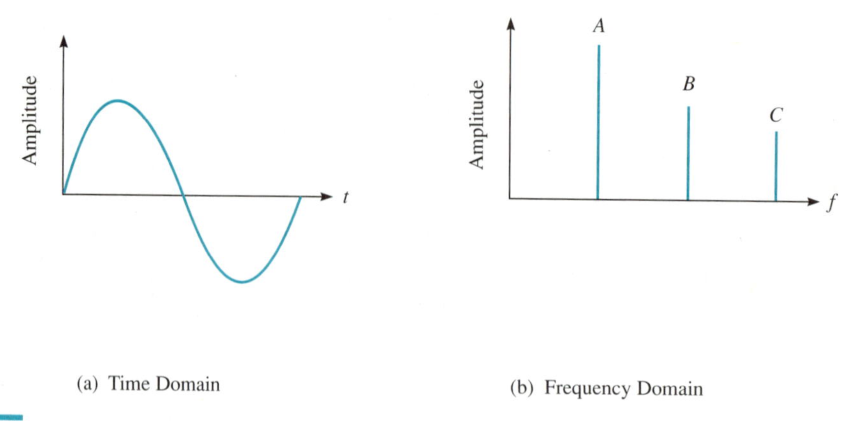

Figure 1.3 Time and frequency domain.

harmonic (point *C*). So it can be seen that a frequency domain presentation has many advantages for communications systems. This is not to say that the time display, or oscilloscope, is an ideal instrument for characterizing amplitude-modulated signals.

Another term that is often used but not fully understood is propagation. Very basically, **propagation** means that an electromagnetic wave "moves" from one point to another through some type of transmission line or medium. This may be through a coaxial cable, a piece of waveguide, or through the air. Whatever the medium, the wave is propagated, or transmitted, from one point to another. A **medium** is the agent used to move electromagnetic waves from one point to another. The medium may be a coaxial cable with its conductors and a dielectric separating them, it may be a low-loss waveguide with air in the center of the guide, or it may be the open air itself through which antennas radiate waves into space. Each of these materials are known as mediums through which the waves are allowed to propagate and are shown in Figure 1.4.

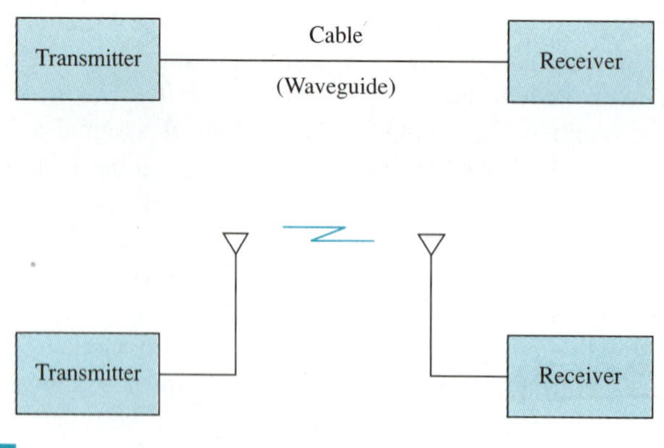

Figure 1.4 Propagation mediums.

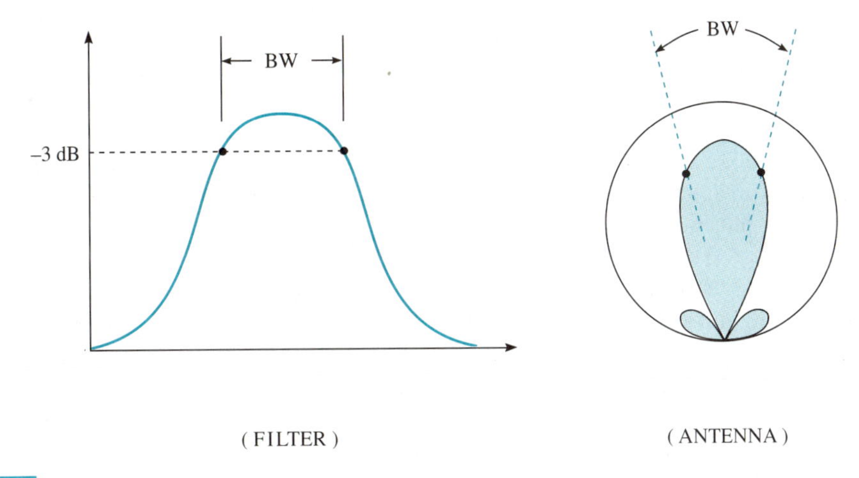

Figure 1.5 Bandwidth representations.

The final term we need to define in this section is bandwidth, but we first need to specify what type of component we are referring to, in other words the bandwidth of "what?" If we are speaking of a filter, the **bandwidth** is the point where the response falls off 3 decibels (dB) above and below the center frequency of the filter. In other words, the bandwidth is the difference between these frequency limits. If we are speaking of an antenna, the bandwidth is the point where the radiation pattern drops off 3 dB in amplitude from the main lobe on both sides. Both of these cases are shown in Figure 1.5. In a communications system we refer to **informational bandwidth**. That is, the total band in which the intelligence of the modulated signal can be found. These bandwidths will be discussed in detail later.

These are a few of the terms that you should be familiar with in order to understand and to speak intelligently about communications systems. Other terms will be defined in detail as they are presented in the text. The ones presented here are just to get you acquainted with some of the basic and more often used ones.

1.3 Basic Communications

As shown in Figure 1.1, a basic communications system consists of an rf signal (carrier), which is modulated with a modulation signal (intelligence), transmitted through a medium, received and demodulated to reacquire the intelligence, processed into a suitable form and then fed to some sort of output. Stated this way, it sounds like a simple task to design a communications system, but let us look at the role each individual block plays in producing the overall system. We will cover each block only briefly, since each will be discussed in greater detail later in this text.

The first block is the rf signal source, or rf carrier. This can consist of a wide variety of sources, depending on the frequency range to be covered, the stability needed, the power output needed, or the temperature range to be encountered by the system, to name a few. Typical sources are voltage-controlled oscillators (VCO), single-frequency crystal-controlled oscillators, sweep oscillators, or frequency synthesizers. Many of these sources may be placed in ovens when a severe temperature range is to be encountered, to ensure that the carrier's frequency and power output will remain constant. The type of source, as mentioned above, will depend on the application.

The next block to be investigated is the one marked "Intelligence." This is the area of the system that will contain the voice, video, tones, or codes that represent the information to be transmitted. The intelligence may be as simple as the frequency produced by a single-tone oscillator or as complex as the information coming from a television camera or a microphone system used to broadcast a symphony orchestra.

The block that ties the first two blocks together is the **modulator**, which combines the rf carrier and the modulating signal (intelligence). The type of modulator depends on the type of modulation, which can be either amplitude modulation (AM) or angle modulation (FM or PM).

An important part of any communications system is the **transmission medium**. This may be a simple transmission line connecting two locations or it may be the air throughout in which there are a series of obstructions with a great deal of distance between them. Regardless of the type of medium, it must be considered when studying even the most basic of communications systems. The topic of propagation between locations will be discussed in Chapter 2 and transmission lines will be covered in Chapter 3.

Once the modulated signal has been sent through the transmission medium, it must go through a process of **demodulation**, or removal of intelligence from the carrier. The method of demodulation depends on the type of modulation used. The goal is to reproduce the original intelligence placed on the carrier back in the transmitter.

The processor block in Figure 1.1 is one that takes the demodulated signal and transforms it into one that can easily be used at the receiving end. The processor may be something as simple as an audio amplifier or as complex as a sophisticated signal-processing unit. What it must do is to translate the information into a form that is useable, a form that will drive a speaker, for example. The processor may also need to drive some sort of indicator or a plotter or printer. The items just mentioned will be those that will occupy the **output** block in the figure.

1.4 Communications Requirements

When speaking of analog communication systems, the following areas must be addressed: transmission frequencies, emission classifications, and bandwidth and information capacity. Transmission frequencies are definite frequency bands. You may have heard of these before but did not really understand what they meant. The frequency bands are designated by the Federal Communications Commission (FCC), are

Table 1.1 Frequency Bands

Frequency*	Designation
30–300 Hz	ELF (extremely low frequency)
0.3–3 kHz	VF (voice frequency)
3–30 kHz	VLF (very low frequency)
30–300 kHz	LF (low frequency)
0.3–3 MHz	MF (medium frequency)
3–30 MHz	HF (high frequency)
30–300 MHz	VHF (very high frequency)
0.3–3 GHz	UHF (ultrahigh frequency)
3–30 GHz	SHF (superhigh frequency)
30–300 GHz	EHF (extremely high frequency)
0.3–3 THz	Infrared light
3–30 THz	Unassigned
30–300 THz	Visible light

* Frequency prefixes are as follows: Hz = hertz, kHz = kilohertz (1×10^3), MHz = megahertz (1×10^6), GHz = gigahertz (1×10^9), THz = terahertz 1×10^{12}).

very broad, and have many different applications. The bands with their frequency designations are shown above in Table 1.1.

There are other frequencies beyond the visible-light designation shown in Table 1.1. These frequency ranges are for ultraviolet light, x-rays, and cosmic rays.

In order to be licensed in the United States, radio transmitters are classified according to bandwidth, type of modulation, and type of intelligence information. This classification takes the form of a combination of letters and numbers as shown in Table 1.2. For example, the designation A3a describes a single-sideband, reduced-carrier, amplitude-modulated signal carrying voice or music information.

Table 1.2 Emission Classification

Type of Modulation		Type of Information		Supplementary Characters	
A	Amplitude	0	Carrier on only	None	Double sideband full carrier
F	Frequency	1	Carrier on–off	a	Single sideband, reduced carrier
P	Pulse	2	Carrier on, keyed on–off	b	Two independent sidebands
		3	Telephony, voice, or music	c	Vestigal sideband
		4	Facsimile, nonmoving, or slowscan TV	d	Pulse amp mod (PAM)
		5	Vestigal sideband, commercial TV	e	Pulse width mod (PWM)
		6	Four-frequency diplex telephony	f	Pulse position mod (PPM)
		7	Multiple sidebands	g	Digital video
		8	Unassigned	h	Single sideband, full carrier
		9	General, all others	i	Single sideband, no carrier

It should be pointed out that the supplemental characters in Table 1.2 consider both analog and digital types of communications. The characters d, e, and f refer to pulse-amplitude modulation (PAM), pulse-width modulation (PWM), and pulse-position modulation (PPM), respectively, which are methods of digitally modulating a signal for transmission. So, it can be said that Table 1.2 truly encompasses all of the communications industry, not only the analog portion that is considered in this text.

We define the bandwidth of a communications system as the **intelligence bandwidth**, or the passband required to propagate the information through the system. **Information capacity** is a measure of how much information can be carried through a system in a given time period. This amount of information, designated by C, is proportional to the product of the system bandwidth B and the time of the transmission T. This principle is called **Hartley's law**, so named for R. Hartley, who in 1928 developed the relationship expressed below:

$$C \propto B \times T \tag{1.1}$$

It can be seen from Equation (1.1) that information capacity, bandwidth, and time of transmission are linearly dependent on one another. If either the bandwidth or the time of transmission changes, there will be a change in the information capacity of the system. Examples of bandwidths in common systems are 3 kHz for voice-quality telephony, 200 kHz for commercial FM transmission, and approximately 6 MHz for commercial television. Thus, it can be that the control of bandwidth and time of transmission are important in maintaining the movement of information (intelligence) throughout the entire communications system.

1.5 Summary

This chapter has served to introduce the student to the basic concepts of electronic communications. The initial section presented the very important terminology vital to understanding the field of electronic communications. Such terms as Analog Communications, Carriers, Modulation and Demodulation, Sidebands, and Propagation were presented and defined.

Time domain and frequency domain were defined and their application to the communications system was discussed. These ideas were then applied to a basic block diagram of a communications system. Finally a section on communications requirements was presented, including a discussion of transmission frequencies (frequency bands), emission characteristics, and bandwidth and information capacity.

Questions

1.2 Terminology

1. Compare the terms **analog** and **digital.**
2. Why is the carrier an rf signal?

3. Distinguish between modulation and demodulation.
4. Distinguish between time domain and frequency domain by means of a definition and a sketch of each.
5. Define medium as it applies to electronic communications.
6. Define **analog communications**. Why is the modulating signal referred to as the intelligence?
7. Define **sidebands**.
8. What is **informational bandwidth**?
9. Define **modulating signal**.

1.3 Basic Communications

10. Draw a block diagram of a basic communications system.
11. Name two (2) types of oscillators that can be used as an rf source for a communications system.
12. What is contained in the intelligence block shown in Figure 1.1?
13. What is the purpose of the modulator in a communications system?
14. What is the purpose of the processor block in the basic communications system block diagram?

1.4 Communications Requirements

15. What frequencies are associated with the following bands: VLF, VHF, SHF, visible light?
16. What emission characteristics are associated with the designation F4c?
17. If a signal has an emission characteristic of A3, will it be received by an FM receiver? Why?
18. Define **information capacity**.
19. What is Hartley's Law?
20. If the bandwidth of a system is increased, what must happen to the time of transmission in order to keep the information capacity the same?

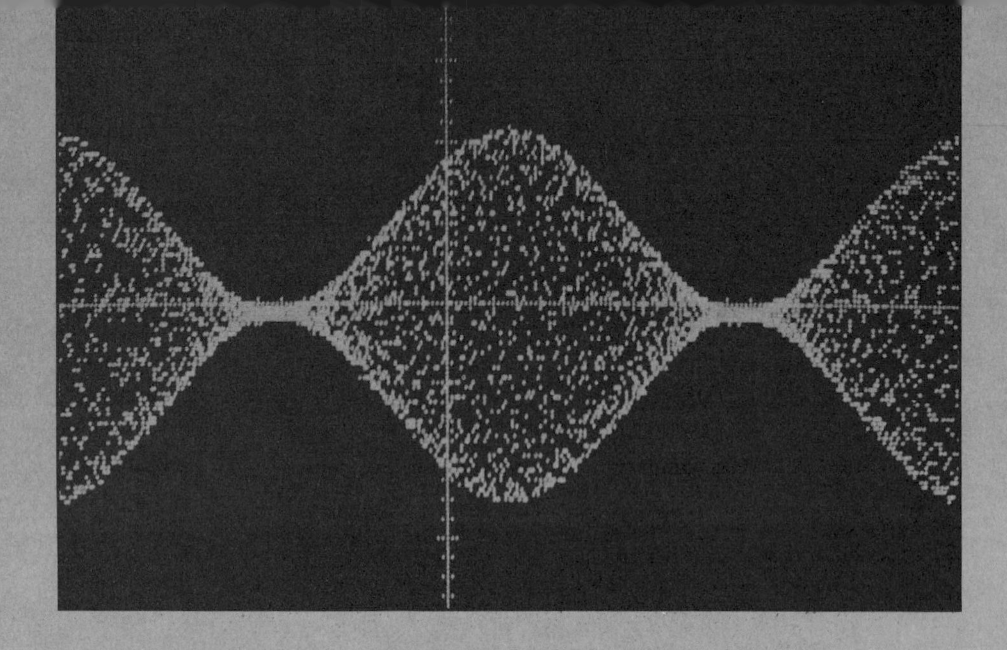

2

Outline

Objectives

- To introduce the student to the concept of electromagnetic waves
- To present the terminology and parameters of an electromagnetic wave
- To present propagation of electromagnetic waves and relate it to the wave itself
- To introduce and discuss the losses that can occur within a system where electromagnetic waves are being propagated

Key terms

propagation
transmission line
transverse
frequency domain
sky wave
wavefront
isotropic source
field intensity
dielectric constant
space wave
permeability
refraction
diffraction
Snell's law
ground wave
transverse electric
transverse magnetic
attenuation
electromagnetic wave
TEM mode
time domain
Huygens' principle
rays
power density
characteristic impedance
wavelength
absorption
permittivity
reflection
interference
incident angle
medium
optical effects
propagation

Propagation

2.1 Introduction to Propagation

We cover the topic of propagation early in this text because the student must have a thorough understanding of just what propagation involves and how it affects different types of transmissions in order to understand the field of electronic communications in any form. You must also understand the different means of propagating energy and which types of signals and the frequencies associated with them are suitable for each method of propagation. Understanding propagation depends on a thorough comprehension of the properties and the characteristics of waves. This is true since the electromagnetic waves used for electronic communications are more than the usual current flowing down a piece of wire and creating voltage drops along the way. It is of great importance that the student develop the ideas that a wave is propagating down a transmission line rather than a current flowing through a wire. We will first investigate the properties and characteristics of electromagnetic waves and see how they make propagation more understandable; then we will examine the area of propagation in more detail.

2.2 Electromagnetic Waves

The idea of waves ''propagating'' in a transmission line is one that the students must get straight in their minds. The concept of trying to understand propagation of a communications signal by relating it to current in a wire will frustrate the student and will result in more memorization rather than learning and reasoning out the topic. If the students think of a communications signal as a moving wave packet, rather than single particles of current, they should have no trouble with the concepts of electronic communications.

2.2.1 TEM Mode

To completely understand waves and how they travel from one point to another, it is necessary, and mandatory, to know the concepts of the **transverse electromagnetic (TEM) mode**. The TEM mode is used for all forms of transmission of electromagnetic energy, except that which is transmitted through waveguide. Waveguide transmission is either by **transverse electric (TE)**, or **transverse magnetic (TM) mode**. However, because waveguide is used in the microwave and millimeter frequency ranges, we will not cover it in this text.

To understand the TEM mode you must first be able to picture how a wave will look as it propagates in this mode. To imagine this we will first discuss the name of the mode. **Transverse** means something that is lying across, or is crossing from side to side. In this case we mean that some component of the wave is across the axis of the actual propagation. That ''something'' is explained by the term **electromagnetic.** *Electromagnetic* means that both an electric and a magnetic field are across the axis of propagation. To help picture a TEM mode, consider a familiar example of an electric field, a 60-Hz ac sine wave (Fig. 2.1). Figure 2.1 is a two-dimensional picture of electricity, with one axis (the y axis) representing amplitude and the other (x axis) representing time. You have seen this kind of display on an oscilloscope many times: it is the **time domain** representation discussed in detail in Chapter 1.

In the TEM mode, we find that not only is the electric field across the direction of propagation, but the magnetic field is also. Figure 2.2 (a) shows a signal that varies

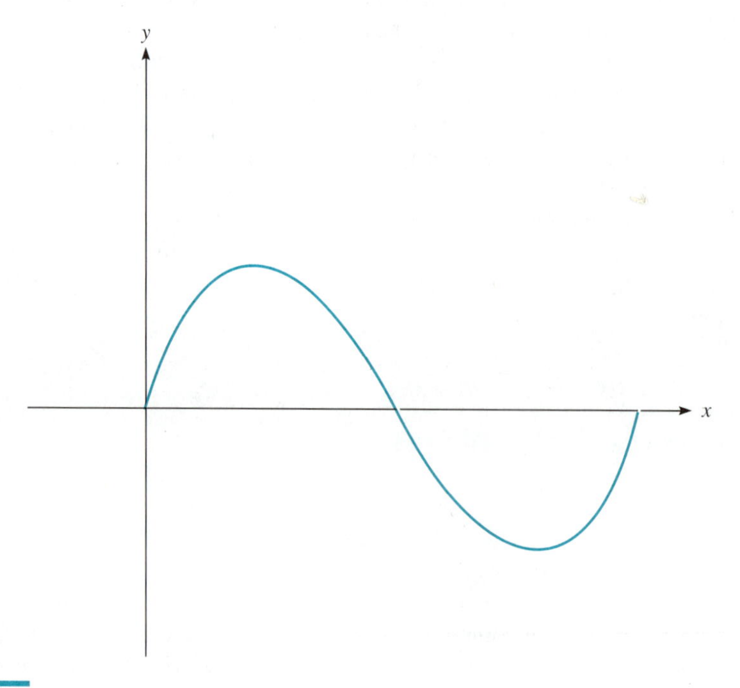

Figure 2.1 60-cycle ac signal.

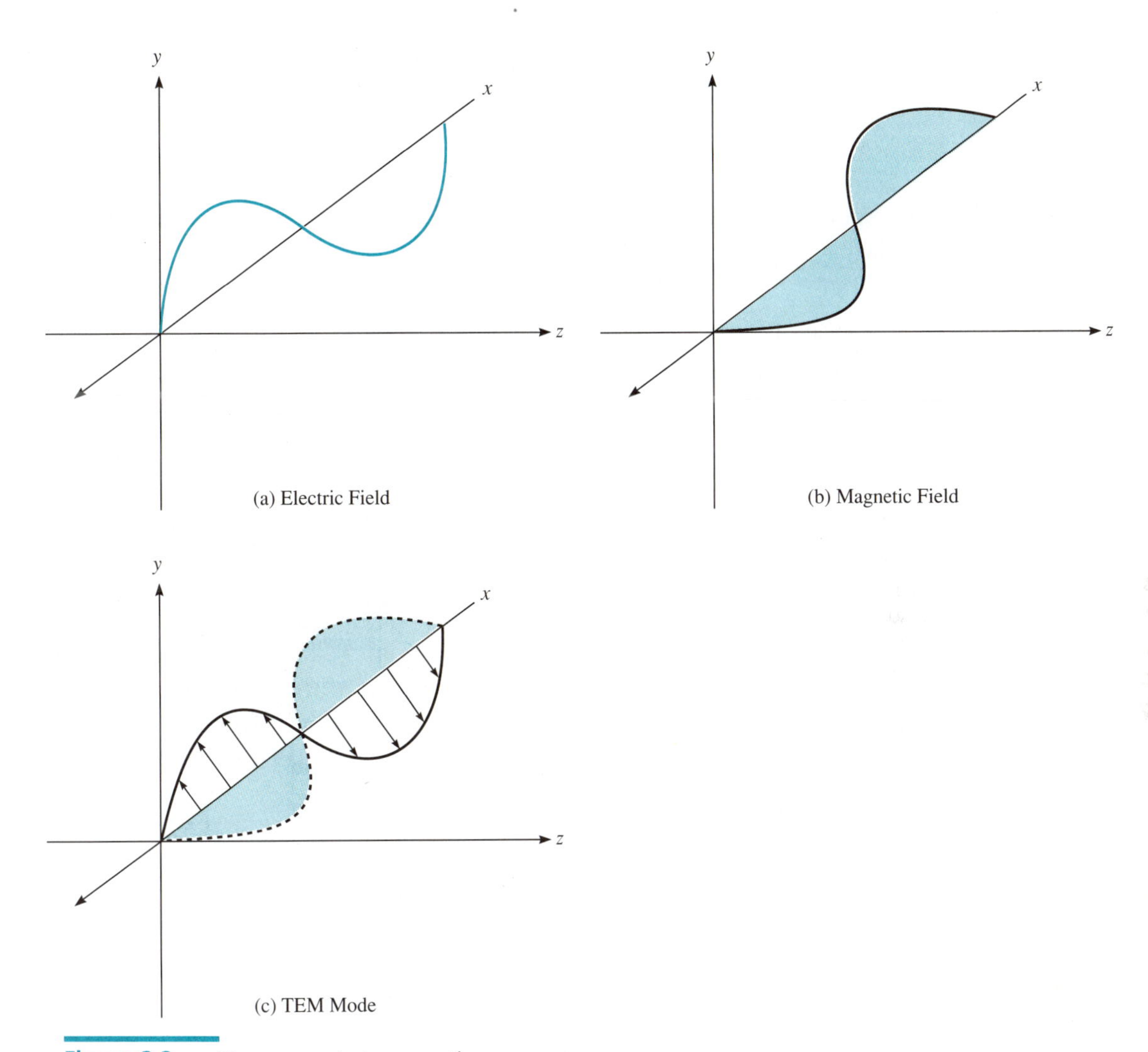

Figure 2.2 Transverse electromagnetic waves.

with time on the x direction and with amplitude in the y direction. The only difference between Figure 2.2(a) and Figure 2.1 is that the time period for Figure 2.2(a) is much shorter since the frequency is in the radio frequency range rather than 60 Hz. Figure 2.2(a) is a representation of the *electric field*. Figure 2.2(b) shows a similar signal, except that the z axis now represents amplitude, and the x axis still displays time (the same time as the signal in Figure 2.2a). Figure 2.2(b) is a representation of a *magnetic field*.

If we now combine the two signals in Figure 2.2 and have the signal propagate in the x direction, we now have a signal composed of both an electric and a magnetic field that vary across the direction of propagation (although the magnetic field is 90 degrees away from the electric field). Thus, we have a transverse *electromagnetic* wave set up (Fig. 2.2c) and are now operating in the TEM mode.

It should now be clearer what we mean when we speak of a wave. The combined signals shown in Figure 2.2c are the wave that is moving and carrying any information that is superimposed on it. If you can picture more than one signal moving together you should be able to understand the nature of a wave and be able to analyze its propagation.

2.2.2 Wavefronts

Now that you understand what a wave is we will discuss a term that is used for many types of transmission, wavefront. A **wavefront** is the surface of constant phase of a wave and is formed when points of equal phase on rays propagated from the same source are joined together. The rays we are speaking of are lines drawn along the direction of propagation of an electromagnetic wave and are used to show the relative direction of the propagation. The rays usually show several waves rather than a single one. Figure 2.3 shows a series of waves from a single source. It contains four rays that are all of constant phase (*a, b, c, d*). The long rectangular box drawn in the wavefront represents the result of all of the rays of this phase being joined together. If we were to look at a **point source** (a single location from which rays propagate equally in all directions, sometimes called an **isotropic source**) we would see rays going out in all directions with the wavefront being around the outer edge of a circle with radius equal to the strength of the rays. Figure 2.4 illustrates a point source.

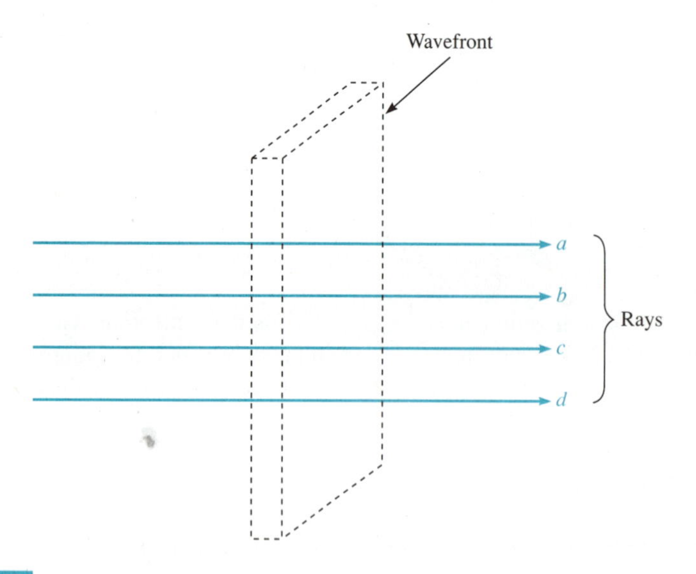

Figure 2.3 Rays and wavefront.

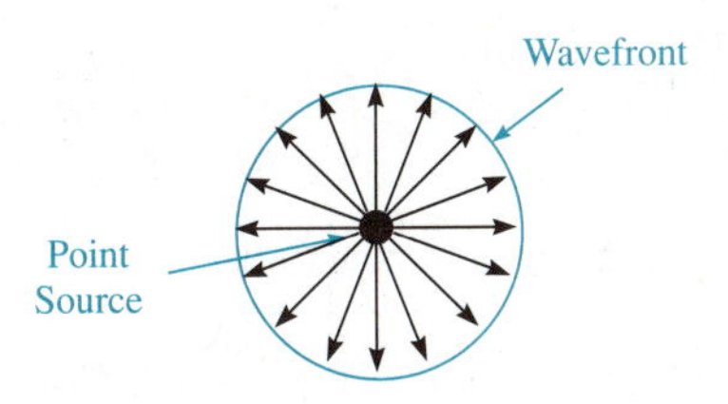

Figure 2.4 Point source.

2.2.3 Power Density

Now that we understand waves and wavefronts, we can better understand how an electromagnetic wave radiates from a source (an antenna, for example). An electromagnetic wave with its TEM characteristics moving down a line to an antenna radiates into space. When this happens, the energy involved passes through a surface (transmission line and antenna) and out into free space. This energy is referred to as the **power density** and is usually expressed in watts per square meter since it is the power density per unit time per unit area. The power density is calculated by knowing the **field intensity**, which is the intensity of both the electric and magnetic fields of the wave. This should now be understandable since we now know that the wave is made up of both an electric field and a magnetic field (TEM). Expressed mathematically we have

$$P = EH \tag{2.1}$$

Where: P = Power density in watts per square meter
E = Root mean square voltage intensity in volts per meter
H = Root mean square magnetic intensity in ampere-turns per meter

Equation (2.1) shows how the electromagnetic wave can be completely accounted for when you know the power density of a signal.

2.2.4 Characteristic Impedance

Another term that is of great importance in considering the radiation characteristics of an electromagnetic wave is its characteristic impedance. As we will see in Chapter 3, characteristic impedance depends heavily on the structure of the transmissions line used and the dielectric constant of the material. However, because we are now discussing only radiation into free space, there will be no structures or dielectric constants to contend with (the dielectric constant of free space is 1.0). Thus, the **characteristic impedance** of the radiated wave in free space is the resistance of free space, which is equal to the square root of the ratio of the free space permeability (μ_0) and the free space permittivity (ϵ_0). Expressed mathematically we have

$$Z_0 = \sqrt{\frac{\mu_0}{\epsilon_0}} \quad \text{(in ohms)} \tag{2.2}$$

Where: Z_0 = Characteristic impedance in ohms (Ω)
μ_0 = Permeability of free space (1.26×10^{-6} H/m: henrys per meter)
ϵ_0 = Permittivity of free space (8.85×10^{-12} F/m: farads per meter)

If we substitute values for μ_0 and ϵ_0 to find the characteristic impedance we have

$$Z_0 = \sqrt{\frac{1.26 \times 10^{-6}}{8.85 \times 10^{-12}}}$$
$$= 377\Omega$$

This value is a widely used standard that we will examine in Chapter 3 when we discuss transmission lines and their characteristic impedances.

Applying Ohm's law to Equations (2.1) and (2.2) we discover

$$P = \frac{E^2}{377} = 377\ H^2 \qquad \text{(watts per meter squared)} \qquad (2.3)$$

$$H = \frac{E}{377} \qquad \text{(ampere-turns per meter)} \qquad (2.4)$$

Another relationship useful for determining power density at a predetermined distance from a radiator is

$$P = \frac{P_r}{4\pi R^2} \qquad (2.5)$$

Where: P_r = Total radiated power (in watts)
R = Radius of a sphere equal to the distance from any point on a surface of the sphere to the source
$4\pi R^2$ = Area of a sphere

We refer to a sphere here because this relationship is derived from the use of an isotropic radiator (point source from previous discussions), and a sphere best illustrates a three-dimensional picture of the radiation from that source. This is a sphere with constant radiation in all directions. Because an isotropic radiator does not actually exist, an omnidirectional antenna is often used to simulate the isotropic radiator.

Example 2.1 An isotropic antenna radiates 200 watts (W) of power; determine

(a) The power density 555 meters from the source.
(b) The distance from the source necessary for a power density of 100 microwatts (μW).

Solution **(a)** Using Equation (2.5) we obtain

$$P = \frac{200}{4\pi(555)^2}$$

$$= \frac{200}{3.87 \times 10^6}$$

$$= 51.7\ \mu W$$

(b) By using the same equation with P given and R unknown, we proceed as follows:

$$R^2 = \frac{P_r}{4\pi P}$$

$$= \frac{200}{4\pi(100 \times 10^{-6})}$$

$$= \frac{200}{1.26 \times 10^{-3}}$$

$$= 158{,}730$$

$$R = 398.41$$

2.3 Properties of Electromagnetic Waves

Such conditions as refraction, reflection, diffraction, or interference will cause electromagnetic waves in free space to change their direction from the ideal. These conditions are sometimes called **optical effects** because of their similarity to conditions affecting light waves. Let us look at each of these and see how they can affect the path of electromagnetic waves.

2.3.1 Refraction

Refraction is the change in direction of the rays of a wave as they pass obliquely from one medium to another. The two media have different dielectric constants and thus there is a *bending* of the rays as they pass through. Figure 2.5 shows the refraction of an electromagnetic wave going from a medium of one density to one that is denser. The bending occurs because the velocity of the wave in the less dense medium is greater than that of the more dense medium. Thus, the energy slows down in the second medium and propagates through at angle 2, which is less than the original incident angle 1 of the wave. The amount of refraction, or bending, that takes place depends on the **refractive index** or **index of refraction**, which is the ratio of the

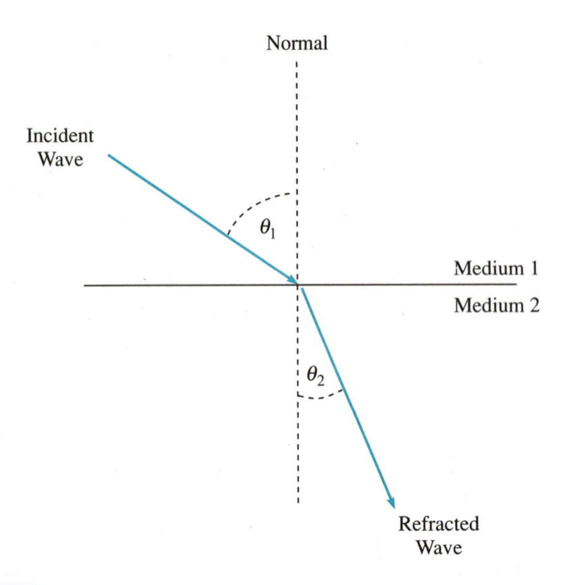

Figure 2.5 Refraction.

velocity of propagation c of the ray in free space to the velocity of propagation v of the ray in the material. Represented mathematically the relationship is

$$n = \frac{c}{v} \tag{2.6}$$

Where: n = Index of refraction
c = Velocity in free space
v = Velocity in a material

Example 2.2 The index of refraction of fused quartz is 1.46. What will the velocity of a signal be through this material?

Solution Using Equation (2.6) we obtain

$$n = \frac{c}{v}$$

$$v = \frac{c}{n} = \frac{3 \times 10^{10}}{1.46}$$

$$v = 2.054 \times 10^{10} \text{ cm/s}$$

The angles formed when waves are refracted when using **Snell's law**, which states

$$n_1 \sin \theta_1 = n_2 \sin \theta_2 \tag{2.7}$$

and

$$\frac{\sin \theta_1}{\sin \theta_2} = \frac{n_2}{n_1}$$

Where: n_1 = Index of refraction of material 1
n_2 = Index of refraction of material 2
θ_1 = Angle of incidence
θ_2 = Angle of refraction

This relationship can also be expressed in terms of the dielectric constants of the two materials as follows:

$$\frac{\sin \theta_1}{\sin \theta_2} = \sqrt{\frac{\epsilon_1}{\epsilon_2}} \tag{2.8}$$

Example 2.3 For material 1 the index of refraction is 1.5. For material 2 it is 1.36. If the angle of incidence is 30° what is the angle of refraction?

Solution Using Equation (2.7) we obtain

$$n_1 \sin \theta_1 = n_2 \sin \theta_2$$

$$\sin \theta_2 = \frac{n_1 \sin \theta_1}{n_2}$$

$$= \frac{1.5 \sin 30}{1.36}$$

$$= 0.5514$$

$$\theta_2 = 33.47°$$

Example 2.4 The angle of incidence of a ray is 35° with an index of refraction in material 1 of 1.55. The angle of refraction in material 2 is 41°. What is the index of refraction of material 2?

Solution

$$n_1 \sin \theta_1 = n_2 \sin \theta_2$$

$$1.55 \sin 35 = n_2 \sin 41$$

$$1.55(0.5735) = n_2(0.656)$$

$$n_2 = \frac{0.8889}{0.656}$$

$$= 1.355$$

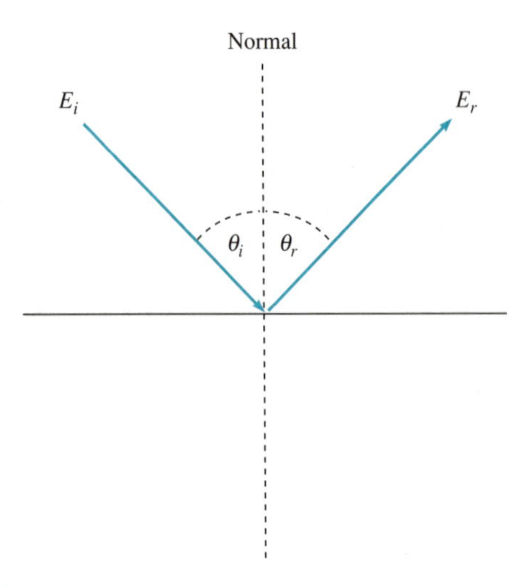

Figure 2.6 Reflection.

2.3.2 Reflection

In the case of refraction, all of the energy that was incident on the more dense medium penetrated that medium. It was then bent at a different angle and continued on its way. With **reflection**, the second medium is so dense that none, or very little, of the incident energy penetrates the medium. Thus, it is reflected back. The angle that the energy has when it is incident on the medium is the **incident angle**. The angle the energy has when it is reflected off the surface is called the **reflection angle**. You will recall from high school physics that the angle of reflection equals the angle of incidence. This is also true of electromagnetic waves as shown in Figure 2.6.

We must consider the principle of reflection where there are objects that will interfere with the transmission of the electromagnetic waves of our communication system. In urban areas reflections can be caused by tall buildings and other structures. Mountains reflect waves, thereby causing a problem in getting transmissions out of low-lying areas. Placing the antenna close to the ground can also be problematic because a certain amount of energy bounces off the ground and interferes with the main transmission. This last case will be covered in more detail later in this chapter. Early consideration of possible terrain problems will make the task of communications transmission much easier and much more efficient.

2.3.3 Diffraction

The terms diffraction and scattering are synonymous. The idea of scattering is probably the best one word definition we could come up with for diffraction. One other definition of diffraction sometimes used is the redistribution within a wavefront when it passes near the edge of an opaque object (one which is impervious to light or is not transparent).

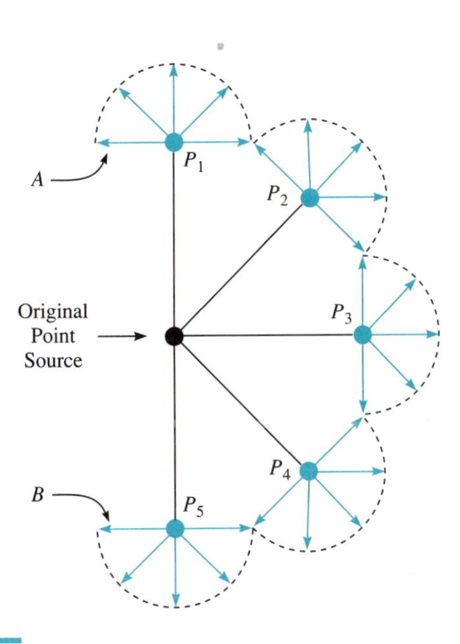

Figure 2.7 Huygens' principle.

In a uniform medium, electromagnetic waves travel in a straight line, but this straight line can be changed by passing the waves through a different medium or by placing obstacles in its way. (Actually, encountering an obstacle is the same as encountering a very dense medium.) Thus, we see that reflection and refraction change the wave's direction, accounting for the phenomenon of bending around some obstacles. This bending, or scattering, of the waves of energy is what we mean by diffraction; it allows electromagnetic waves to "peek around the corner" of an object. This concept is explained by **Huygens' principle**, which states that every point on a given spherical wavefront can be considered to be a secondary point source of electromagnetic waves from which other secondary waves are radiated outward.

Figure 2.7 is a basic illustration of Huygens' principle. Note that the original point source has five beams moving out from it. To reduce the complexity of the drawing, we limit the beams to five here, but in actuality there are an infinite number of beams coming from the point sources. Off the five beams of the original point source there are now five new point sources P_1, P_2, P_3, P_4, P_5, which now have five beams of their own radiating from them. This process goes on and on. As you can see in areas *A* and *B* of Figure 2.7, the electromagnetic energy has bent, or scattered, so that there is now energy around the corners from the original plane. This verifies our statement of the energy "peeking around the corner."

2.3.4 Interference

Interference probably can be described with one word: *opposition*. Interference is anything that obstructs and opposes the movement or proper operation of the original item.

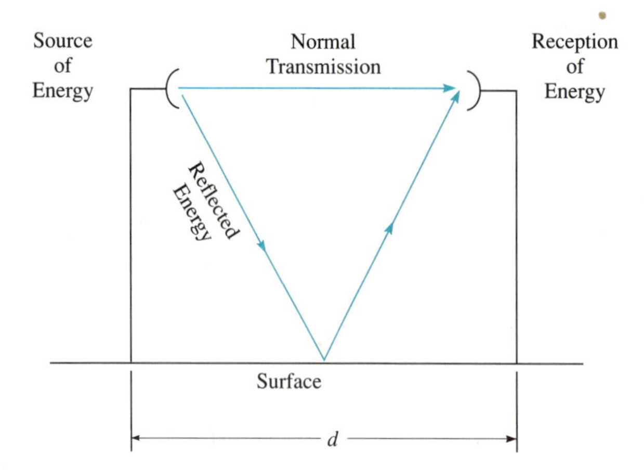

Figure 2.8 Interference.

Some simple sports analogies can illustrate the concept. Consider, for example, pass interference in a football game, which occurs when one player obstructs the ability of another to catch the ball. The call of interference is made in baseball when a batter steps in front of the catcher when the catcher is trying to throw a runner out at second base. In both these cases something gets in the way and opposes the original person's movement or effectiveness. There should be no argument that these are interference.

This notion of opposition holds true for interference with electromagnetic waves in communications systems. Interference with the original signal causes a problem in the signal's efficient transmission. We briefly mentioned the concept of interference when we discussed reflection in Section 2.3.2. Figure 2.8 illustrates the interference of waves. We can see that there is an energy source (a transmitting antenna) and an energy receiver (a receiving antenna) separated by a certain distance *d*. The figure shows that there is a normal (or ideal) path that the energy will take from the source to the receiver. If this were the only path of propagation we would have very little trouble with our systems. However, an alternative path exists in the reflected energy which bounces off the ground and arrives at the receiving antenna also. It is this signal that can cause problems with your system. What becomes important is just how far this second signal travels before it gets to the receiver. If the difference in distance between the original signal and the reflected signal is an odd multiple of one-half wavelengths, the two signals will add together to form a larger signal at the receiver. If the difference in distance is an even multiple of one-half wavelengths, they will totally cancel and nothing will get to the receiving antenna. Usually the difference in distance between the two signals is somewhere in between these two cases and only a partial cancellation occurs. The areas where wave interference becomes a great concern is the UHF (0.3–3 GHz) band and above. With these frequencies a half wavelength is so small that we frequently get even multiples and the cancellation associated with them.

We have presented four properties of electromagnetic waves that should be considered when developing an electronic communications system. By knowing the properties of refraction, reflection, diffraction, and interference, the student will be more familiar with some of the problems that can occur and will be able to achieve better performance in his system.

2.4 Propagation of Waves

With the exception of the times that electromagnetic waves are sent through transmission lines, the most widely used medium for propagation is free space. To efficiently move waves from one point to another you must employ the proper type of propagation for the frequency range being used. For free-space wave propagation three methods are used: **ground-wave propagation**, **space-wave propagation**, and **sky-wave propagation**. We will look at all three of these methods and examine the advantages and disadvantages of each to enable you to accurately judge the best type of propagation to use.

2.4.1 Ground-Wave Propagation

A **ground wave** is an electromagnetic wave that travels along the surface of the earth. A representation of a communications system using ground waves is shown in Figure 2.9. One problem with ground-wave propagation is that the earth has resistance and dielectric losses, meaning that the earth will weaken the signal as it propagates from source to receiver. This weakening, known as **attenuation**, is an exponential decrease in a wave's amplitude with distance. In order to efficiently use the ground wave we must use a much higher level of output power in order to overcome these losses. The only case where ground-wave propagation is excellent is over salt water because it is a good conductor. An example of a very poor conductor, and thus a poor propagator of ground waves, is a dry desert area. Ground-wave losses also increase rapidly as a function of frequency and are therefore never used above 2 MHz.

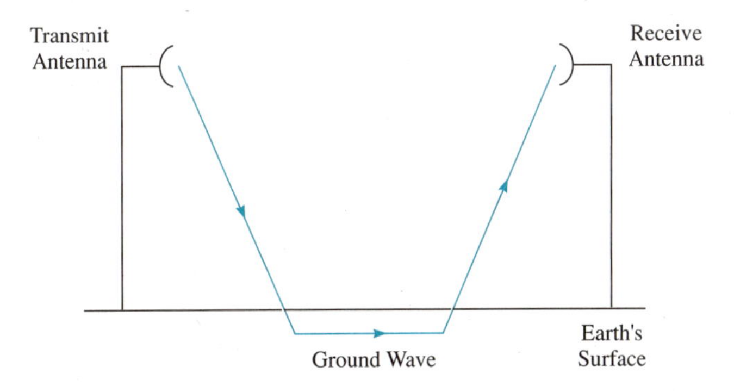

Figure 2.9 Ground-wave propagation.

Typical applications of ground waves are ship-to-shore transmissions, radio navigation, and maritime mobile communications. The advantages and disadvantages of ground waves are summarized below:

ADVANTAGES

1. Communication is possible between points anywhere in the world, provided there is enough power from the transmitter.
2. Ground waves are unaffected by atmospheric conditions because they do not rely on the atmosphere for propagation.

DISADVANTAGES

1. Ground losses are very high and increase with frequency.
2. Ground waves require relatively high transmitter powers.
3. They are limited to the low frequency (LF) and very low frequency (VLF) bands for their proper operation.
4. Because of the low-frequency limitation, the antennas used for ground waves are large.

2.4.2 Space-Wave Propagation

Space-wave propagation is probably the most widely used and most recognizable form of propagation for electromagnetic waves. It consists of two methods. The first is the one most often thought of, the direct line-of-sight (LOS) path, which involves a transmitter and receiver with a direct path between them with basically nothing between to obstruct the signal (see Fig. 2.10). The second path is the reflected wave and moves from the transmitter (source), down to the ground, and back up to the receiver antenna. This is shown in Figure 2.10 as the reflected wave. As you can see, the curvature of the earth is a limitation of the space-wave method. However, as Figure 2.11 shows, we can increase the range somewhat beyond the curvature of the earth by increasing the height of the transmitter antenna, the receiver antenna, or both. The line-of-sight horizon for a single antenna is given as

$$d = \sqrt{2h} \tag{2.9}$$

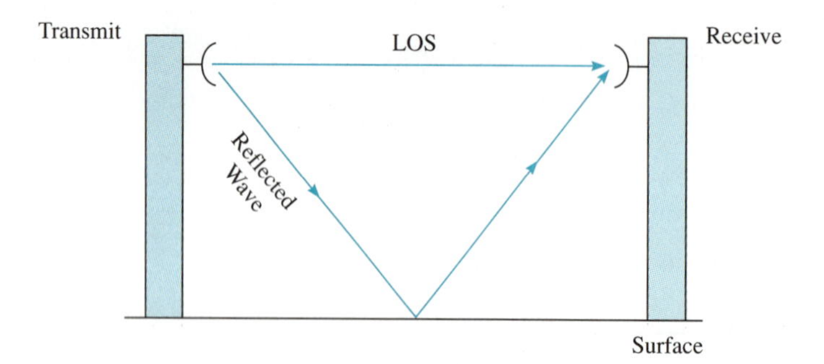

Figure 2.10 Space-wave propagation.

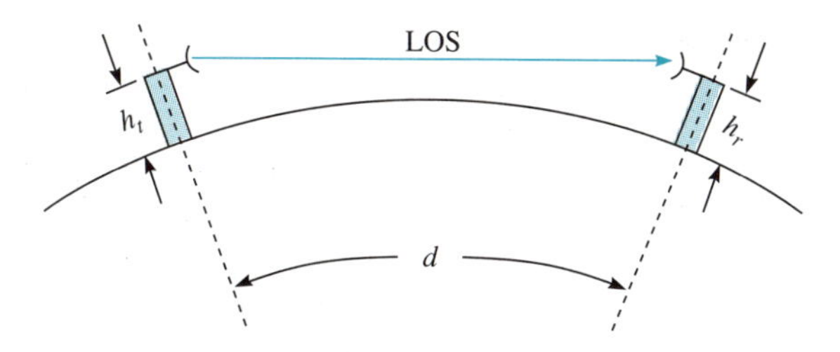

Figure 2.11 Space wave and the horizon.

Where: d = Distance to the horizon in miles
h = Height of the antenna in feet

If we now consider both the transmitter antenna and its height (h_t) and the receiver antenna and its height (h_r), we have the total distance d between the antennas as

$$d = \sqrt{2h_t} + \sqrt{2h_r} \tag{2.10}$$

We therefore see how the curvature of the earth can be extended a bit by adjusting antenna heights.

Example 2.5 If the transmitter antenna height is 35 ft, how high must the receiver antenna be to obtain a signal from 15 miles away?

Solution Using Equation (2.10) we obtain

$$d = \sqrt{2h_t} + \sqrt{2h_r}$$

$$15 = \sqrt{2(30)} + \sqrt{2(h_r)}$$

$$15 = 7.74 + \sqrt{2(h_r)}$$

$$7.26 = \sqrt{2(h_r)}$$

$$52.7 = 2(h_r)$$

$$h_r = 26.35 \text{ ft}$$

The space-wave propagation method is used for many of the higher frequency transmissions, including the familiar commercial FM. FM stations use the space-wave method with a line-of-sight transmission. This becomes apparent when you tune in an FM station during a long car trip you notice very quickly that you lose the original station as you move farther from it. This is why some people set their radios to AM stations on long trips because of the much longer range (due to their lower frequency) of AM stations.

2.4.3 Sky-Wave Propagation

The final type of wave propagation we will discuss is sky wave. This type of propagation is very much affected by atmospheric conditions and by the time of day of your transmission. Sky waves are radiated above the horizon and into the sky where they are either reflected or refracted back to Earth. As we mentioned previously, when an electromagnetic wave comes in contact with a denser medium than the medium one from which it originated, the wave is either reflected or refracted. If the medium is very dense, the wave is reflected; none of the energy is absorbed into the atmosphere. It is all reflected back to earth. If the medium the energy is entering is only slightly denser, a certain amount of the energy is absorbed and the remainder is refracted back to Earth.

The medium for sky waves is the ionosphere—the upper portion of the earth's atmosphere, located approximately 30 to 250 miles (50 to 400 km) above the earth's surface. This region absorbs a great deal of the sun's energy and ionizes the air. This ionization changes the velocity of electrons in the ionosphere, thereby altering its dielectric constant and consequently changing its density. This change in density causes reflection and refraction.

There are three layers in the ionosphere of relevance to the propagation of electromagnetic sky waves: the D, E, and F layers. The location of these layers changes with the time of year and the time of day. Let us examine each of these layers to see how they affect the transmission of energy.

D Layer

This, the lowest layer of the ionosphere, is located between 30 and 60 miles (50 to 100 km) above the Earth. Very little ionization occurs in it since it is the farthest from the sun, and it therefore has little effect on energy as it is radiated into space. Because it is so far from the sun and relies on the sun for its very existence, it disappears at night.

E Layer

This layer, located between 60 and 85 miles (100 to 140 km) above the Earth, has a maximum density at about 50 miles at noon. The upper portion of this layer is sometimes considered separately and is called the *sporadic E layer*. It is caused by solar flares and sunspot activity, it greatly improves long-distance transmission when it appears. Similar to the D layer, the E layer also virtually disappears at night.

F Layer

This layer is actually made up of two separate layers—F_1 and F_2—as shown in Figure 2.12. During the day the F_1 layer is 85 to 155 miles (140 to 250 km) above the earth, and the F_2 layer is 85 to 185 miles (140 to 300 km) above the earth in winter and 155 to 220 miles (250 to 350 km) in summer. At night the layers change just as the D and E layers do. The difference is that the F layers do not disappear, they simply combine.

Figure 2.13 shows the relationship between the different methods of propagation and the frequency bands associated with them. Note that the standard AM broadcast

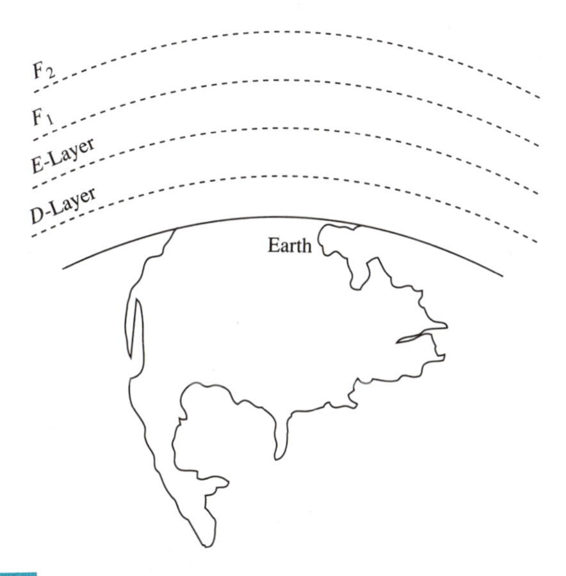

Figure 2.12 Ionosphere layers.

band (550 to 1600 kHz) uses ground waves and the sky wave at night, and that the FM broadcast band and the TV channels use line of sight only.

2.5 Free Space Transmission Losses

The two areas to be investigated with free space transmission losses are attenuation and absorption. Both of these phenomena will cause a transmitted signal to decrease in amplitude, thereby decreasing the range of the communications system, sometimes drastically. It is vitally important, therefore, that the student understand these concepts in order to avoid them wherever possible.

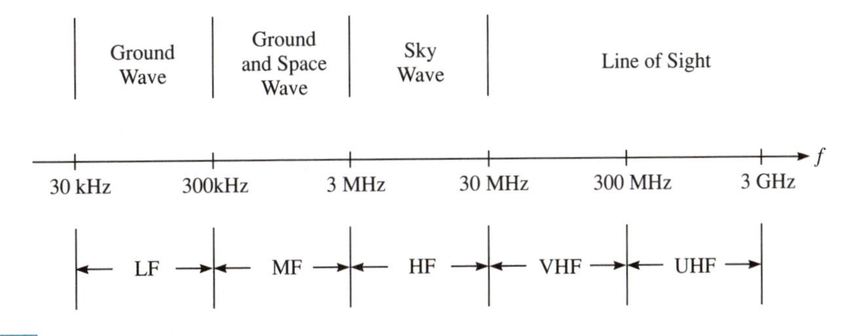

Figure 2.13 Wave propagation vs. frequency.

2.5.1 Attenuation

Basically, **attenuation** is a property that decreases the value of something. Attenuation in communications systems takes place after the electromagnetic wave leaves the source, or antenna, and proceeds out into space on its way to the receiver. It occurs because the waves continually spread out as the energy gets farther and farther from the source. This causes a reduction in the power density, which we discussed in the previous sections. The reduction is actually a power loss called **wave attenuation**, or **space attenuation**. Mathematically, it is

$$L_s = 37 \text{ dB} + 20 \log f + 20 \log d \qquad (2.11)$$

Where: L_s = Space attenuation in decibels
f = Frequency in megahertz
d = Distance in miles
37 dB is a constant.

To convert to results using kilometers, change the 37 dB in Equation (2.11) to 32.4 dB and proceed as before.

Example 2.6 What is the attenuation of a signal at a distance of 21 miles operating at a frequency of 55 MHz?

Solution Using Equation (2.11), we obtain

$$L_s = 37 + 20 \log f + 20 \log d$$

$$= 37 + 20 \log (55) + 20 \log (21)$$

$$= 37 + 20\,(1.74) + 20\,(1.322)$$

$$= 37 + 34.8 + 26.44$$

$$L_s = 98.24 \text{ dB}$$

2.5.2 Absorption

The concept of absorption in the atmosphere is based on the idea that the atmosphere is not a vacuum. Some people believe that everything above the earth should be a vacuum. The atmosphere in reality consists of a variety of gases, liquids, and solids, many of which are capable of absorbing electromagnetic energy and causing severe attenuation of signals traveling through them.

Absorption occurring in the atmosphere is dependent on the frequency of the transmission. The effects of absorption are relatively insignificant until the frequency of operation approaches 10 GHz, as can be seen in Figure 2.14 where the attenuation in decibels per kilometer is plotted as a function of frequency. Thus, unless you are working well into the microwave region of the electromagnetic spectrum, the factors of absorption do not become a consideration.

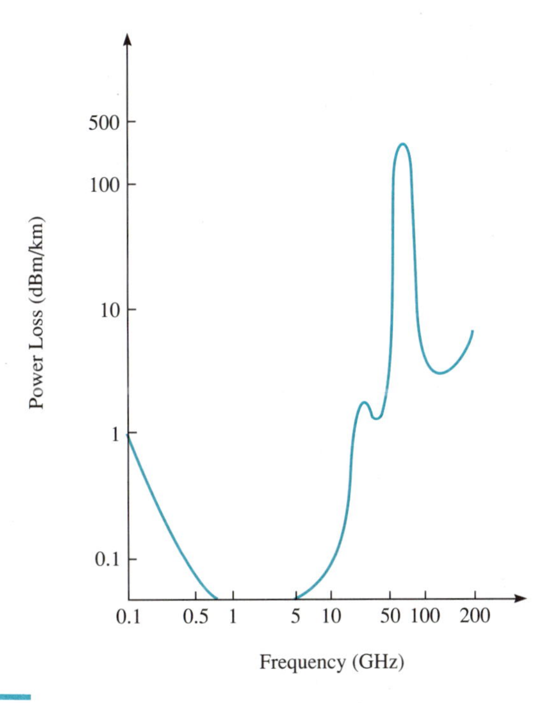

Figure 2.14 Absorption curves.

It is important to note that wave attenuation due to absorption does not depend on distance from the radiating source but on the total distance that the wave propagates through the atmosphere. This is analogous to a vehicle going through a patch of ground fog. It does not make any difference how far the vehicle is from home, the only thing that matters is the distance that the vehicle has to travel through the fog. The amount of obstruction of vision is the same just as you enter the fog as it is 1 mile down the road. Similarly, the electromagnetic wave entering the atmosphere where absorption can occur has the exact same amount of attenuation during the first mile of propagation as it does for the last mile it encounters.

2.6 Summary

This chapter has introduced the student to propagation of electromagnetic waves used for communications systems. A discussion of electromagnetic waves and the TEM mode were presented in order to familiarize the student with the concept of a wave rather than a current traveling through a wire.

The electromagnetic properties such as wavefront, power density, and characteristic impedance were then discussed, as were the properties of refraction, reflection, diffraction, and interference. The propagation of the waves using either ground-, space-, or sky-wave techniques were then explored.

We ended the chapter by discussing the losses that can occur within a propagation scheme due to attenuation and absorption.

Knowing what propagation is and how it is accomplished, you should now be able to look at a communications system with a more informed eye.

Questions

2.2 Electromagnetic Waves

1. Define **wavefront**.
2. Describe a TEM mode.
3. What is the relation of field intensity and power density?
4. What is free-space characteristic impedance?
5. Define **isotropic radiator**.
6. What is the mathematical expression for free space characteristic impedance?
7. If the distance from a radiator is decreased, what must be done to the total radiated power to keep the received power constant?

2.3 Properties of Electromagnetic Waves

8. What is **refraction**?
9. What is another term for refraction?
10. What term is used to describe the amount of refraction?
11. Define **reflection**.
12. Define **diffraction**.
13. Explain how diffraction is used to make energy ''peek around the corner'' of an object.
14. For what condition does interference cancel the transmitted signal?
15. Define **incidence angle**.
16. What is the result of a reflected signal arriving at a different time than that of the line of sight signal?

2.4 Propagation of Waves

17. What frequencies propagate best using **ground waves**?
18. Show the two paths for a **space wave**.
19. Name the layers of the ionosphere used for **sky waves**.
20. What is one advantage of using ground waves?
21. What is the major disadvantage of using ground waves?
22. What is a limitation of space waves? What does the transmission distance for space waves depend on?
23. Where does space-wave transmission work best?
24. How does the density of the atmosphere effect sky-wave propagation?

2.5 Free Space Transmission Losses

25. How does **wave attenuation** occur?
26. If we have a transmission at 3 GHz and are using sky waves, should we worry about absorption? Why?
27. If the frequency of operation is doubled and the distance from the transmission is cut in half, what happens to the space attenuation?

28. Would the space attenuation be greater at 3 MHz or 3 GHz, assuming the same distance from the source? Explain.
29. Describe the idea of **absorption**.
30. What is the main parameter that determines the amount of absorption?

Problems

2.2 Electromagnetic Waves

1. Determine the power density for a radiated power of 2500 W at a distance of 50 km from an isotropic antenna.
2. What happens to the power density of a signal if we double the distance at which we measure the density?
3. Determine the voltage intensity for the conditions in Problem 1.
4. The power density of an isotropic antenna is measured and found to be 65.3 microwatts (μW). This measurement was made 700 m from the antenna. What is the radiated power?
5. For the power density in Problem 4, what would be the distance from the source if the radiated power were raised to 500 W?

2.3 Properties of Electromagnetic Waves

6. Given a dielectric ratio $\sqrt{\epsilon_2/\epsilon_1} = 0.75$, and the angle of incidence $\theta_i = 32°$, determine the angle of refraction θ_r.
7. Material 1 has an index of refraction of 1.8, material 2 has an index of refraction of 1.24. If the angle of refraction is 26°, what is the angle of incidence?
8. If the angle of refraction in a system is 33° and the angle of incidence is 43°, what is the ratio of indices of refraction (n_1/n_2) of the two materials used?
9. When initial tests were run a system had an angle of incidence of 28°, index of refraction of material 1 of 1.4, index of refraction of material 2 of 1.22, and the angle of refraction was 32.6°. When the system was checked again the angle of refraction was found to have changed to 41.2°. What is the new angle of incidence?

2.4 Propagation of Waves

10. Determine the distance to the radio horizon for a receiver antenna 40 ft high and a transmitter 30 ft high.
11. If a transmitter antenna is erected at a height of 36 ft, what is the line-of-sight horizon distance for this antenna?
12. The distance between a transmitter and receiver is 7.5 miles. If the transmitter antenna is 12 ft high, what must be the height of the receive antenna?

2.5 Free Space Transmission Losses

13. What is the space attenuation value for a signal at 250 MHz after a distance of 3 miles?
14. For the frequency shown in Problem 13, how far must we go from the source to encounter a space attenuation of 125 dB?
15. A system is to be designed that must operate over a distance of 12 miles and have no more than 95 dB of loss. At what frequency should this system operate?

3

Outline

Objectives

- To introduce transmission lines
- To discuss a variety of transmission lines and their characteristics
- To present parameters of transmission lines such as characteristic impedance, standing waves, reflection coefficient, and return loss
- To present the Smith chart and provide examples of its use

Key Terms

transmission line
resistance
inductance
balanced line
BALUN
shield
capacitance
VSWR (voltage/standing-wave ratio)
CW (continuous wave)
return loss
coaxial
conductance
unbalanced line
dielectric
D/*d* ratio
velocity of propagation
reflection coefficient
Smith chart

Transmission Lines

3.1 Introduction

A transmission line can be defined as a component of a communications system that moves energy from one point to another. This means that a transmission line could be used to transfer an rf carrier with a modulation applied to it at point A to point B. This definition, however, may be too broad, because it might lead us to think that a clip lead could act as a transmission line for our rf signal. This is obviously not possible. We, therefore, need a different definition that takes into account all the conditions we may encounter in communication systems. Thus, we will define a **transmission line** as a component that moves energy from one point to another *efficiently*.

3.2 Types of Transmission Lines

There are a wide variety of transmission lines that can be used for many applications. There are open wire, twin lead, twisted pair, shielded cable, and coaxial, and for higher frequencies there are strip transmission lines (strip line), microstrip, and waveguides. The type of transmission line is determined by the application, but to be able to make an educated decision about the type of cable needed for a particular application, you will need to know the properties of each of these transmission lines. We will discuss the first five listed above; the remaining three types are primarily microwave transmission lines and will not be covered here.

3.2.1 General Transmission Line

Before we get into the specifics of transmission lines it is important to understand the general properties that apply to all transmission lines. Figure 3.1 shows a general equivalent circuit for all of the transmission lines we have listed above. It can be seen that four parameters are used to describe the lines: **resistance** (*R*), **inductance** (*L*), **capacitance** (*C*), and **conductance** (*G*).

Note that the equivalent circuit of Figure 3.1 shows two conductors associated with the transmission lines, and these two conductors can be of two types: one may be the conductor of signal energy and the other the ground, or both may be conductors of the main energy of the signal. It is from these two conductors that the four parameters are derived.

The first parameter to be considered is resistance, which occurs in every cable because of the metallic conductor present in every transmission line. Similarly, the center conductor, which is metallic, has a current flowing (or a wave propagating) through it. As you remember from earlier course work on circuits, this flow of current (or energy) will result in the creation of a magnetic field. This is also the case with transmission lines in communication systems. Because of this magnetic field there will be an inductance set up in every center conductor of the transmission line. This is the inductance shown in Figure 3.1.

The last two parameters of the transmission-line structure, the capacitance and conductance, are associated with the dielectric constant (ϵ) and the spacings between the two conductors. As you will recall from previous circuit courses, we defined **capacitor** as an element that has two conductor plates separated by a dielectric material. In Figure 3.1, we can see that the top conductor is one plate, the bottom conductor is another plate, and they are separated by a dielectric material. Thus, we have an excellent capacitor. Capacitance, then, is a function of the dielectric used for the transmission line and the spacing between the two conductors.

In most cases, the dielectric discussed above is assumed to be perfect; that is, the resistance between the top conductor and the bottom conductor is infinite—there is no conduction between the two conductors. In reality, however, there is no such thing as a perfect dielectric. There is always some conduction between the conductors. For this reason the conductance parameter must be considered when evaluating transmission lines for a particular application.

Now that we have outlined the parameters for general transmission lines, we can proceed to specific types of transmission lines and relate them to the general case.

Transmission lines can generally be classified as either **balanced** or **unbalanced** with the following characteristics:

Balanced

Both conductors carry current with the current being 180° out of phase.

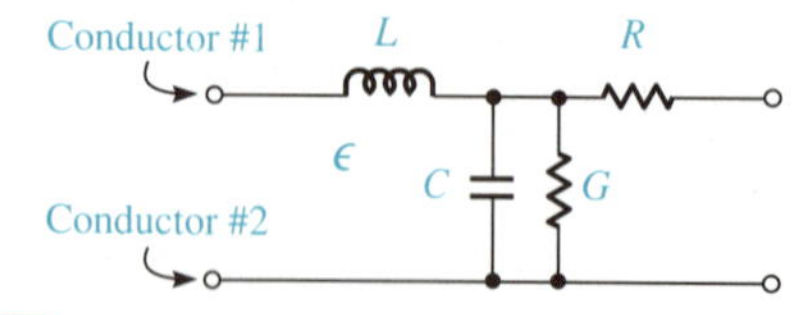

Figure 3.1 Transmission line equivalent circuit.

Unbalanced

One wire is at ground potential, and the other carries current.

Both conductors in a balanced line carry current of equal magnitude with respect to electrical ground but traveling in opposite directions. Currents that flow in opposite directions in a balanced wire pair are called **metallic-circuit currents**. Currents that flow in the same direction are called **longitudinal currents**. A balanced pair has the advantage that most of the noise interference is induced equally in both wires, producing longitudinal currents that cancel in the load. Balanced lines can be connected to balanced lines or they may be connected to unbalanced lines through a special transformer called a **BALUN** (BALanced to UNbalanced).

3.2.2 Open-Wire Transmission Line

The **open-wire transmission line** presented here is more an illustration than a practical application of communications transmission lines. Its structure is good to show how the four parameters discussed previously are distributed in transmission lines. As can be seen from Figure 3.2, the line consists of two parallel conductors, 1 and 2, separated by an air dielectric. The spacers shown are nonconductive material placed at periodic intervals for support of the two conductors. The distance maintained by these spacers (anywhere from 2 to 6 in.) is the reason the open-wire transmission line is presented more for illustration of concepts than for practical applications. This distance makes the line unruly, because it necessitates an overall line of great length. The type of structure also makes for a line with less structural integrity. Finally, the line has no shielding properties to keep signals separate from each other, and it is susceptible to noise pickup.

The negative features of the open-wire transmission line notwithstanding, it is a good structure to show the four parameters associated with every transmission line. Look again at Figure 3.2 and notice its relationship to the parameters shown in Figure 3.1. There will be resistance in each of the conductors, as there is in every wire that we use for circuits; there will be inductance in these conductors because of current flowing through them; there will be a value of capacitance between the two conductors of the transmission line because of their size, the spacing between the conductors,

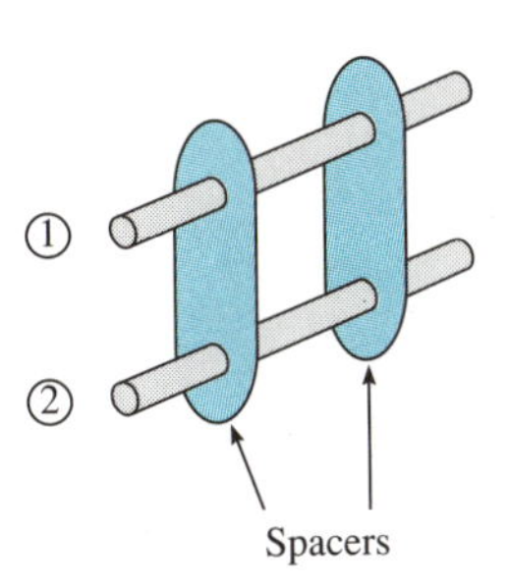

Figure 3.2 Open-wire transmission line.

and the air dielectric between the conductors; and there will be some conductance between the conductors since dielectrics are not perfect. The conductance may be the most difficult to see since we are talking about air as a dielectric and it is difficult to imagine conductance between conductors separated by air. Some minute conduction does occur, however, and it must be accounted for in some applications. Thus, we see that the open-wire line is an excellent example to begin our discussion of transmission lines for communication systems.

3.2.3 Twin-Lead Transmission Line

The **twin-lead transmission line**, more commonly known as **ribbon cable**, is one that you may recognize more readily than the open-wire transmission line. This is the type of line used to connect a TV antenna on the roof of your house to the television set. Such arrangements are rather uncommon these days, however, because of cable television and satellite dishes.

The twin-lead resembles the open-wire transmission line, except that the dielectric is a solid instead of air as can be seen in Fig. 3.3. This is a much more compact, and flexible transmission line than the open-wire type. The dielectric in the twin-lead line is usually made of Teflon or polyethylene. This means that the velocity of the electromagnetic waves passing through this cable are slowed by a factor determined by the dielectric constant (ϵ) of either Teflon or polyethylene.

The twin lead is a much more uniform type of transmission line, and its characteristics are more useable than those of the open wire. Moreover, the impedance of the line is a function of the distance between the conductors, the radius of the conductor, and the dielectric constant. Expressed mathematically the impedance is

$$Z_0 = \frac{276}{\sqrt{\epsilon}} \log \left(\frac{D}{r}\right) \tag{3.1}$$

Where: ϵ = Dielectric constant of the material
D = Distance between the conductors
r = Radius of the conductor

This relationship is the same as that used for general two-wire transmission lines, except that the dielectric constant is taken into account. In most of the two-wire calculations, an air dielectric ($\epsilon = 1.0$) is used.

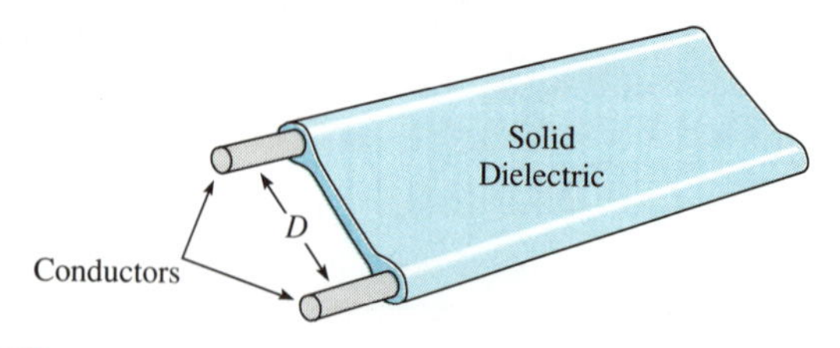

Figure 3.3 Twin lead.

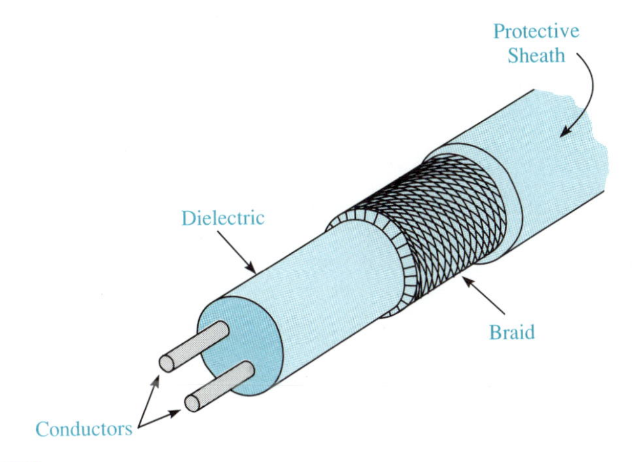

Figure 3.4 Shielded pair.

3.2.4 Twisted-Pair Transmission Line

The **twisted pair**, fabricated by twisting two insulated conductors together, is used where there is a danger of interaction between transmission lines. The twisting action causes any signals to be out of phase and, thus, cancel one another if they are present on the wrong line. If more than one twisted pair is needed within a system, the pairs are twisted with a different pitch. The **pitch** of a twisted pair is the twist length, or how tight the pair of conductors is twisted. The change in pitch is usually accomplished by applying more or less pressure to the ends of the lines as they are twisted. To understand this, consider a pair of conductors placed in a drill press with their other ends held straight down. The drill press is turned on for a short time, twisting the pair of wires to some degree. If the conductors are held tighter and the drill press is not turned on as long, wires will not be as twisted—the pitch will be less. The pitch will be greater if the drill press is turned on longer. Thus, the pitch, and the phasing of the signals can be changed by changing the pitch of the twist.

3.2.5 Shielded Cable

Figure 3.4 represents a **shielded-pair transmission line**, showing a balanced parallel two-wire transmission line enclosed in a conductive braid. This arrangement reduces radiation losses from the transmission line as well as interference. The braid on this line is connected to ground and also acts as a shield. As stated, the shielded cable transmission line is a balanced configuration which was described earlier in this chapter. (Recall that a balanced configuration was one where both conductors carry current with the currents being 180° out of phase.)

3.2.6 Coaxial Transmission Line

A **coaxial transmission line** is the one most often used in high-frequency systems, or when there is a need for a well-shielded type of transmission line for a commu-

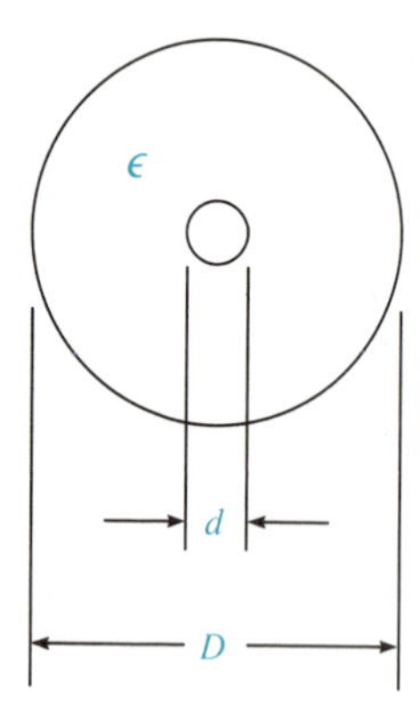

Figure 3.5 Coaxial cable.

nications system. Figure 3.5 shows an end view of a coaxial cable. The term **coaxial** refers to one conductor surrounded by another, as shown in the figure.

In Figure 3.5 the center conductor d is considered the outside diameter of the inner conductor. The second conductor D is the outside shield and is called the inside diameter of the outer conductor. These dimensions are very important to the operation of the coaxial cable because they form what is called the ***D*/*d* ratio**, which is used in the determination of a variety of parameters for the cable. The other term shown in the figure is ϵ, the dielectric constant of the material separating the two conductors. This material is usually made of Teflon or polyethylene ($\epsilon = 2.1$ and $\epsilon = 2.26$, respectively). The D/d ratio is also used for many cable parameters, the most important one being the characteristic impedance of the cable. Characteristic impedance is calculated as follows:

$$Z_0 = \frac{138}{\sqrt{\epsilon}} \log \left(\frac{D}{d} \right) \tag{3.2}$$

Equation (3.2) shows that the characteristic impedance of a coaxial cable can be adjusted by varying the ratio of the two diameters. This principle is reinforced when you realize that coaxial cables for 50, 75, and 93 Ω (ohms) all have approximately the same outside dimensions. (The 50 Ω cable has an outside diameter of 0.195 in., the 75 Ω and 93 Ω cable are 0.242 in.)

Example 3.1 A piece of coaxial cable is found with the marking RG-214 A/U on it. It can be seen that the dielectric is Teflon ($\epsilon = 2.1$). When measurements are taken we find that $D = 0.285$ in. and $d = 0.087$ in. Can this cable be used in a 75 Ω system?

Solution To find the answer we must determine the characteristic impedance Z_0 of the cable. Using Equation (3.2) we obtain

$$Z_0 = \frac{138}{\sqrt{\epsilon}} \log (D/d)$$

$$= \frac{138}{\sqrt{2.1}} \log (0.285/0.087)$$

$$= \frac{138}{\sqrt{1.44}} \log (3.275)$$

$$= 95.83\ (0.5153)$$

$$Z_0 = 49.38\ \Omega$$

Thus, the cable cannot be used in a 75 Ω system because it is a 50 Ω cable.

The characteristic impedance is one of many specifications of a coaxial cable that can be determined from the given parameters. Below is a list of possible additional specifications that may be considered when selecting a coaxial cable:

- voltage/standing-wave ratio (VSWR)
- attenuation
- velocity of propagation
- capacitance and inductance
- CW power rating
- shielding
- flexibility
- operating temperature range

The **voltage standing-wave ratio (VSWR)** of a coaxial cable is basically a measure of any reflections that may be on the cable. The concept of standing waves will be discussed in detail in later sections of this text.

The value of attenuation in a cable is especially important in a system with very long transmission lines. The value of attenuation in some cables may be so great as to cause the system to either operate poorly or not at all. Thus, you must be concerned with the attenuation of all coaxial cables. The mathematical expression for the attenuation of a coaxial cable is

$$\alpha = 4.34 \left(\frac{R_t}{Z_0}\right) + 2.78 \sqrt{\epsilon}(f) \tan d \tag{3.3}$$

Where: α = attentuation in dB per 100 ft

$$R_t = \left(\frac{1}{D} + \frac{1}{d}\right) \sqrt{f}$$

f = frequency in MHz

$\tan d$ = dissipation factor

(The *Dissipation factor* is the loss encountered within a dielectric material when energy is passed through it. This value is given on dielectric data sheets)

As mentioned previously, the velocity of propagation within a cable is determined primarily by the dielectric constant of the dielectric material separating the two conductors. It is expressed as a percentage of the velocity of light in free space, which is 3×10^{10} cm/sec. It is calculated as follows:

$$\text{Percentage} = \frac{1}{\sqrt{\epsilon}} \times 100 \tag{3.4}$$

Equation (3.4) tells us that the velocity of the wave being propagated changes by a factor that is the reciprocal of the square root of the dielectric constant. The velocity of propagation of Teflon ($\epsilon = 2.1$) will be 69% of the speed of light. Similarly, the velocity of a wave in a polyethylene medium ($\epsilon = 2.26$) is 66.5% of the speed of light. This factor is used many times when making calculations at high frequencies.

The capacitance and inductance of a cable are expressed in the following mathematical relationships:

$$C = \frac{7.354\epsilon}{\log (D/d)} \tag{3.5}$$

Where: C = Capacitance in picofarads per foot

$$L = 0.1404 \log (D/d) \tag{3.6}$$

Where: L = Inductance in microhenrys per foot

Example 3.2 For the cable presented in Example 3.1 (RG-214 A/U) determine the capacitance and inductance per foot of the cable.

Solution Using Equation (3.5) to find the capacitance per foot, we obtain

$$\begin{aligned} C &= \frac{7.354\epsilon}{\log (D/d)} \\ &= \frac{7.354(2.1)}{\log (0.285/0.087)} \\ &= \frac{15.443}{0.5153} \\ &= 29.96 \text{ pF/ft} \end{aligned}$$

Equation (3.6) is used to find the inductance per foot.

$$L = 0.1404 \log\left(\frac{D}{d}\right)$$
$$= 0.1404 \log\left(\frac{0.285}{0.0878}\right)$$
$$= 0.1404(0.5153)$$
$$= 0.072\ \mu\text{H/ft}$$

The **continuous-wave** (**CW**) power rating of a cable is the measure of how much continuous power it is capable of handling. This is not the figure that would be used for a pulsed system, it is only for use when considering the continuous flow of power. The rating is dependent on such parameters as frequency, temperature, altitude, and the VSWR of the transmission line.

Coaxial cable is shielded for two reasons (1) to keep outside interference from upsetting the rf energy inside the cable, and (2) to prevent the internal energy of the cable from interfering with any outside circuitry.

Depending on the application, coaxial cable may be shielded in one of the following ways:

Single Braid
: May be bare, tinned, or silver-plated wires

Double Braid
: Two single braids with no insulation between them

Triaxial
: Two single braids with a layer of insulation between them

Strip Braids
: Flat strips of copper, rather than round wires

Solid Sheath
: Solid aluminum or copper tubing

The flexibility of a transmission line is of vital importance when assembling large communications systems. Because there is almost never a case where every component in a system is lined up in a straight line with the output of one circuit exactly in front of the input to the next, it is always necessary to bend the transmission lines around corners and to weave them through the entire system. Other applications, such as laboratory testing also require that coaxial transmission lines be able to withstand repeated flexing. This is probably the most severe of all applications for transmission lines. In these cases the cables are subjected to many repeated flexes. Standard braided constructions of coaxial cable can withstand over 1000 flexes through 180° if they are bent to a radius equal to 20 times the outside diameter. This cable is usually stored on reels with hubs ten times the outside diameter of the cable, and installation in a permanent system should not be less than five times the outside diameter of the cable.

The operating temperature of a coaxial transmission line is determined primarily by the operating range of the dielectric and the jacket material. The dielectrics most commonly used for coaxial cables are Teflon (also termed *polytetrafluorethylene*) and polyethylene, whose operating temperature range is −250 to 250°C and −65 to +80°C, respectively. The jackets for the cables may be of a variety of materials; common materials are fluorinated ethylene propylene (FEP), polyvinylchloride (PVC), and silicon rubber. Operating temperatures for these materials are −70 to 200°C, −50 to 105°C, and −70 to 200°C, respectively. Because the ranges presented here are well beyond the temperatures at which the system will be operating, there is usually not much concern with the operating temperatures of coaxial transmission lines. However, you should always be aware of the temperature demands placed on the system, no matter what the application.

3.3 Transmission Line Characteristics

A transmission line is much more than a straight piece of bus wire through which a current will flow. It is itself a circuit, with its own characteristics and parameters that must be taken into consideration. We will cover four of the important characteristics of transmission lines in this section: characteristic impedance, standing waves, reflection coefficient, and return loss. In this section we wish to acquaint the student with the concept of waves propagating down a transmission line and causing certain phenomena to occur.

3.3.1 Characteristic Impedance (Z_0)

In the previous section, we mentioned the term characteristic impedance, a vital component to understanding transmission lines and how they differ. You will recall from courses on ac theory that there is a maximum transfer of power when the load impedance equals the source impedance. The source impedance usually was that of the generator. For the cases we will be discussing in this text, source impedance is defined as the characteristic impedance (Z_0) of the transmission lines being used to feed the load.

Basically, the **characteristic impedance** of a transmission line is a complex ac quantity

$$R \pm jX$$

Where: R = The resistive part of the impedance
X = The reactive part of the impedance

This value of impedance is independent of length and frequency and is a quantity that cannot be measured directly. The actual impedances that are present are usually only the real part of the complex value because the complex portion is cancelled out within the cable and only the real portion remains. Thus, Z_0 of a cable is actually $50 + j0$ Ω, which is usually only designated as a 50 Ω cable.

As previously stated, the value of Z_0 cannot be measured directly. An ohmmeter placed across the conductors of a cable will read an open circuit, and not any value of impedance, because impedance is an ac quantity resulting from an electromagnetic wave propagating down the line. It is actually the impedance you would see looking into the end of a transmission line if the line were infinitely long. This results in a very consistent impedance along a line; that is, if the impedance at the input is 50 Ω, it is 50 Ω 1 ft away, 50 Ω 50 ft away, and so on.

To see how impedance works, consider the following example:

Example 3.3 Consider the circuit shown below.

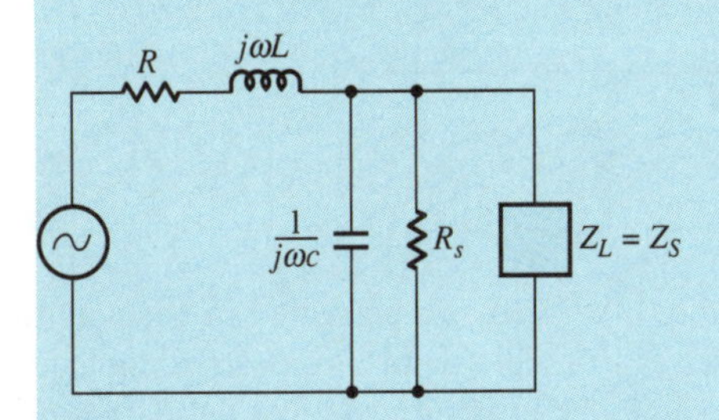

The impedance looking into such a circuit or line containing n sections is determined by

$$Z_0^2 = Z_1 Z_2 + \frac{Z_L^2}{n} \tag{3.7}$$

For an infinite number of sections, Z_L^2/n approaches 0.

$$\lim_{n\to\infty} \frac{Z_L^2}{n} = 0$$

Then:

$$Z_0 = \sqrt{Z_1 Z_2}$$

$$Z_1 = R + j\omega L$$

$$\frac{1}{Z_2} = \frac{1}{R_s} + \frac{1}{1/j\omega C} = G + j\omega C$$

$$Z_2 = \frac{1}{G} + j\omega C$$

$$Z_0 = \sqrt{\frac{R + j\omega L}{G + j\omega C}} \tag{3.8}$$

For low frequencies: $Z_0 = \sqrt{\frac{R}{G}}$ (3.9)

For high frequencies: $Z_0 = \sqrt{\frac{L}{C}}$ (3.10)

From these equations it can be seen that Z_0 is independent of frequency and length, has a phase angle of 0° and is purely resistive.

To illustrate the independence of Z_0 with regard to length, consider the following example.

Example 3.4

Given the following resistive network, with $R = 10\ \Omega$ and $R_s = 110\ \Omega$, determine the input impedance of this network.

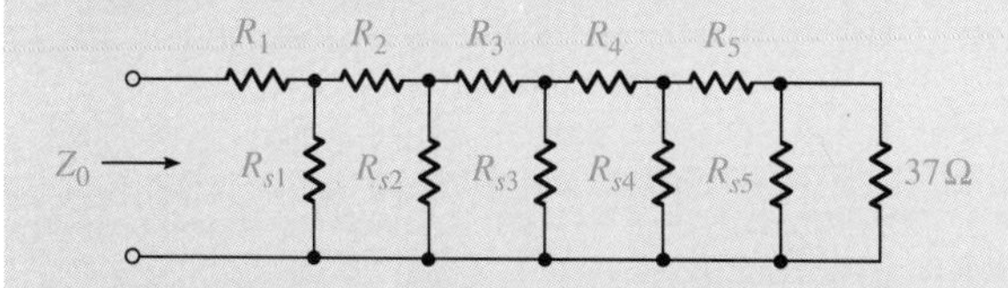

Solution

This network is similar to any other network presented in basic ac theory courses. The parallel and series–parallel combinations are combined and a resultant impedance is found. If, for example, we take the output resistance (37 Ω) and the R_{s5} value (110 Ω) and combine them in a parallel combination we have

$$R_{eq} = \frac{37(110)}{37 + 110}$$

$$= \frac{4070}{147}$$

$$= 26.7\ \Omega$$

This value is now added to R_5 and we obtain 36.7 Ω. This sum is now combined in parallel with R_{s4}, and so on until the final value of Z_0 is found. The final value in this case is 38.4 Ω. To show Z_0's independence of length, add three more sections of R and R_s to the circuit shown above. Now calculate the new value of Z_0.

Solution

This problem is treated similarly to the previous one, with the load resistance (37 Ω) put in parallel with the last R_s value, added to the last R value, and proceeding on until the input is reached. The answer of 38.5 Ω indicates that the impedance of the line is virtually independent of length.

3.3.2 Standing Waves

In Chapter 2 we discussed the concept of waves and how they propagate down a transmission line, concentrating at that time on the forward, or **incident, wave**. We did not mention that there may be another wave, called a reverse, or **reflected, wave**, coming back down the line in the opposite direction. The combination of these two waves form a wave referred to as the **standing wave** on the line. Figure 3.6 shows incident and reflected waves. This concept will now be discussed to show how the standing waves affect the overall properties of the transmission lines being used.

As mentioned previously, there is a maximum transfer of power from a transmission line to a load when the impedance of the transmission line Z_0 equals the impedance of the load Z_L. Under this condition all of the power propagated down the transmission line is absorbed into the load. There are no reflections from the load because there is no mismatch. This means that there is no standing wave on the line, and the VSWR is 1.0:1, the best ratio possible on a transmission line. The 1.0:1 ratio is a theoretical value since a perfect match is not actually possible. A more practical value that can be considered a near perfect match is a VSWR between 1.02:1 and 1.05:1. However, the most common type of "good" match is a VSWR in the range of 1.2:1 to 1.5:1, and these are typical values for communications components and solid-state devices used in communications systems.

To illustrate how standing waves are formed, what effect they have on a transmission line, and how they indicate the type of load, let us now look at two extreme cases: the open circuit and the short circuit. Amplifier manufacturers often test their circuits at these two extremes and deduce from the data that the circuits will operate under any load condition if they pass these two tests. Each of these conditions will be placed at the end of a transmission line to illustrate the effects of such a load on the line. The two conditions are chosen since they are the two extremes of conditions

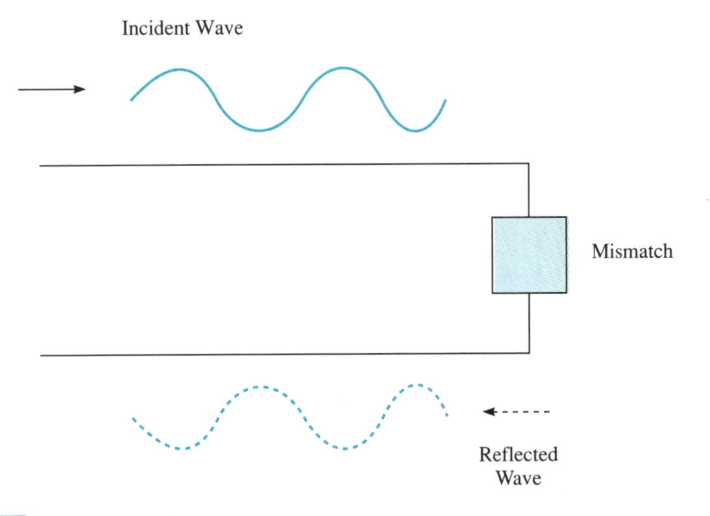

Figure 3.6 Voltage standing wave ratio.

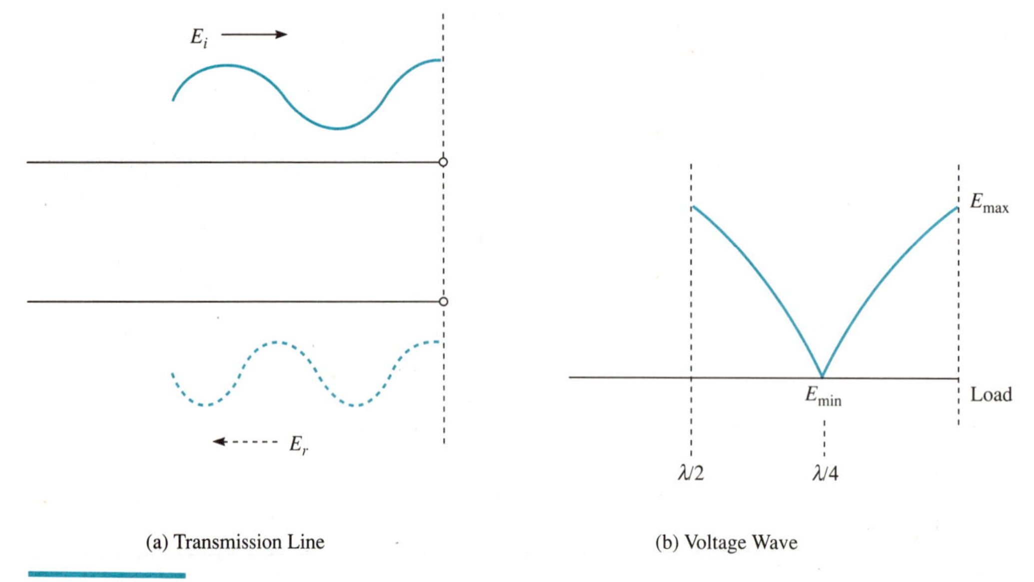

Figure 3.7 Open circuit.

that can occur at the output of a transmission line. Everything else that can occur is in between these two conditions.

The first condition is the open circuit at the end of a transmission line. You will remember that an open circuit has a maximum voltage across it and zero (or minimum) current—conditions that are the same for an open-circuited transmission line. Figure 3.7(a) is a representation of an open-circuited transmission line. The incident voltage E_i propagating forward down the line gets to the open circuit and is a voltage maximum. At this point there is no place for the energy to go so it is all reflected back toward the generator. Figure 3.7(b) shows the resulting detected standing wave ratio now on the transmission line.

To determine the VSWR of a system in which the values of the voltage maximum and voltage minimum are known, the following relationship is used

$$\text{VSWR} = \frac{E_{max}}{E_{min}} \quad (3.11)$$

Where: E_{max} = The peak value of voltage of the standing wave.

E_{min} = The lowest value of voltage of the standing wave.

With Equation (3.11) in mind let us now go back to Figure 3.7 and substitute values to determine the VSWR of the line. We can see from Figure 3.7 that E_{max} equals a value that we will call, simply, E_{max}. The value is not of any real significance because it is relative to the incident voltage. However, E_{min} is significant: it is 0 V (volts) because it is right on the zero line. Using Equation (3.11) we find a VSWR of

$$\text{VSWR} = \frac{E_{\max}}{E_{\min}}$$
$$= \frac{E_{\max}}{0}$$
$$= \infty$$

In a similar manner we can look at a transmission line with a short circuit at the output (See Fig. 3.8(a)). You will recall that a short circuit has a maximum current and a zero voltage across it (See Fig. 3.8(b)). As the figure shows, the voltage at the load (short circuit) is at zero and increases to an $E_{\max}$ one-quarter wavelength from the load. Using Equation (3.11), we find that the VSWR is again infinite.

$$\text{VSWR} = \frac{E_{\max}}{E_{\min}}$$
$$= \frac{E_{\max}}{0}$$
$$= \infty$$

Thus, we see that the VSWR of a short circuit is exactly the same as for the open circuit. The only difference is that the two waves are 180° out of phase. To understand this compare Figures 3.7 and 3.8 and notice the phase relationship of the signals for both the open and the short circuit. In the case of the open circuit, the voltage is a

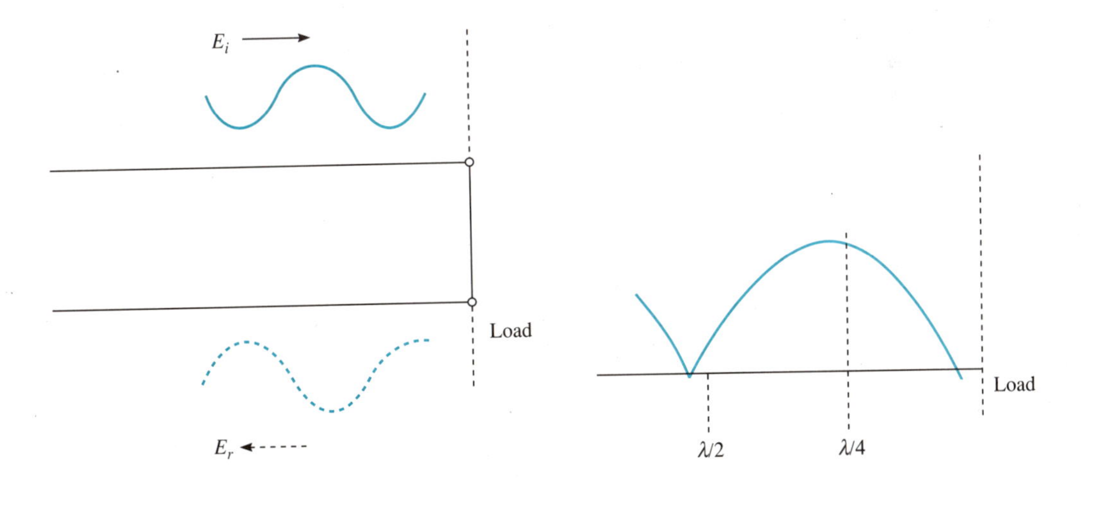

(a) Transmission Line

(b) Voltage Wave

Figure 3.8 Short circuit.

maximum at the load and at every one-half wavelength; in the case of the short circuit, the voltage is zero, or minimum, at the load and at every one-half wavelength: a difference in phase of 180°.

We will next examine the case of a resistive load that is greater than the characteristic impedance Z_0. If, for example, we have a transmission line with a characteristic impedance of 50 Ω, and place a load at the end of the line that is 150 Ω, what would the VSWR of such a line be?

In Figure 3.9(a) you can see that the 50 Ω line is terminated in an impedance of 150 Ω. Figure 3.9(b) shows that the voltage relationship of the resulting standing wave is a combination of the incident and reflected waves. By using the values in Figure 3.9(b), we can calculate the value of VSWR for this condition.

$$\begin{aligned} \text{VSWR} &= \frac{E_{\text{max}}}{E_{\text{min}}} \\ &= \frac{1.5}{0.5} \\ &= 3.0{:}1 \end{aligned}$$

It is interesting to compare the example above with a case where $Z_L < Z_0$, that is, where the characteristic impedance is still 50 Ω but the load impedance is now

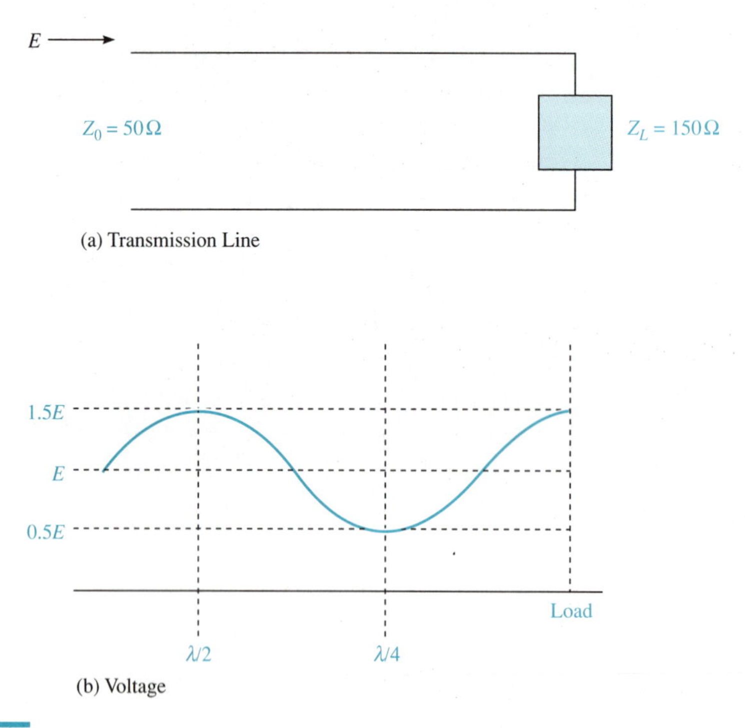

Figure 3.9 Transmission line with $Z_L > Z_0$.

16.666 Ω, or $\frac{1}{3}$ Z_0. This condition is shown in Figure 3.10. If we use the voltage values in Figure 3.10(b), we can calculate the VSWR of the line as follows:

$$\text{VSWR} = \frac{E_{\text{max}}}{E_{\text{min}}}$$

$$= \frac{1.5}{0.5}$$

$$= 3.0{:}1$$

This is the same value we obtained when calculating the VSWR for a 150 Ω termination. We would find that this was the case if we had a 100 Ω or a 25 Ω termination (VSWR = 2:1), a 200 Ω or a 12.5 Ω (VSWR = 4:1), or a 75 Ω or a 33.33 Ω (VSWR = 1.5:1) termination. These results show that you cannot tell if the load impedance is greater or less than the characteristic impedance simply by knowing the VSWR; all you can tell is that you have a mismatch that may be greater or less than Z_0. Further investigation will need to be conducted to find out the exact impedance of the load.

One final area should be covered before we leave the topic of VSWR. When the transmission line is terminated with a reactive component (an inductor or capac-

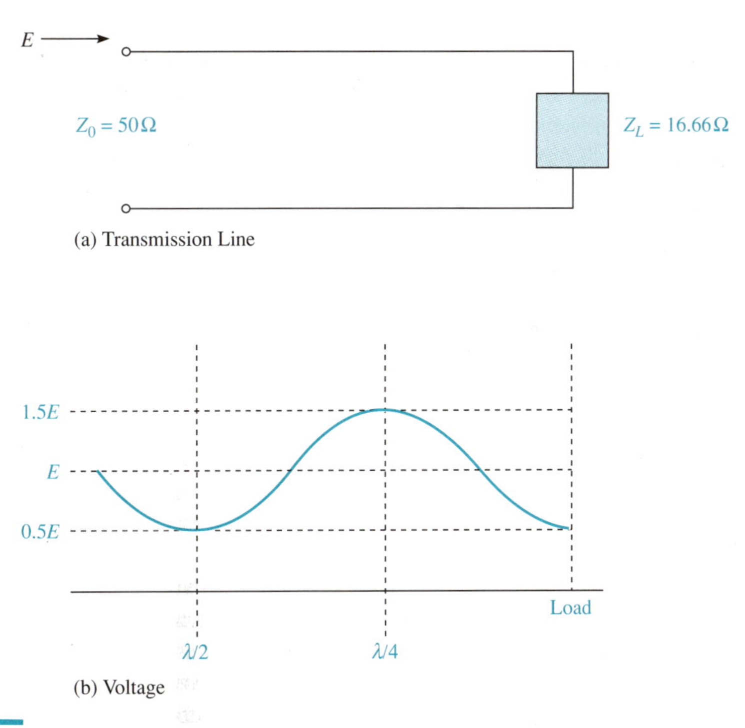

Figure 3.10 Transmission line with $Z_L < Z_0$.

itor), the VSWR is once again infinite because all of the energy is reflected back toward the source. To understand this consider that only a resistive portion of a component, not a reactive component, can dissipate power. Since none of the power is absorbed by the reactive load, it all must be reflected back, and the VSWR is therefore infinite.

Example 3.5 A representation of a signal shows that there is a maximum voltage of 2.3 V and a minimum of 0.3 V. What is the VSWR of this signal?

Solution

$$\begin{aligned} \text{VSWR} &= \frac{E_{\text{max}}}{E_{\text{min}}} \\ &= \frac{2.3}{0.3} \\ &= 7.66{:}1 \end{aligned}$$

3.3.3 Reflection Coefficient

The **reflection coefficient** is the percentage of incident, or input, signal being reflected back from a load. It is a voltage reflection coefficient that ranges between 0 and 1, with $\Gamma = 0$ signifying no reflected power (0%) and a VSWR of 1.0:1, and $\Gamma = 1.0$ signifying that all of the power (100%) is reflected and a VSWR of infinity. So it can be seen that the reflection coefficient and VSWR are related. Let us now see how the examples from Section 3.3.2 can be related to reflection coefficient.

In Section 3.3.2 we examined an open and a short circuit at the output of a transmission line, and in each of these cases the VSWR was equal to infinity. If we take this information and place it into the following reflection coefficient-equation, we will see that the percentage of reflected power is

$$\begin{aligned} \Gamma &= \frac{\text{VSWR} - 1}{\text{VSWR} + 1} \\ &= \frac{\infty - 1}{\infty + 1} \\ &= 1.0 \end{aligned} \tag{3.12}$$

A reflection coefficient of 1.0 tells us that all of the power sent down the transmission line will be reflected back. This agrees with our previous statements that both open and short circuits totally reflect the energy sent down a transmission line.

Our example using a characteristic impedance of 50 Ω and a load impedance of 150 Ω resulted in a VSWR of 3:1. If we now use Equation 3.12, we find the value of reflection coefficient for the VSWR of 3.

$$\Gamma = \frac{\text{VSWR} - 1}{\text{VSWR} + 1}$$
$$= \frac{3 - 1}{3 + 1}$$
$$= \frac{2}{4}$$
$$= 0.5$$

Thus, it can be seen that one-half of the power is being reflected back when we have a VSWR of 3:1 on a transmission line.

Our last example was that of a reactive load. Recall that with a reactive load all of the power was reflected because a reactive load cannot dissipate power. The VSWR is, therefore, infinite, the reflection coefficient is 1.0 for reactive loads.

The reflection coefficient may also be found when only the characteristic impedance and the load impedance are known. This relationship is shown below:

$$\Gamma = \frac{Z_0 - Z_L}{Z_0 + Z_L} \tag{3.13}$$

Where: Z_0 = Characteristic impedance (ohms)
Z_L = Load impedance (ohms)

Example 3.6

Given: $Z_L = 77\ \Omega$
$Z_0 = 50\ \Omega$

Find: The reflection coefficient Γ.

Solution

$$\Gamma = \frac{Z_0 - Z_L}{Z_0 + Z_L}$$
$$= \frac{50 - 70}{50 + 70}$$
$$= \frac{|50 - 70|}{|50 + 70|} \quad \text{Absolute value used when } Z_0 < Z_L$$
$$= \frac{20}{120}$$
$$= 0.1666$$

3.3.4 Return Loss

Return loss is probably one of the most difficult parameters of transmission lines that the student is asked to understand because for a VSWR of 1.0 (perfect match) the return loss is infinite. Similarly, for a VSWR of infinity (open or short circuit), the return loss is 0 dB. This seems to be a direct contradiction of everything we have said to this point. When we discuss it further, however, you will see that there really is no contradiction but rather a reinforcement of our previous statements.

It will be somewhat easier to grasp the idea of return loss if you consider the loss as a negative number. It may also help to picture return loss as the reflected signal that is "returning" from a mismatch and being a certain number of decibels below the incident signal. Thus, there is a loss in the signal that is returning, or a **return loss**.

To further illustrate the idea of return loss, once again let us consider the transmission line with an open or short circuit at the output, as well as the case where we have a load impedance that is resistive and greater or less than the characteristic impedance. These examples will show how return loss varies with the match and also how it relates to reflection coefficient and VSWR.

In the examples of the open or short circuit, we recall that the VSWR was infinite and the reflection coefficient was 1.0. To determine the return loss, we use Equation (3.14).

$$\text{Return loss} = -20 \log \Gamma \tag{3.14}$$

Where: Γ = Reflection coefficient

Substituting the reflection coefficient for the open or short circuit, which is 1.0, we obtain a return loss of 0 dB, since the log of 1 is 0. This means that the level of

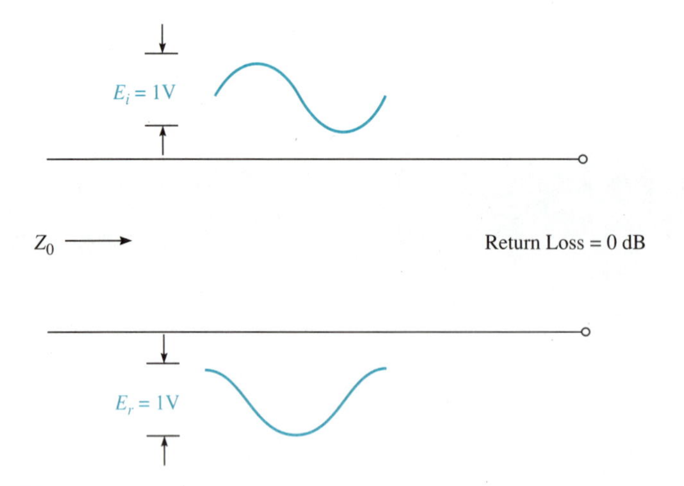

Figure 3.11 Return loss for an open circuit.

the return, or reflected, signal is the same as the level of the incident signal. This is shown in Figure 3.11. You can see that the incident, or forward, signal (E_i) has a magnitude of 1 V and the reflected, or reverse, signal (E_r) is also 1 V. This results in a 0 dB difference, or return loss, for this case. This 0 dB difference can be verified if we plug the two signal levels into the decibel formula below:

$$\text{dB (power)} = 10 \log \frac{P_1}{P_2}$$

$$\text{dB (voltage)} = 20 \log \frac{E_1}{E_2}$$

Using the voltage formula we obtain

$$\begin{aligned}\text{dB (voltage)} &= 20 \log \frac{1.0}{1.0} \\ &= 20 \log 1.0 \\ &= 20(0) \\ &= 0 \text{ dB}\end{aligned}$$

In the case of a characteristic impedance of 50 Ω and a load impedance of 150 Ω, we have found that this produces a VSWR of 3:1 and a reflection coefficient of 0.5. If we substitute the reflection coefficient into Equation (3.13), we have the following:

$$\begin{aligned}\text{Return loss} &= -20 \log \Gamma \\ &= -20 \log (0.5) \\ &= -20(-0.3) \\ &= 6.02 \text{ dB}\end{aligned}$$

Thus, it can be seen that a 3:1 VSWR will produce a return loss of 6 dB. That is, the level of the reflected signal will be 6 dB lower than the original incident signal.

To tie the previous discussion together, consider the following example.

Example 3.7

Given: Characteristic impedance Z_0 is 50 Ω
Load Impedance Z_L is 500 Ω.

Find: **(a)** Reflection coefficient
(b) Return loss
(c) VSWR

Solution

(a) Using Equation (3.13), we have

$$\Gamma = \frac{Z_0 - Z_L}{Z_0 + Z_L}$$

$$= \frac{50 - 500}{50 + 500}$$

$$= \frac{450}{550} \quad \text{(Absolute value)}$$

$$= 0.82$$

(b) The return loss is calculated by using Equation (3.14).

$$\text{Return Loss} = -20 \log \Gamma$$

$$= -20 \log (0.82)$$

$$= -20(-0.09)$$

$$= 1.72 \text{ dB}$$

(c) By changing Equation (3.12) and solving for VSWR, we obtain

$$\text{VSWR} = \frac{1 + \Gamma}{1 - \Gamma}$$

$$= \frac{1 + 0.82}{1 - 0.82}$$

$$= \frac{1.82}{0.18}$$

$$= 10.11{:}1$$

3.4 Smith Chart

In the January, 1939 issue of *Electronics* magazine, Phillip H. Smith published an article entitled, "Transmission Line Calculator." Five years later (January, 1944) the same magazine published the "Improved Transmission Line Calculator." These two articles were the beginning of what has become the most useful tool ever derived for working with transmission lines—this tool now bears the name of the man who designed it, the **Smith chart**. Phillip Smith passed away in 1987, but the Smith chart will be used for many years to solve virtually any transmission line problem imaginable.

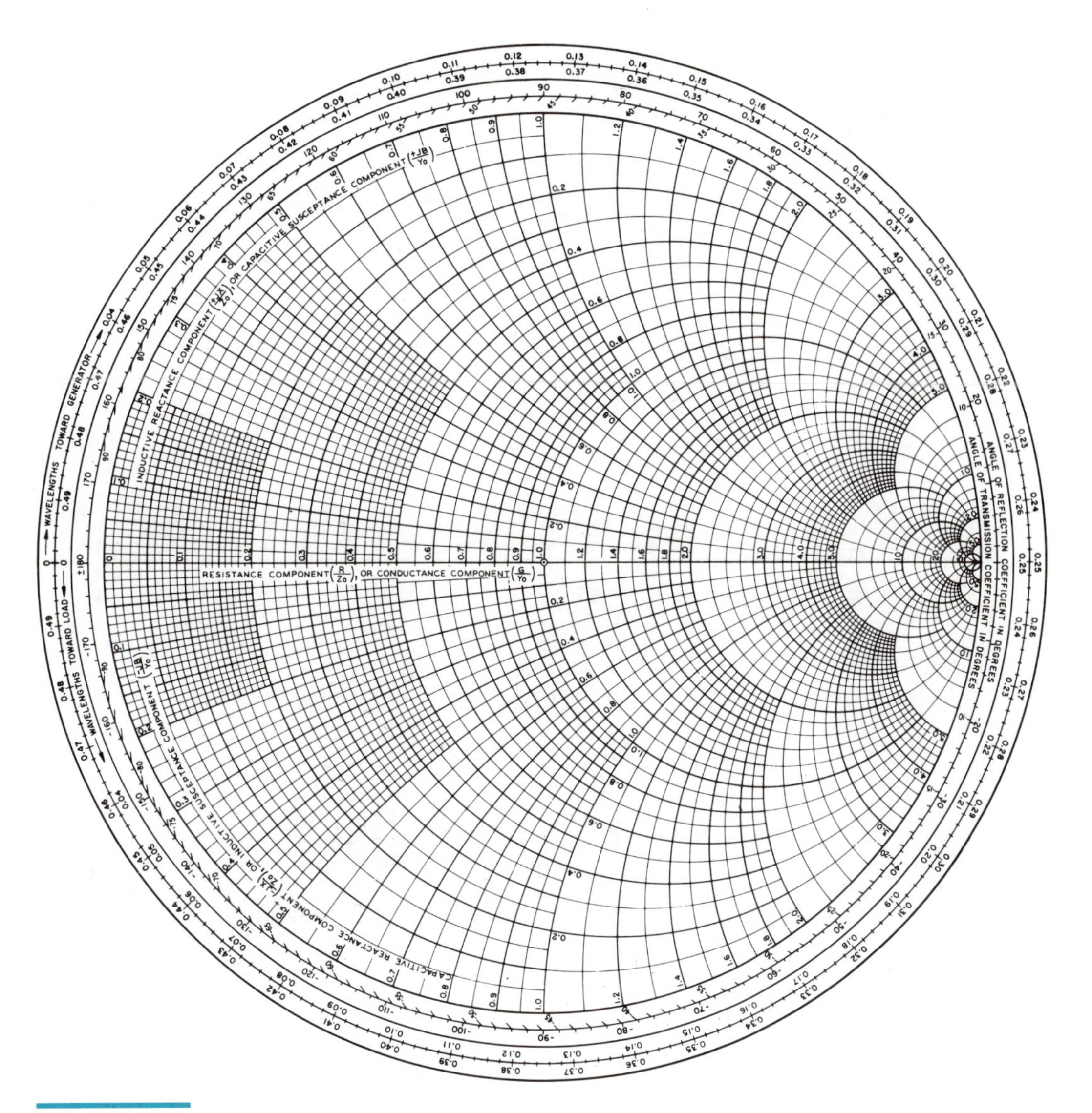

Figure 3.12 Smith chart.

As shown in Figure 3.12, the Smith chart is a normalized chart for plotting any impedance that you can imagine. This is a broad statement to make, but the student will see how this is possible very shortly. At first glance the chart appears to be an overwhelming group of circles and lines that go nowhere, but to understand it you need to realize that impedance is actually made up of two components: a real part (resistive) and an imaginary part (reactance). Expressed mathematically, it is

$$Z = R \pm jX \tag{3.15}$$

It is in this type of **complex impedance** that the Smith chart specializes.

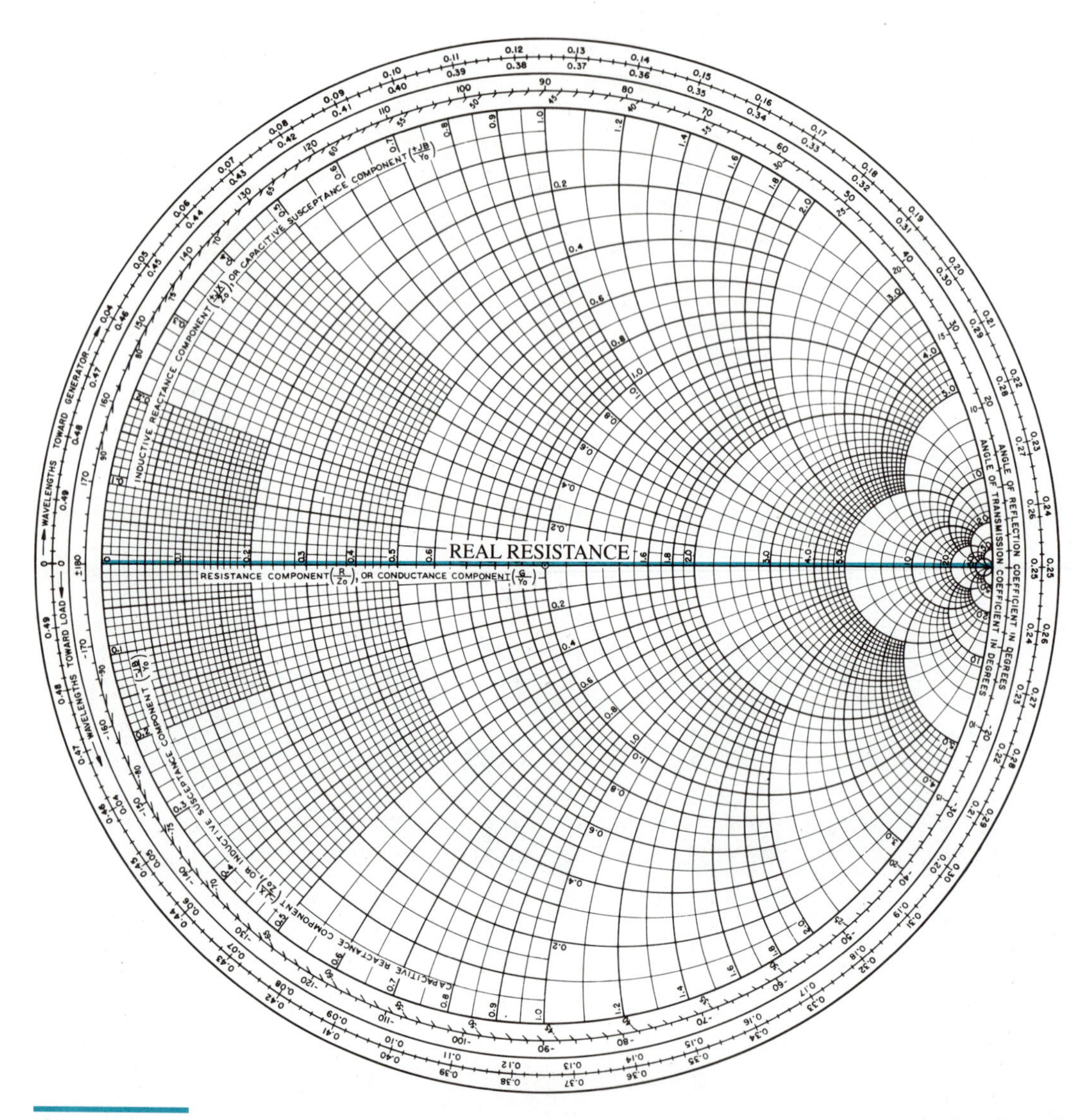

Figure 3.13 Real resistance line.

To find an impedance on the chart, first consider that all impedances along the center line of the chart are real values only; there is no reactive value at all along this line. Therefore, if a real impedance is needed (100 Ω), it would be plotted along this line (Fig. 3.13). Similarly, all the impedances around the inside edge of the chart are imaginary. The top half of the chart is inductive reactance and the bottom half is capacitive reactance (Fig. 3.14).

It follows then, that if all the real values of impedance are on the center axis of the chart and all the imaginary values are around the inside edge of the chart, the combination of the two will be in between these two, or on the chart itself.

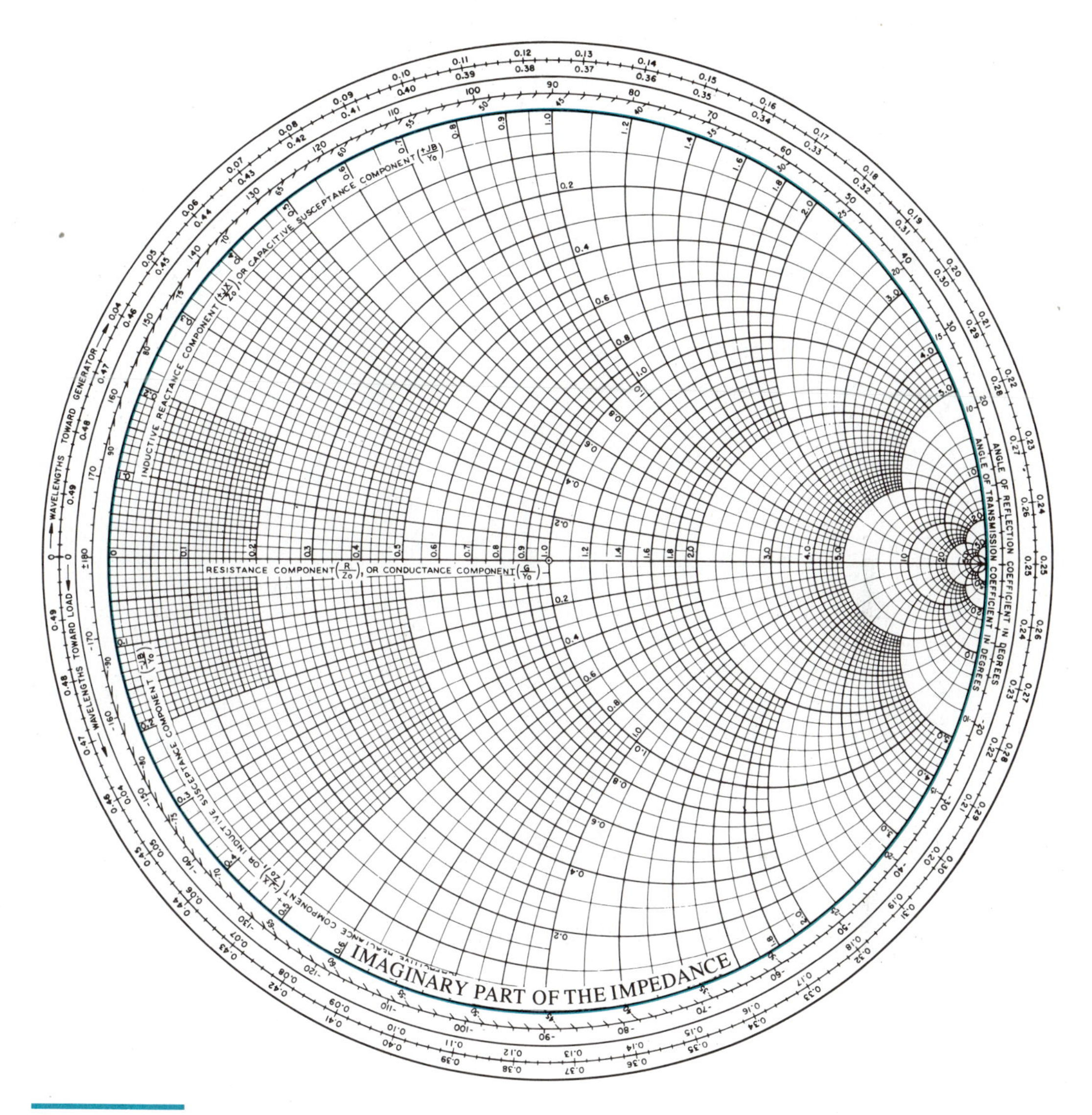

Figure 3.14 Imaginary part of the impedance.

In the beginning of this section we said that Figure 3.12 was a normalized chart. This means that the value in the center of the chart is 1.0. This value is obtained by dividing each component of the complex impedance by the characteristic impedance (Z_0). That is,

$$Z_n = \frac{R}{Z_0} + \frac{X}{Z_0} \tag{3.16}$$

Where: Z_n = Normalized impedance

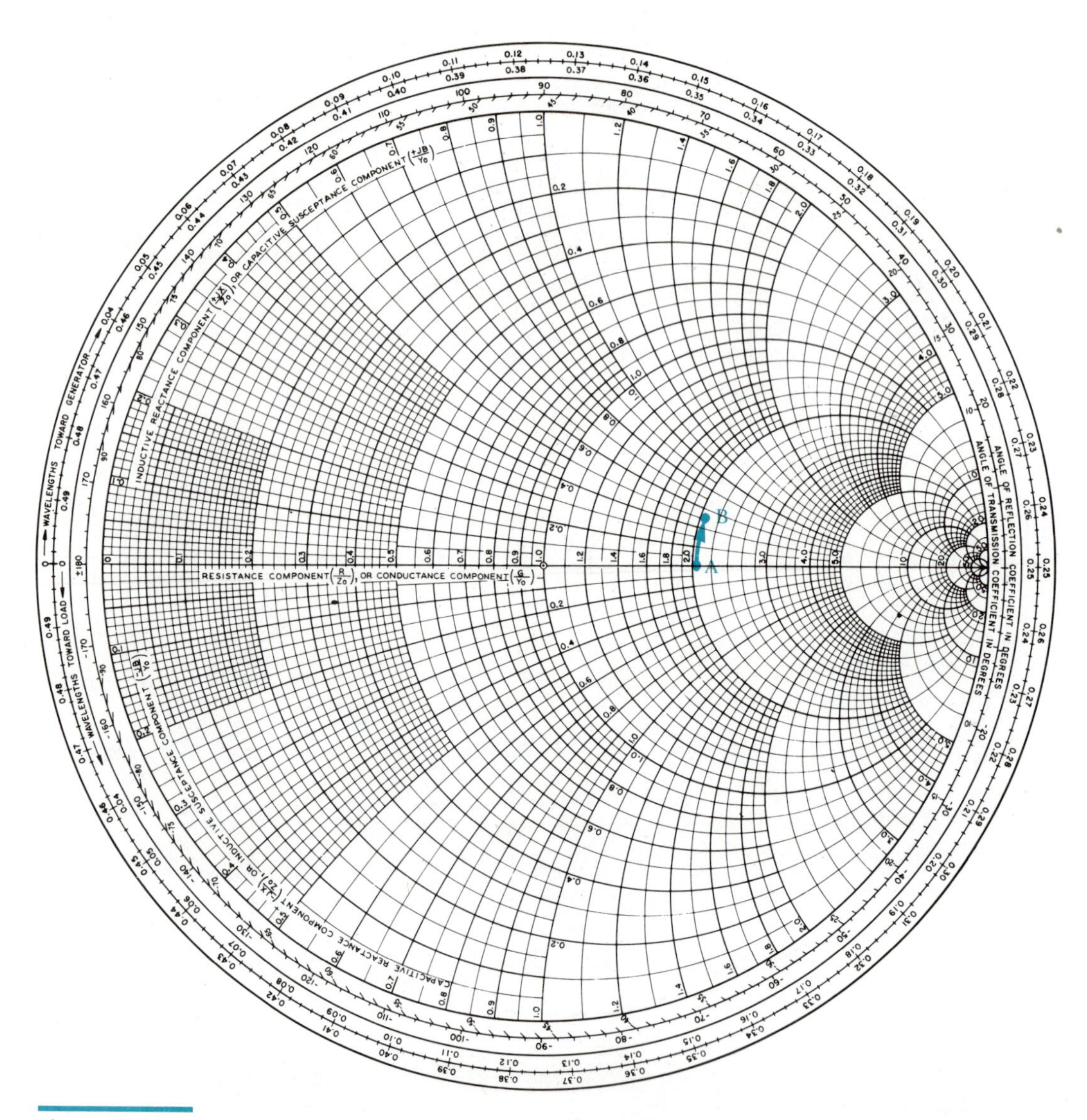

Figure 3.15 Impedance example.

As an example of how an impedance is normalized and plotted, consider $Z_0 = 100 + j25\ \Omega$ as an impedance that must be plotted on the Smith chart. The first step is to normalize the impedance. To do this the characteristic impedance must be known. For this case, let it be 50 Ω. The normalized impedance now becomes $2 + j0.5\ \Omega$. To plot this value on the chart, first find the 2.0 point on the horizontal axis of the chart (point *A* in Fig. 3.15). Now move up the 2.0 circle (since the impedance is $+j$, or inductive) until the 0.5 line is intersected (point *B*). This is the point $2 + j0.5$.

Now that we know how to plot impedance on the chart, we need to examine how we move along a transmission line and how it relates to the Smith chart. Recall from earlier discussions and courses that when we speak of sinusoidal variations of signals there is a reversal every quarter-wavelength, and a repetition of characteristics every half-wavelength. That is, if we take a sine wave (Fig. 3.16) and start at zero time, we have a maximum amplitude at 90°, or one-quarter cycle, away from this initial zero point. If we keep going, the signal will return to zero in another 90°, or 180° total (one-half wavelength). It is this characteristic that is employed in the Smith chart: one revolution around the chart results in one-half wavelength of travel on a transmission line. This half-wavelength is broken up into increments all the way around the chart on the two outer scales. Clockwise movement around the chart is termed **wavelengths toward the generator**; counterclockwise movement is **wavelengths toward the load**. These points are *A* and *B*, respectively, in Figure 3.17.

To see how the Smith chart can be used to calculate impedances on a transmission line, consider the following example.

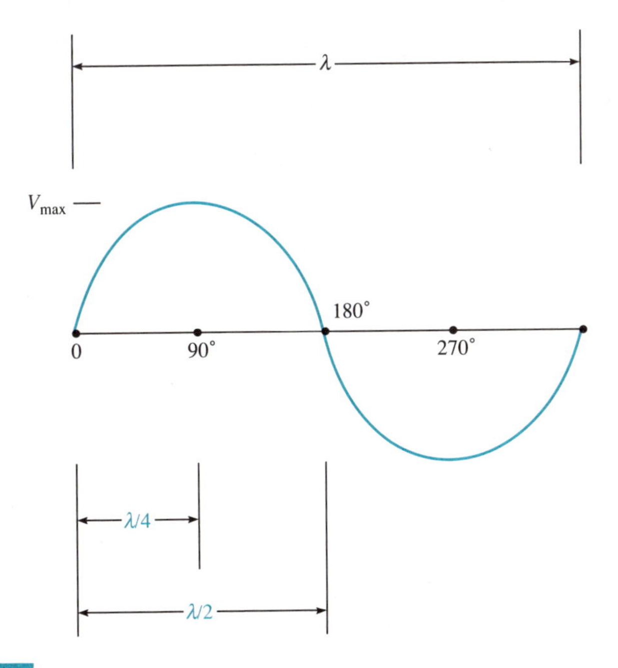

Figure 3.16 Wavelength relationship of a wave.

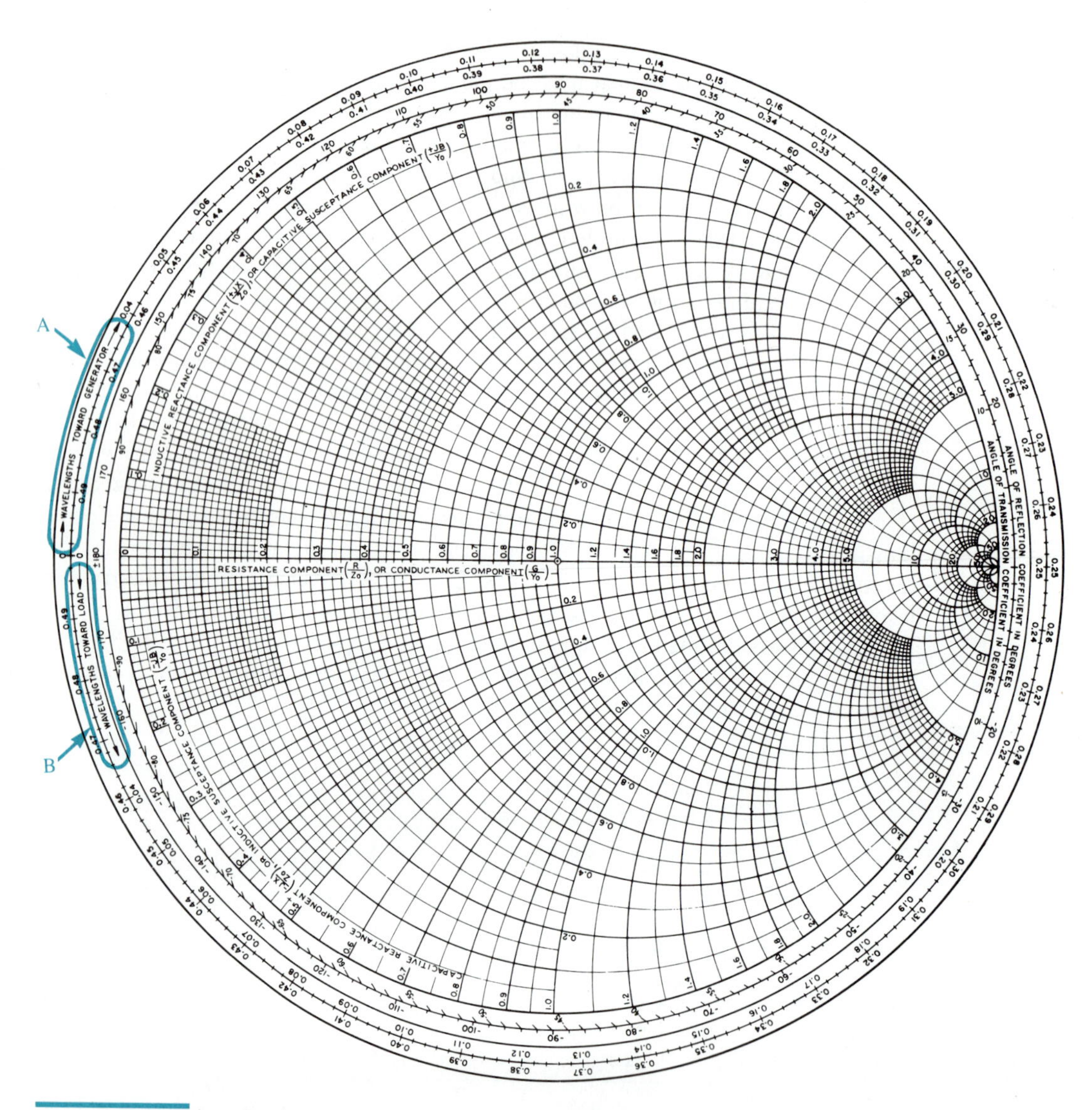

Figure 3.17 Wavelengths toward the generator and the load.

Example 3.8

Given: Load impedance Z_L of $25 - j50\ \Omega$
Characteristic impedance $Z_0 = 50\ \Omega$

Find: The impedance which is 0.1 wavelengths away from the load.

Solution

1. First, normalize the impedance

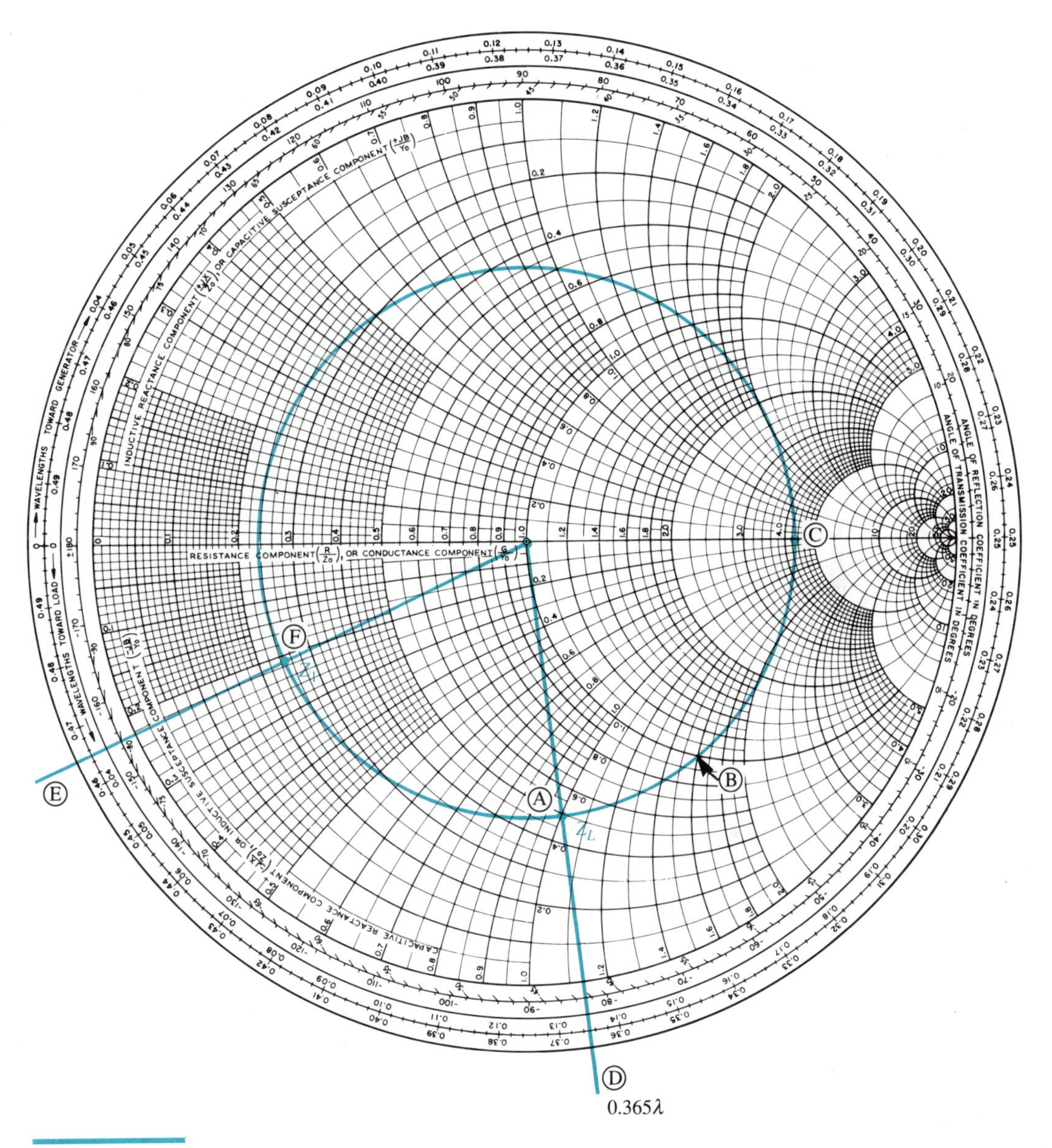

Figure 3.18 Impedance problem.

$$Z_n = \frac{25}{50} - j\frac{50}{50}$$

$$= 0.5 - j1.0\ \Omega$$

(This is plotted in Fig. 3.18 at point A)

2. Place the point of a compass in the center of the chart and the pencil on Z_L and draw a circle around the chart as shown at point *B* of Figure 3.18. This is the VSWR circle. The VSWR can be read directly from the chart by reading where the VSWR circle crosses the real axis on the right side of the chart (point *C*). This is read as a 4.3:1 VSWR.
3. Draw a line from the center of the chart through Z_L and read the outside scale. This should say, 0.365 wavelengths (point *D*). This is the reference point to find the new impedance.
4. Find the spot on the scale that says 0.465 wavelengths (point *E*) and draw a line from the center of the chart to point *E*. Point *F*, the desired impedance, is where the line crosses the VSWR circle.
5. The desired impedance is now read as $0.24 - j0.2\ \Omega$. This is a normalized value. To obtain the actual value multiply the normalized value by the characteristic impedance. Thus, the impedance is $12 - j10\ \Omega$.

In this example we have taken a load impedance, transformed it down a transmission line (which is equivalent to moving it around the VSWR circle), and found a new impedance a certain fraction of a wavelength away from the load.

A widely used method of matching two transmission lines employs a quarter-wave transformer. This arrangement is shown in Figure 3.19. The transformer section is the quarter-wave section of transmission line between the load impedance Z_L and the characteristic impedance Z_0. To use the quarter-wave transformer to match transmission lines, it is necessary to convert the load impedance (which is a complex impedance) to a real value of impedance. Recall from the early discussions of the Smith chart that real values of impedance can be found on the horizontal axis of the chart. Thus, it is necessary to move the load impedance around to the real axis.

To understand what needs to be done, consider the following example. We have a transmission line with a characteristic impedance of 50 Ω and want to match a load impedance of $75 - j50\ \Omega$ to it with a quarter-wave transformer. What needs to be done?

To begin the process we must first normalize the load impedance, making it $1.5 - j1$. This value is now plotted on the Smith chart (point *A* in Fig. 3.20). A line is now drawn from the center of the chart through point *A* to determine the reference point of 0.308 wavelengths. The VSWR circle is now drawn using point *A* and the

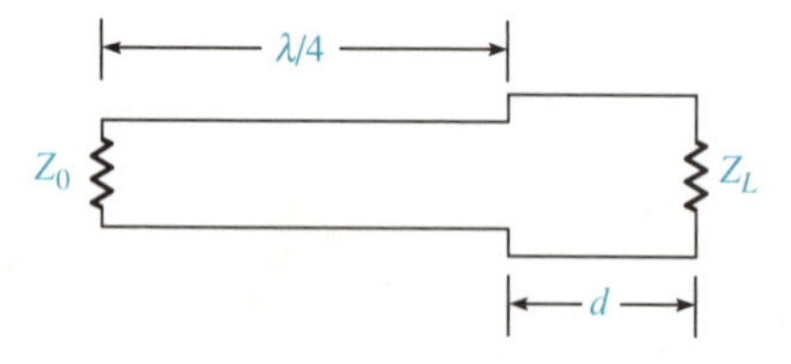

Figure 3.19 Quarter-wave transformer.

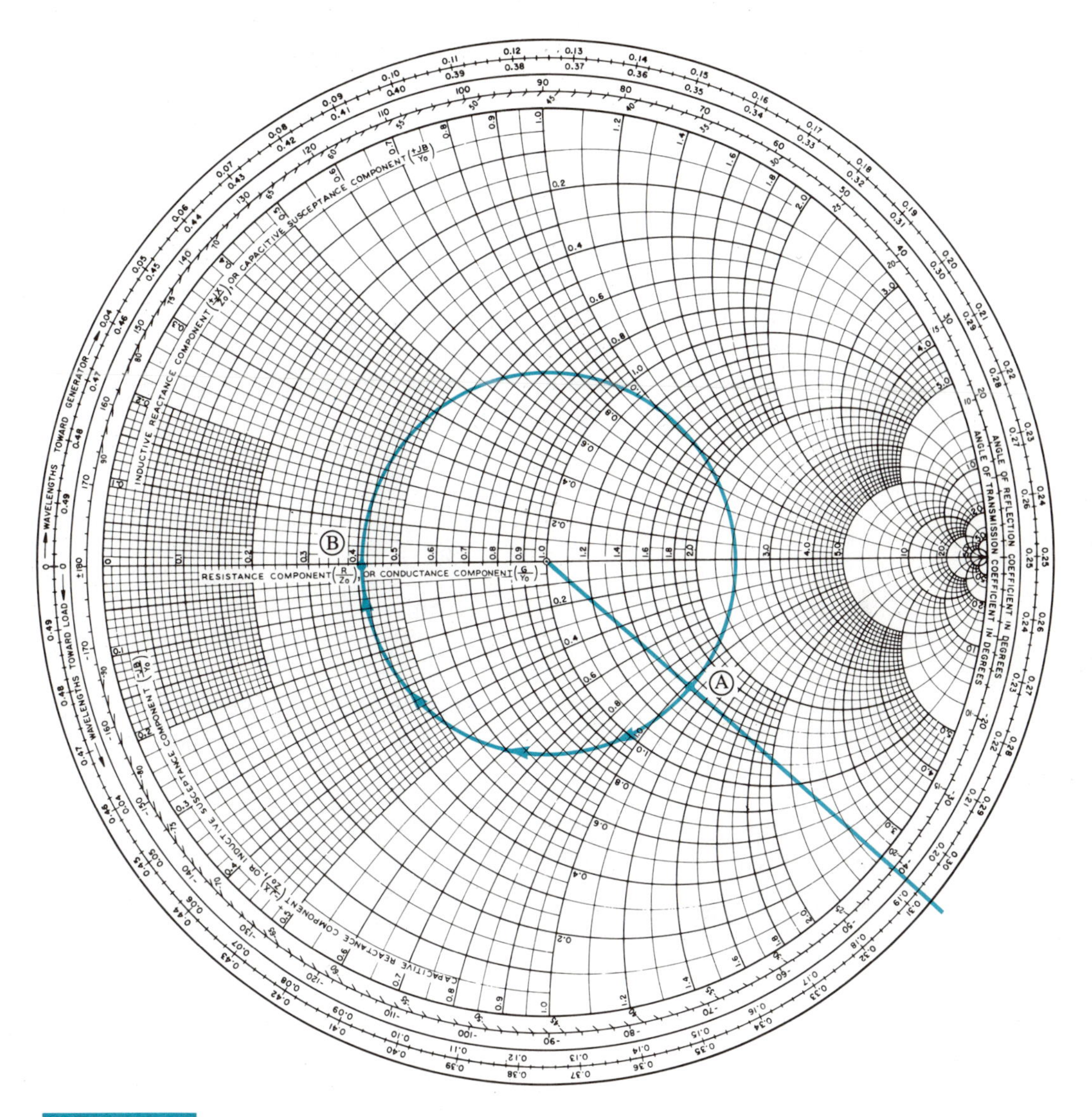

Figure 3.20 Quarter-wave transformer example.

center of the chart. To get to the real axis in order to be able to calculate the transformer section point A is moved clockwise (toward the generator) until we reach the real axis (point B). This value is 0.42, which is $0.42(50) = 21\ \Omega$. This is the impedance that the transformer section has to match to 50 Ω.

To get to point B from point A we had to move down the transmission line, but how far did we have to move? Point A was at 0.308 wavelengths and point B is at 0.500 wavelengths. We have, therefore moved 0.192 wavelengths to get to a point where the impedance is 21 Ω. This is the d dimension shown in Figure 3.19.

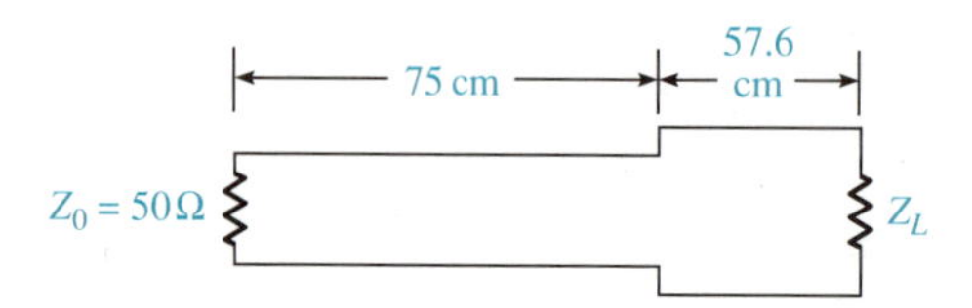

Figure 3.21 Final quarter-wave transformer.

With the new impedance determined it is simply a matter of mathematics to find the impedance of the transformer section. If we designate the 21 Ω impedance as Z_1, and the characteristic impedance as Z_0, the matching impedance will be Z_{01}. To calculate this impedance, use the following equation:

$$Z_{01} = \sqrt{Z_0 Z_1} \quad (3.17)$$

Using Equation (3.17) in the example, we find

$$Z_{01} = \sqrt{(50)(21)}$$
$$= \sqrt{(1050)}$$
$$= 34.2\ \Omega$$

The length of the transformer is, as the name implies, one-quarter wavelength at the frequency of operation. The wavelength is determined by the equation

$$\text{Wavelength} = \frac{c}{f} \quad (3.18)$$

Where: c = Velocity of light (3×10^{10} cm per s)

f = Frequency in hertz

Thus, if we include a frequency of 100 MHz in our example, the final transformer would have a length of $3 \times 10^{10}/100 \times 10^6$, or 300 cm divided by 4, or 75 cm. The dimension d is 0.192(300 cm) or 57.6 cm. The final circuit with dimensions is shown in Figure 3.21.

Thus, we see how useful the Smith chart is for finding impedances on transmission lines and for matching transmission lines to one another. These are only two of the many tasks that the Smith chart can perform for communications systems.

3.5 Summary

This chapter has presented the theory of the most useful components used in communications systems—transmission lines. These devices determine a system's effectiveness by either exhibiting very low loss matched characteristics, or by introducing high losses and intolerable standing-wave ratios.

We introduced types of transmission lines such as open wire, twin lead, twisted pair, shielded, and coaxial, describing each so that the student may choose the best type of transmission line for a particular application.

The chapter concluded by describing the transmission line characteristics and parameters: characteristic impedance, standing waves, reflection coefficient, and return loss. We then introduced the Smith chart to calculate these parameters.

Questions

3.1 Introduction

1. Define **transmission line**.

3.2 Types of Transmission Lines

2. Name and describe the four parameters used for an equivalent circuit for transmission lines.
3. What is the difference between **balanced** and **unbalanced** transmission lines?
4. What are the disadvantages of an open-wire transmission line?
5. What is another name for twin lead?
6. What is the advantage of a twisted-pair transmission line?
7. What term is used to determine the majority of the parameters for coaxial lines?
8. List four parameters of a coaxial cable that are important to its operation.
9. Define **CW power rating**.
10. What does PTFE stand for?
11. Which parameters of a transmission line equivalent circuit are associated with the dielectric?
12. Which parameters of a transmission line equivalent circuit are associated with the conductor of the line?
13. Name two (2) advantages of an open wire transmission line.
14. If the distance between the conductors on a twin lead transmission line is decreased, what effect will this have on the characteristic impedance?
15. How is a twisted pair transmission line constructed?
16. What is **pitch** in a twisted pair transmission line?
17. What is a good property of the shielded cable?
18. Sketch and label the construction of a coaxial cable.
19. What is the D/d ratio of a coaxial cable?
20. What are four factors that affect the amount of attenuation in a coaxial cable?
21. If the dielectric constant increases, what is the effect on the velocity of propagation?
22. What is the common term used to calculate both the capacitance and the inductance of a coaxial cable?
23. Define, in your own words, **shielding** in a coaxial cable.

3.3 Transmission Line Characteristics

24. Define **characteristic impedance**.
25. Define **standing wave**.
26. Relate reflection coefficient to VSWR.
27. What are two components that make up the characteristic impedance?

28. If the impedance attached to a transmission line is the same as the impedance of the line, what is the amplitude of the reflected signal? What is this called?
29. What is the relationship of the VSWR, return loss, reflection coefficient, and phase between an open circuit and a short circuit at the end of a transmission line?
30. Using the equations given in this chapter, is it possible to tell if the load impedance on a transmission line is more or less than the characteristic impedance of the line?
31. What effect does a reactive load have on a transmission line?
32. Define **reflection coefficient**.
33. Which mismatch has a higher reflection coefficient, an open circuit or a short circuit?
34. Define **return loss**.
35. Which mismatch has a higher return loss, VSWR = 5:1 or VSWR = 7:1?

3.4 Smith Chart

36. What is another name for the Smith chart?
37. What does the line across the center of the chart indicate?
38. What is on the outside edge of the Smith chart?
39. Which direction do we move on the Smith chart to move toward the generator?
40. Where would an impedance of $R + jX$ be found on a Smith chart?
41. When using a quarter-wave transformer, why is it necessary to move to the real resistance line on the Smith chart?
42. When finding impedance along a transmission line with the Smith chart, what line do you move around?

Problems

3.2 Types of Transmission Lines

1. Given a coaxial cable with a dielectric of polyethylene ($\epsilon_r = 2.26$), center conductor diameter of 0.087 in., outer conductor of 0.285 in., find the characteristic impedance.
2. Given a coaxial cable with a dielectric constant of 2.1, center conductor of 0.025 in., outer conductor of 0.146 in., find Z_0.
3. We have a 50 Ω coaxial cable with $\epsilon_r = 2.1$, $D = 0.034$ in., $d = 0.012$ in., frequency = 6 GHz, and $\tan d = 0.0015$. Find the attenuation for 30 ft of cable.
4. Find the velocity of propagation for cables with dielectric constants of 2.1, 2.26, 3.78, and 10.2.
5. Find the value of capacitance and inductance for the conditions given in Problems 2 and 3.
6. A 75 Ω coaxial cable is to be used for a system which will be 275 ft long. If $D =$ 0.146 in., $d = 0.083$ in., $\epsilon = 2.39$, dissipation factor = 0.0001, and frequency of operation is 100 MHz, what will be the attenuation of this section of cable?
7. For the 75 Ω cable described in Problem 6, what is the total value of capacitance and inductance for the 275 ft length?

3.3 Transmission Line Characteristics

8. For Example 3.2, add five more sections to the transmission line and calculate Z_0.
9. We have a circuit with a VSWR of 2.6:1, what is the reflection coefficient and return loss?
10. Given a load impedance of 176 Ω and $Z_0 = 50$ Ω, what is the reflection coefficient?

11. Given that $Z_0 = 50\ \Omega$ and the load impedance is 200 Ω, find the reflection coefficient, return loss, and VSWR.

3.4 Smith Chart

12. Given $Z_L = 80 + j60\ \Omega$ and $Z_0 = 50\ \Omega$, find Z_{in} of a line 4 cm long. Also find the VSWR of the system. (frequency = 2 GHz, $\epsilon = 2.1$)
13. A load impedance of $65 - j30\ \Omega$ is to be matched to a 60 Ω characteristic impedance using a quarter-wave transformer. Calculate the length and impedance of the transformer and the distance from the load that the transformer must be placed (frequency = 275 MHz, $\epsilon = 2.5$).
14. An impedance of $45 - j22\ \Omega$ is read on a transmission line. It is determined that this impedance is 13 cm from the load. If the frequency of operation is 450 MHz and the dielectric constant is 2.2, find the value of load impedance ($Z_0 = 50\ \Omega$).

4

Outline

Objectives

- To introduce you to antennas and their operation
- To familiarize you with the terminology used with antennas
- To present various types of antennas such as dipoles, arrays, log periodic, and helical antennas
- To discuss the parabolic reflector used with many types of antennas

Key Terms

antenna
Marconi antenna
radiation pattern
power gain
near field
radiation resistance
EIRP (effective isotropic-radiated power)
beam width
array
log-periodic antenna
helix
dipole
Hertz antenna
directional gain
directivity
far field
antenna efficiency
polarization
bandwidth
director element
parabolic antenna

Antennas

4.1 Introduction

When the term **antenna** is mentioned, many people think of the antenna on their car or the piece of wire imbedded into their windshield that allows them to hear their favorite AM or FM station. This, of course, is one application of antennas, but there is much more to antennas than a wire sticking up into the air. In this chapter, we will look at some basics of antennas, antenna terminology, and some typical antennas used in electronic communications.

4.2 Antenna Basics

The simplest, and probably the most understandable, way to explain antennas is as a transmission line with its output end left open. You will recall from our discussions of open-circuited transmission lines that the voltage at this point is a maximum value, and there will, therefore, be radiation from this open-ended line. That is something that occurs even if you are not planning on the transmission line being an antenna. Any open-circuited transmission line, including microstrip transmission lines for microwave frequencies, will radiate and act as an antenna. The open-circuited transmission line is shown in Figure 4.1. Although, as we have said, a certain amount of electromagnetic energy radiates from all open-circuited transmission lines, the distance for which this radiation is of any consequence is minimal. We must, therefore, find some means of directing this radiation so that it will cover a much greater range. This can be done, as shown in Figure 4.2, by flaring out the end of the transmission line to allow the energy to be propagated out into the air. This flared structure has the familiar name of dipole. By **dipole** we mean that the antenna has two poles associated with it: one is on the top conductor and the other is on the bottom. The combination of the two poles makes a very economical and efficient antenna.

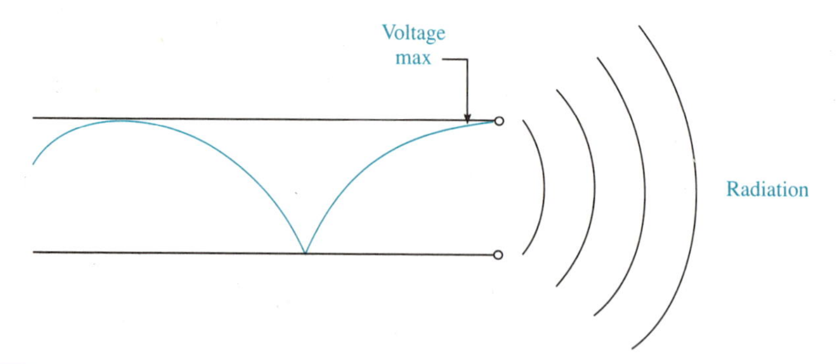

Figure 4.1 Basic antenna radiation.

A more common, or more recognizable, picture of a dipole may be the one shown in Figure 4.3. In this figure we have taken the dipole of Figure 4.2 and continued to spread out the transmission-line conductors until the total distance between them is one-quarter wavelength. This type of antenna is called a **quarter-wave dipole**, or **Marconi antenna**.

Probably the most widely used dipole antenna is the one shown in Figure 4.4. This is the same type of antenna we have been speaking of except that we have continued to spread out the transmission line until each side is one-quarter wavelength long, making the antenna a **half-wave dipole**, or what is commonly termed a **Hertz antenna**.

When we actually construct a half-wave dipole antenna, however, we usually find that the length of the antenna should be approximately 5% less than the theoretical one-half wavelength. (A *wavelength* is defined as the speed of light c, divided by the frequency of operation f.) We actually end up with an antenna that is in reality .95 times the wavelength over 2, or .48 wavelengths. If you use this ratio when constructing a Hertz antenna, you will have a very high degree of success and will produce an antenna that will operate properly 99% of the time.

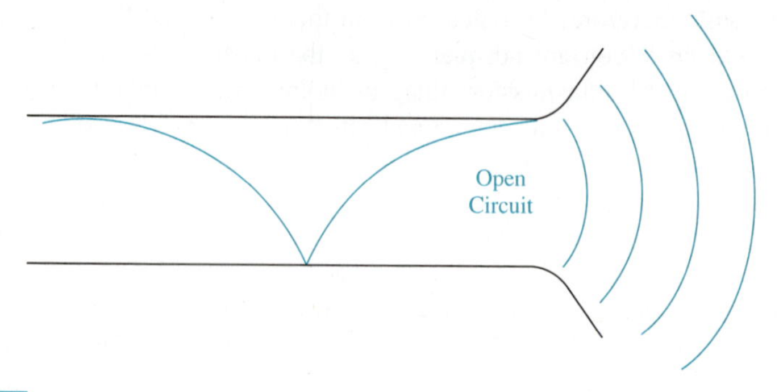

Figure 4.2 Dipole antenna.

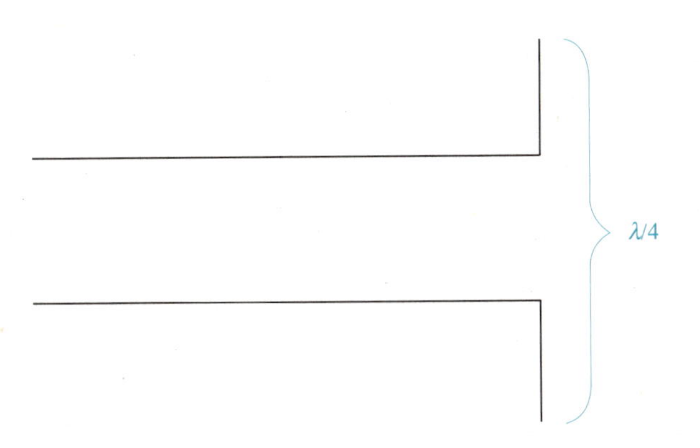

Figure 4.3 Quarter-wave (λ/4) antenna.

Now that we understand the basic idea of how a transmission line can be changed into a radiating antenna, let us look a little deeper into the concept of antennas. The first thing to realize about antennas is that they are **reciprocal devices**, meaning that electromagnetic energy can be both transmitted and received from the same antenna. This should be obvious if you own a CB radio for your car, truck, or van. When you talk to someone on the road you use the antenna mounted on the fender, mirror, or trunk of your vehicle to transmit your message. You use the same antenna to receive incoming replies. The antenna is, therefore, a versatile piece of equipment in any communication system.

It is also important to realize that an antenna is actually an impedance-matching device. Most people think of the antenna as little more than a means of transmitting the power-amplifier output into the air or a means of receiving the small signal from a far-off transmitter. Although the antenna performs these tasks, it does so efficiently only when it acts as an impedance-matching device. To understand what an impedance-matching device is, recall that earlier in the text we calculated free-space impedance to be 377 Ω. Also recall that characteristic impedance is generally in the range

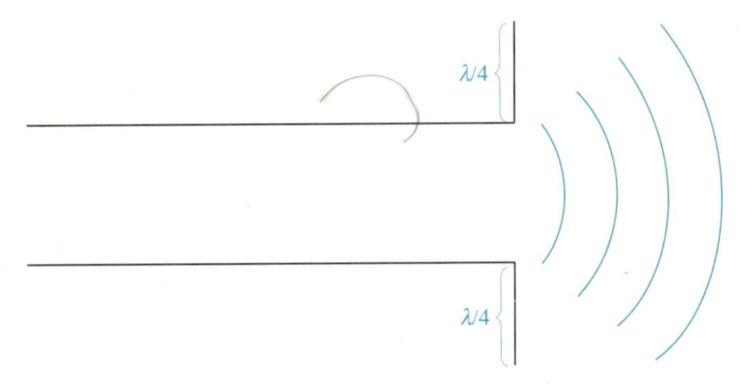

Figure 4.4 Hertz antenna.

of 50 Ω. If we now assume that the output impedance of our power amplifier is 50 Ω and the input to our receiver is also 50 Ω, we have to have some means of matching the transmitter's 50 Ω to the 377 Ω of the air in order to transmit as much energy into the air as possible. Similarly, if we want to receive as much signal from the air into the receiver as we can, we must also be able to match the 377 Ω impedance of the air to the 50 Ω impedance of the receiver. Thus, it can be seen that the antenna serves a much more vital function to the communication system than merely providing a means of radiating a carrier and intelligence into the air. It is actually a very efficient impedance-matching device.

4.3 Antenna Terminology

To understand fully the operation of antennas it is necessary to be familiar with the terminology used to describe and specify them for use in specific communications applications.

4.3.1 Radiation Pattern

When we discussed the basic antenna as shown in Figure 4.1, we were not concerned with how the open transmission line radiated the energy into space; our primary concern was that the configuration would radiate. Although this is a good way to develop basic theory, it is not the actual way antennas are characterized. Every antenna radiates, in a specific way, the electromagnetic energy applied to it. The way that the energy is radiated from an antenna is called its **radiation pattern**.

Figure 4.5 shows a typical radiation pattern for an antenna. It is a plot of the field strength of the antenna as a function of the angle around the antenna. It actually indicates where the major portion of the electromagnetic energy is being directed when a transmitter is connected to the antenna. The patterns plotted for an antenna are relative plots, meaning that the magnitude of the presentation is a relative number with the maximum value at the outside edge of the plot and every other level being determined with respect to the largest level. Generally the plots are in decibels so that other levels are easier to distinguish and to determine. If for example, you wanted to know what the level of the power was at an angle of 22.5°, we could look at Figure 4.5 and determine that it would be approximately 1 dB down from the maximum level. This means that if we were transmitting 1 W from the antenna (+30 dBm), we would have a level of +29 dBm or a 794-mW power level at the 22.5° area.

Another radiation pattern that you should be able to recognize is shown in Figure 4.6. Although this pattern looks like nothing has been done to the piece of paper, it is actually a plot of an **omnidirectional antenna**: That is, an antenna that transmits virtually the same level in all directions. It can be seen from Figure 4.6 that there are some minute variations in the level, but nothing like the other patterns shown in Figure 4.5, which is a more directional type of antenna.

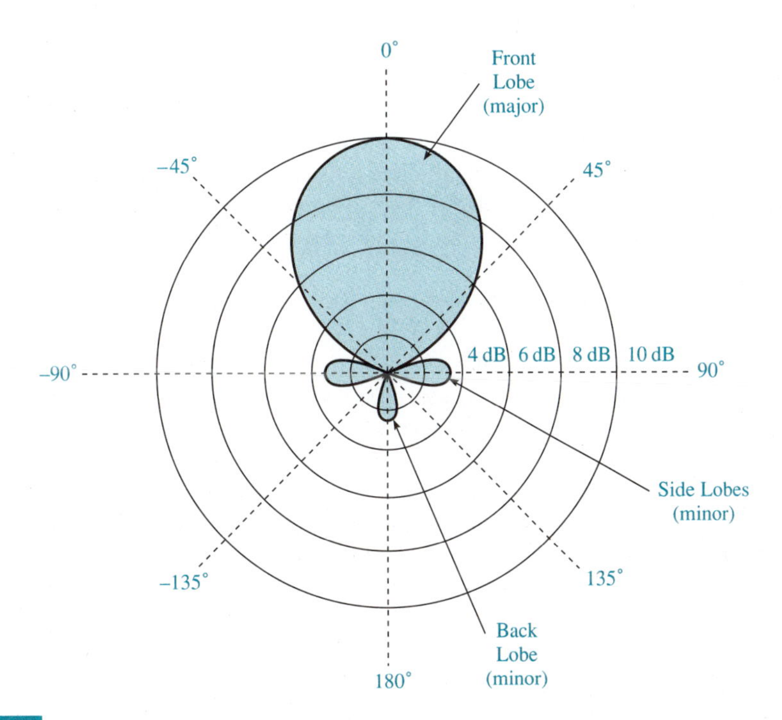

Figure 4.5 Antenna radiation pattern.

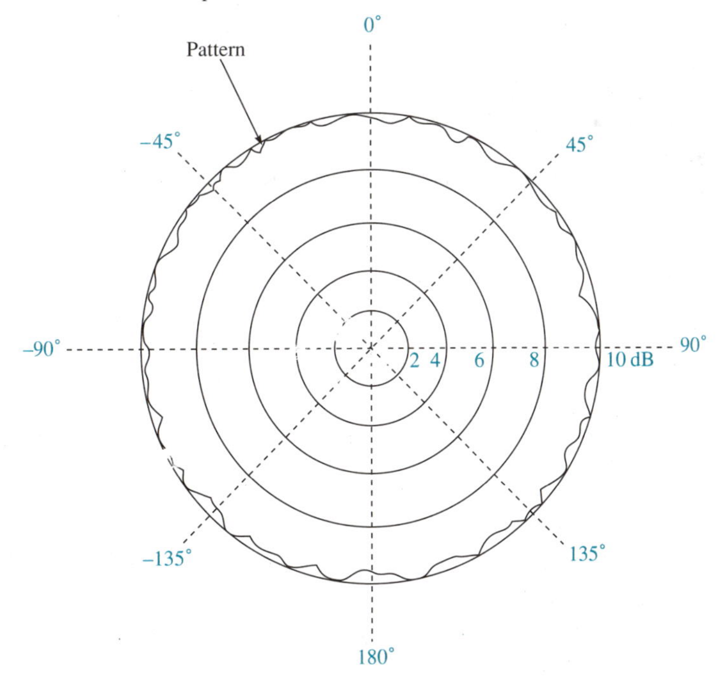

Figure 4.6 Omnidirectional radiation pattern.

4.3.2 Directive Gain

The first radiation pattern mentioned above was one for a directive antenna. The term **directive** means that for both the radiation pattern and the gain, some antennnas will radiate more energy in a specific direction out from the antenna and, consequently, will have more gain in that direction. **Directional gain** is the ratio of the power radiated in a specific direction to the power radiated to the same point by a reference antenna. The reference antenna we refer to here is generally an isotropic antenna. You will recall that we previously discussed the term *isotropic* when dealing with the propagation of electromagnetic waves. To refresh your memory, we called an **isotropic source** one that radiates equally in all directions. Similarly, an **isotropic antenna** is one that has the same gain in all directions. Thus, when we use this type of antenna and compare its performance with that of an actual antenna, we have a reliable and repeatable standard to go by.

The maximum directional gain of an antenna is termed its **directivity**; that is, basically, how well an antenna will send energy out in a specific direction and only that direction. If we express the directivity mathematically, we have

$$D = \frac{P}{P_{\text{ref}}} \tag{4.1}$$

Where: D = Directivity

P = Power of a given point for the test antenna

P_{ref} = Power of same point for the reference antenna

4.3.3 Power Gain

Power gain is very similar to directive gain, except the total power fed to the antenna is used. For this type of calculation, the antenna efficiency is used; it is also assumed that the antennas have the same input power and that the efficiency of the reference antenna is 100% (**lossless**). If we call the power gain A_p and use the directivity D and the antenna efficiency η, we now have Equation (4.2).

$$A_p = D\eta \tag{4.2}$$

If the antenna is lossless, then it radiates 100% of the input power. Under these conditions the power gain A_p will equal the directive gain D. To obtain the power gain in decibels, simply take the log of A_p and multiply it by 10.

Example 4.1 We have an antenna efficiency of 55%, power of a point for a test antenna of 60 W, and power for a reference antenna of 6 W. Find the power gain.

Solution Using Equation (4.1), we find the directivity

$$D = \frac{P}{P_{\text{ref}}}$$

$$= \frac{60}{6}$$

$$= 10(10 \text{ dB})$$

Now use Equation (4.2) to find the power gain.

$$A_p = D\eta$$

$$= 10(.55)$$

$$= 5.5(7.4 \text{ dB})$$

4.3.4 Near and Far Fields

The concept of near and far fields is one you need to understand in order to determine what is happening at the antenna and well out from the antenna. It is often of great importance to know the characteristics of an antenna some distance away from the actual device, whereas at other times you must characterize it close in. This is because the fields are not the same close to the antenna and far away from it.

By **near field**, we mean the fields generated by the antenna that are very close to the antenna. This type of field is also sometimes called the **induction field** because of the induction action that takes place close to the antenna. During the first half of a cycle, the power is radiated from the antenna as you would expect, and some of this power is temporarily stored in the near field. During the second half of the cycle, the stored power is returned to the antenna. The action is very similar to an inductor storing and releasing energy, as mentioned previously.

The **far field** is actually anything beyond the near-field area. It is, however, generally the area where the energy is a great distance away from the antenna. Once the energy reaches the far field, it continues to radiate out into space for some distance, and for this reason this field is often called the **radiation field**. The energy in the radiation (or far) field never returns to the antenna, as some of the energy in the near field does; it just keeps going.

As a measure of where the fields may be, consider that the near field is defined as the area within a distance equal to D/λ, where D is the antenna diameter, and λ is the wavelength of the frequency being used. Units of these parameters must be the same. If, for example, we had a 3-ft antenna operating at 500 MHz, we would have a near-field distance of 1.52 ft. (The wavelength at 500 MHz is 1.97 ft.) Thus, any energy that is beyond approximately 2 ft will be considered to be in the far field, or radiation field, and will radiate out into the air, never to return to the antenna. Some of the energy within approximately 2 ft may return to the antenna by induction.

4.3.5 Radiation Resistance

As is the case with every electronic circuit or component, not all of the power supplied to a unit is used. This is the case with antennas also. Not all the power applied to the input is radiated out into space; there is a certain amount that is converted into heat. For the power to change to heat, there must be a resistance that satisfies the basic relationship of $P = I^2 \times R$.

The resistance that produces the heat is called the **radiation resistance**. It is a "dynamic" resistance in that it cannot be measured with a meter. It is equal to the ratio of the power radiated by the antenna to the square of the current at the antenna feed point. Expressed mathematically, it is

$$R_r = \frac{P}{I^2} \tag{4.3}$$

Where: R_r = Radiation resistance

P = Power radiated at the antenna (rms)

I = Antenna current at the feed point (rms)

Another way to characterize radiation resistance is to say that it is the value of resistance that, if it replaced the antenna, would dissipate the exact amount of power that the antenna would radiate.

4.3.6 Antenna Efficiency

An antenna, like any other device, is not perfect. As we saw in our discussion on radiation resistance, there is a certain amount of input power that does not get radiated. This power is dissipated and lost. The ratio of radiated power to the total input power is a measure of the **antenna efficiency** η. Expressed mathematically, we have

$$\eta = \frac{P_r}{P_r + P_d} \times 100 \tag{4.4}$$

Where: η = Antenna efficiency (percentage)

P_r = Radiated power

P_d = Dissipated power

If we were to express the efficiency only in terms of the resistances of the antenna (R_r and R_{dc}) it would be

$$\eta = \frac{R_r}{R_r + R_{dc}} \tag{4.5}$$

Where: R_r = Radiation resistance

R_{dc} = dc antenna resistance (dissipation)

Example 4.2

Given: Radiated power from an antenna is 150 W, dissipated power is 1.8 W, and radiation resistance is 10 Ω.

Find: The dc antenna resistance.

Solution

1. The first step is to find the antenna efficiency (Equation 4.4)

$$\eta = \frac{P_r}{P_r + P_d} \times 100$$

$$= \frac{150}{150 + 1.8} \times 100$$

$$= 0.988 \times 100$$

$$= 98.8\%$$

2. Next, use Equation (4.5) to find R_{dc}.

$$\eta = \frac{R_r}{P_r + R_{dc}}$$

$$0.988 = \frac{10}{10 + R_{dc}}$$

$$R_{dc} = 0.012\ \Omega$$

4.3.7 Effective Isotropic-Radiated Power (EIRP)

The **effective isotropic-radiated power** (EIRP) figure given for communications systems is the equivalent transmit power that an isotropic antenna would have to radiate to result in the same power density in a certain direction and at a given point as a practical antenna. As you will recall, isotropic antennas send energy out in all directions at the same level. This is the standard we use to characterize all other antennas. If we use an isotropic antenna as the standard and then measure an actual antenna, we will then get out EIRP value. If, for example, we have an antenna with a power gain of 15, the measured antenna's power is 15 times greater than it would be for an isotropic antenna. In order for the isotropic antenna to duplicate the performance and power density of the measured antenna, it would have to radiate 15 times as much power in the required direction. EIRP may also be referred to simply as **effective radiated power** (**ERP**).

This equivalent power can be expressed as

$$\text{EIRP} = P_r A_t \qquad (4.6)$$

Where

P_r = Total radiated power

A_t = Antenna directive gain (This is the gain in a specific direction.)

To find the EIRP in decibels above 1 mW, we use the following equation:

$$\text{EIRP} = 10 \log \frac{P_r}{0.001} + 10 \log A_t \quad (4.7)$$

Example 4.3

Using the radiated power P_r from Example 4.2, and a directive gain A_t of 795, find the EIRP in decibels above 1 mW.

Solution

Using Equation (4.7)

$$\begin{aligned} \text{EIRP} &= 10 \log \frac{P_r}{.001} + 10 \log A_t \\ &= 10 \log \frac{150}{.001} + 10 \log 795 \\ &= 10 \log 150000 + 10 \log 795 \\ &= 10(5.176) + 10(2.9) \\ &= 51.76 + 29 \\ \text{EIRP} &= 80.76 \text{ dBm} \approx 119 \text{ kW} \end{aligned}$$

Equation (4.8) will provide the power density at a given point to enable you to compare the antenna with an isotropic antenna.

$$P = \frac{P_r A_t}{4\pi R^2} \quad (4.8)$$

Where: R = Distance away from the radiator

Since an antenna is a reciprocal device, as we mentioned previously, it will have the same power gain and directivity in the receiving direction as it has in the transmitting direction. Thus, the power that is received, or captured, at a receiver antenna is the product of the power density in a specific area and the antenna gain. Therefore, we can apply Equation (4.8) to a **capture condition**.

$$c = \frac{P_r A_t A_r}{4\pi R^2} \quad (4.9)$$

4.3.8 Polarization

The type of **polarization** associated with an antenna simply refers to how the electric field that is radiated is oriented. If it is oriented in a vertical or horizontal position, it is a *vertically* or *horizontally polarized* antenna, respectively. If the electric field is oriented in a circular direction, it is *circularly polarized*, and if the electric field seems to be in a circular direction but the vertical and horizontal excursions are not the same, it is *elliptically polarized*. Each of these types of polarization have specific applications and should be used accordingly. It is very important to remember that in order to receive a signal from a specific antenna, the receiving antenna must be the same polarization as the transmitting antenna; otherwise, they will not be compatible. Some antennas can switch to different types of polarization. One typical example is changing polarization to eliminate the effects of raindrops on electromagnetic energy being returned to aircraft radar.

4.3.9 Beam Width/Bandwidth

As we discussed many times in this chapter, for antennas to be directional they must radiate energy in a small area and in a specific direction. This small area is the **beam**, or **beam width**, of the electromagnetic energy radiated out into space, or received back from a transmitter. The beam width of an antenna is expressed at 3 dB, as shown in Figure 4.7. The figure is a portion of the radiation pattern for an antenna. You can see that the maximum power is at the outer edge of the pattern, as it always is, and that there are steps of 0 dB and −3 dB that go around the pattern. Where the radiation pattern crosses the 3 dB line is where the beam width is measured. The actual beam width is the difference between points *A* and *B* in Figure 4.7.

Beam width is important in an antenna if you are planning a long-range system. Consider the fact that there is only a certain amount of power being radiated, and if that power is spread over a very wide beam width, the overall power level will be rather low. If you concentrate as much of that radiated power as possible into a very narrow beam width, the result will be an overall increase in power in the narrow direction you require and, consequently, a much higher power level radiated out.

Bandwidth, a term used along with beam width, is defined in some texts as the frequency range over which the antenna will operate. This, however, is a vague and unsatisfactory definition; it is actually the frequency band over which an antenna will

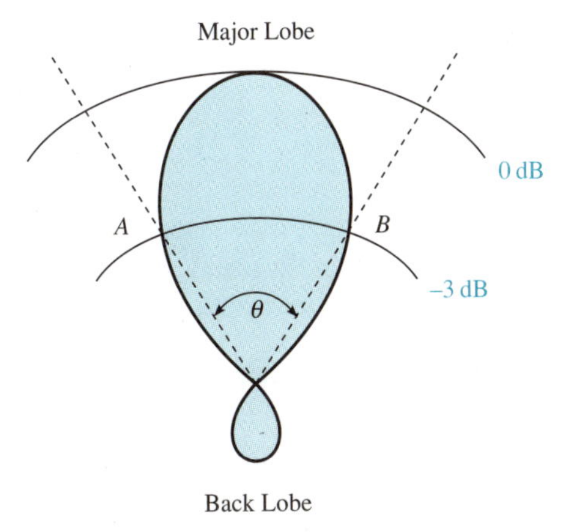

Figure 4.7 Antenna beam width.

operate properly. That is, we find the frequency where the maximum radiation occurs, decrease the frequency until the power falls off by 3 dB, and read the frequency (f_1). We then repeat the procedure for the upper frequency 3 dB point (f_2) and take the difference. This results in $f_2 - f_1$ as the antenna bandwidth.

4.3.10 Input Impedance

Antenna input resistance is actually the feed point of the antenna. It presents an ac load to the transmission line that feeds the antenna. Thus, the same conditions that occur to a load for any other transmission line also apply when attaching an antenna: the impedance (resistance) of the load (the antenna) must equal the impedance of the source (the transmission line Z_0) for a maximum transfer of power to occur. The antenna input impedance can be expressed as follows:

$$Z_{\text{in}} = \frac{E_i}{I_i} \tag{4.10}$$

Where: Z_{in} = Input impedance

E_i = Antenna input voltage

I_i = Antenna input current

The input impedance is generally a complex number ($R + jX$). However, if the impedance is only a resistive number, it can be expressed as the sum of the radiation resistance R_r and the dc resistance R_{dc}.

4.4 Antenna Types

Now that we have defined a wide variety of terms that apply to antennas, it is time to look at some of the typical antennas used for communications systems and discuss their operations and applications.

4.4.1 Half-Wave Dipoles

In Section 4.2 we introduced the half-wave dipole, or Hertz antenna, and used this device to show the basic principles of antennas. In reality, it is many times the basic unit that makes up a much more complex antenna array, as we shall see later in this text.

As mentioned previously, the half-wave antenna is an open-circuited transmission line with two ends that are each one-quarter wavelength long. In Figure 4.8, which shows the construction of the half-wave dipole, you can see that the open transmission line has two segments (or poles) that are each a quarter-wavelength long, with the feed point to the antenna at the intersection of these two poles. The poles of this type of antenna are usually made of a hollow tubing cut to the desired length.

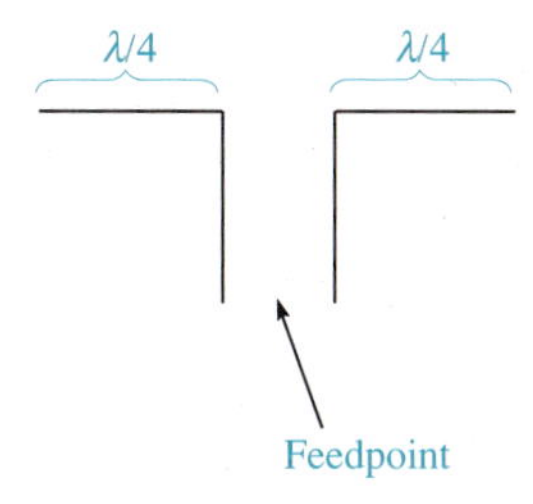

Figure 4.8 Half-wave dipole.

Theoretically, the length should be a total of one-half wavelength for both of the poles; however, as we stated before, the actual length should be 5% less than this, or 0.95 times the wavelength over 2. This now becomes 0.48 rather than 0.5λ.

Figure 4.9 shows the radiation pattern for a half-wave dipole antenna. Figure 4.9(a) is a picture of a vertically mounted dipole, showing the pattern as we look

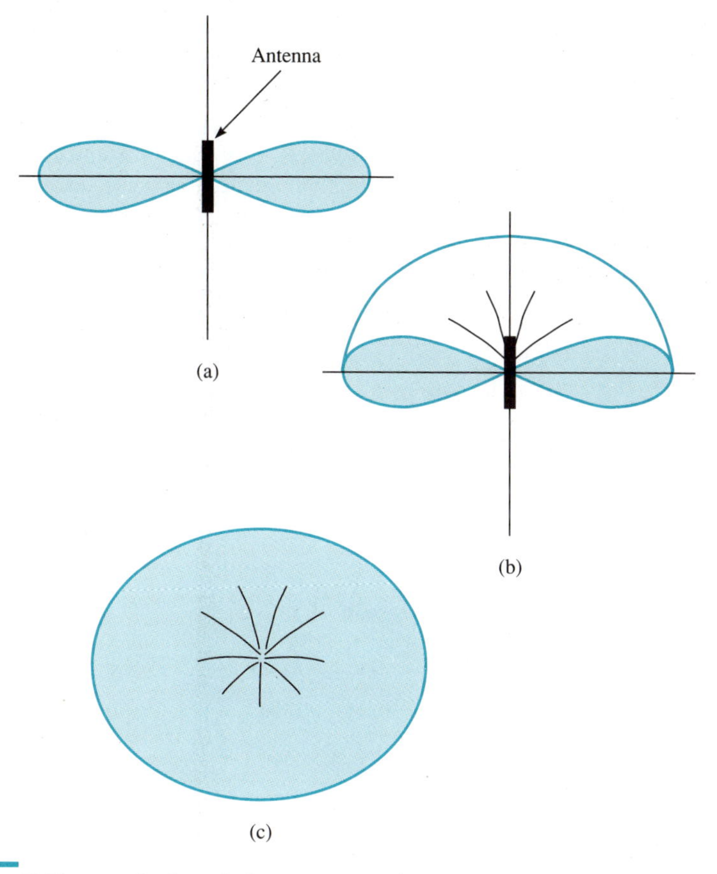

Figure 4.9 Half-wave dipole radiation patterns.

directly into the antenna. It looks like a figure 8 with minimum amplitude in the center and maximum amplitude at each of the ends. Figure 4.9(b) shows a three-dimensional view of the vertically mounted antenna which takes on the familiar ''doughnut'' pattern that most people regard as the pattern for simple antennas. Figure 4.9(c) makes this concept even clearer by looking at the antenna pattern from the top of the display; it looks at Figure 4.9(b) from the top rather than from the side.

The patterns shown in Figure 4.9 are for free space. When in the earth's atmosphere, these patterns are affected by how the antenna is mounted, general atmospheric conditions, and such ground effects we spoke of previously, such as reflections. You will recall that reflections cause two signals to arrive at a receiving antenna, cancelling some signals, while reinforcing others. The problem of reflections can be eliminated by mounting the antenna high enough off the ground, or by having the transmitting and receiving antenna sufficiently far apart to reduce the effects significantly. Unfortunately, these measures are not always possible. You must therefore, live with what happens and compensate for the effects. An illustration of what happens to a half-wave dipole antenna when it is close to the ground is shown in Figure 4.10. This is a vertical radiation pattern for a horizontal dipole one-half wavelength above the ground. This is one-half of the pattern and would be similar to using one-half of the pattern in Figure 4.9. The pattern in Figure 4.9 is what the antenna should look like. The pattern in Figure 4.10 actually shows two lobes in the pattern where it has been split with the direction of maximum radiation now being at about 30° rather than the ideal 90° as would be shown in Figure 4.9 if we rotate the antenna to a horizontal placement. Although this effect is generally thought of as being undesirable, there are times when the angle of maximum radiation may need to be something other than the ideal. At these times the ground effects may be put to good use to adjust for the angle of radiation that is best for your application.

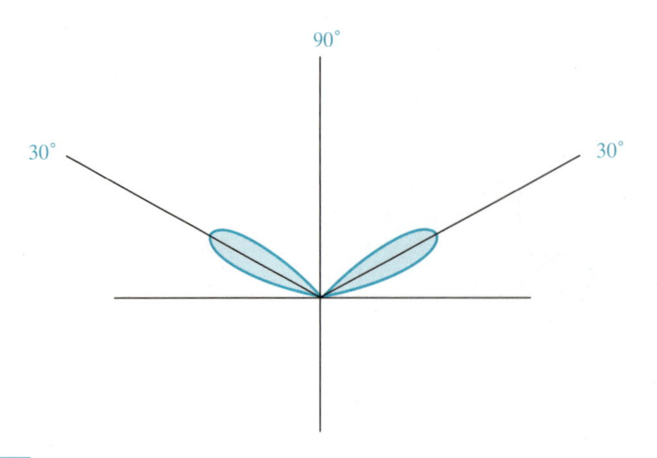

Figure 4.10 Radiation pattern.

4.4.2 Antenna Arrays

Generally an antenna has more than one set of elements visible on it, because most often a single-element antenna will not be adequate to do the task. This means that a series of antenna elements, or an array, must be used. An **array** is a series of antenna elements combined to form a single antenna unit. The basic array consists of a driven element, such as the Hertz antenna; a reflector element; and a director element. These three sections are shown in Figure 4.11.

The main purpose of an array is to increase the directivity of the antenna and concentrate the power in a certain direction for accuracy and range. To accomplish this task, there must first be some sort of driven element as the basic antenna for the array. Driven elements are those that are connected to the transmission line and receive an input power from some source such as a power amplifier. As we have said, this driven element is typically a Hertz (half-wave dipole) antenna, which you will see often in very complex arrays because of its effectiveness.

The other two basic elements of the array, the reflector and director, are termed **parasitic elements**, because they are not connected to a transmission line but receive energy through a mutual induction process with the driven element. The **reflector element** is longer than the driven element, and it acts as a mirror, reflecting the energy from the driven element back in the desired direction. The reflector takes care of the energy that is radiated back from the driven element so that it is not lost in the back of the array. Figure 4.12 shows the relationship between the driven element and the reflector element and how the reflector acts as a reflective mirror.

The **director element** is shorter than the driven element, and it focuses the energy from the driven element into a beam that is in the direction of interest. Thus, it acts as a convex lens. You recall from high-school physics, that a convex lens takes a broad spectrum of light and focuses it down at the focal point of the lens (See Fig. 4.13). The same is true for electromagnetic energy: a spectrum of energy is applied

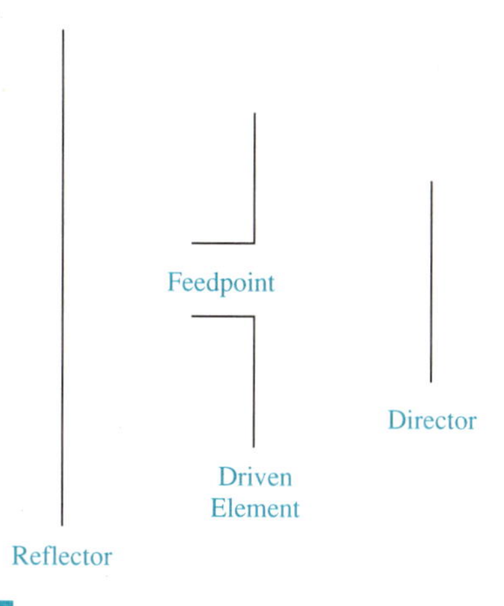

Figure 4.11 Antenna array.

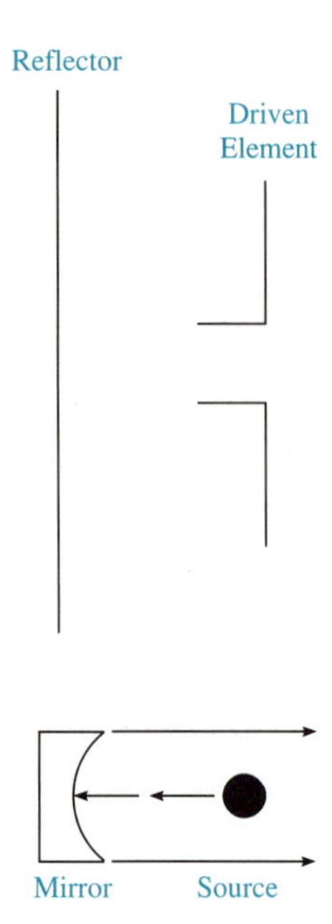

Figure 4.12 Driven and reflector elements.

to the effective lens and is focused into a single beam in a specific direction. Figure 4.14, which shows the relationship of the driven element and the directive element, illustrates how a diverse energy beam is focused into a narrow beam in the specified direction. Thus, we now have a system that takes the energy radiated by a Hertz antenna, for example, focuses it by means of a reflecting element behind the antenna and a directive element in front of the antenna, and finally radiates a much narrower beam in a very selective direction. The directivity of these arrays can be increased greatly by adding more elements to the existing array.

Two types of arrays are shown in Figure 4.15. The first (a) is a **broadside array**, which consists of several dipoles connected in parallel, with each of the elements fed in phase from the same driving source. Radiation is at right angles to the plane of the array as the antenna pattern shows in Figure 4.15(a). There is actually very little

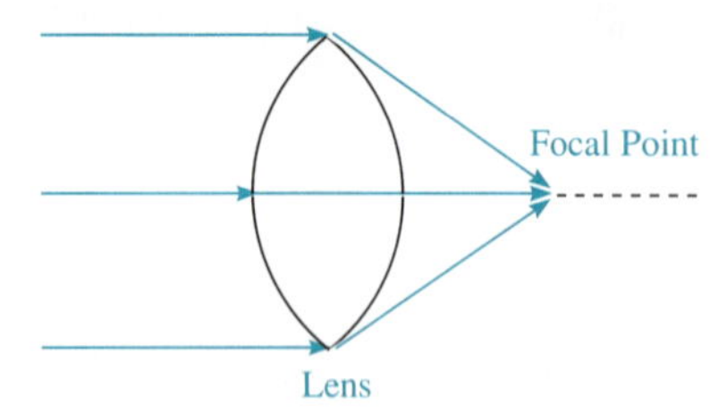

Figure 4.13 Convex lens.

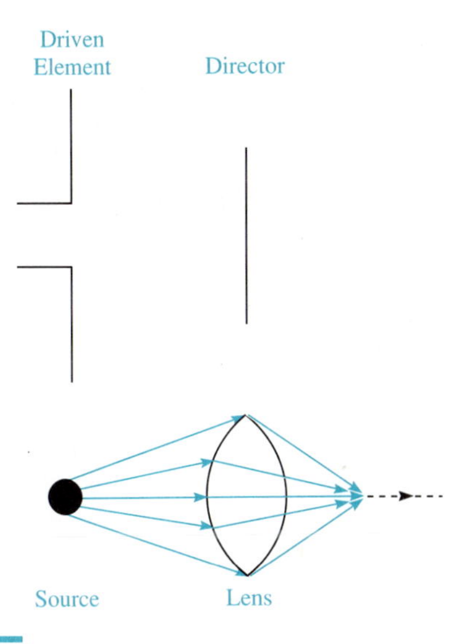

Figure 4.14 Driven element and director.

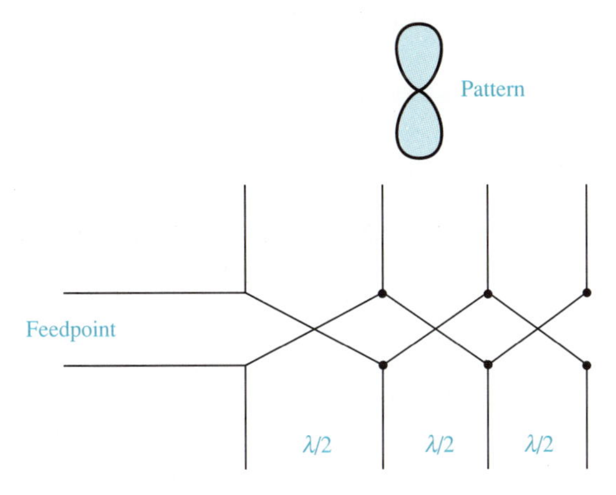

(a) Broadside Array

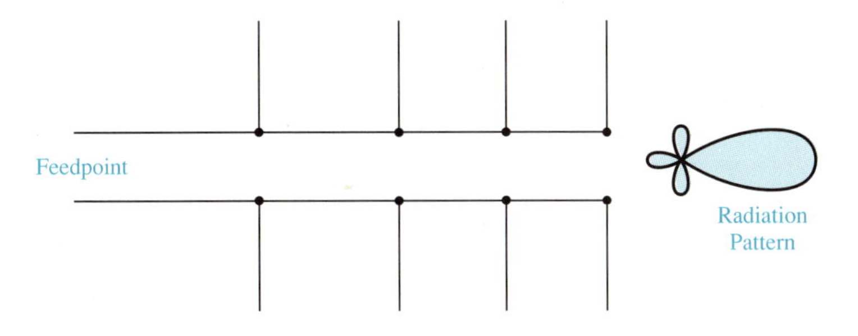

(b) End-Fire Array

Figure 4.15 Antenna arrays.

radiation in the actual plane of the antenna. This is where the name broadside array is derived. The in-phase characteristics of a broadside array come from the fact that the source radiates energy into the first antenna, the energy travels one-half wavelength (180° phase shift), then is again shifted by 180° by the crossing of the input connections to the next element. This produces an in-phase condition for all of the antenna elements. When all of the elements are added together the energy is added at right angles to the array and acts to increase the level of the signal being transmitted. If more elements are added, the directivity of the array is increased.

The second type of array shown in Figure 4.15(b) is the **end-fire array**. This array is basically the same as the broadside array, with the exception that the inputs to the different elements are not reversed. This results in the fields being additive in the plane of the array, as seen in the radiation pattern shown in Figure 4.15(b).

4.4.3 Log-Periodic Antenna

The log-periodic antenna is a good example of how arrays can enhance the performance of an antenna. This type of antenna is a special case of an array that finds many applications in communications systems, and it should be rather familiar to you since it is seen on many roofs in many cities. Figure 4.16 shows the basic log-periodic structure.

The **log-periodic antenna** is one whose bandwidth ratio is many times in excess of 10:1 due to its radiation resistance and pattern being independent of frequency. (By *bandwidth ratio* we mean the ratio of the antenna's highest frequency to its lowest frequency.) This type of antenna can be either unidirectional or bidirectional and can have a low to moderate gain value. These antennas can have a higher gain value if they are used as an element in a more complex array. Figure 4.16 shows that the

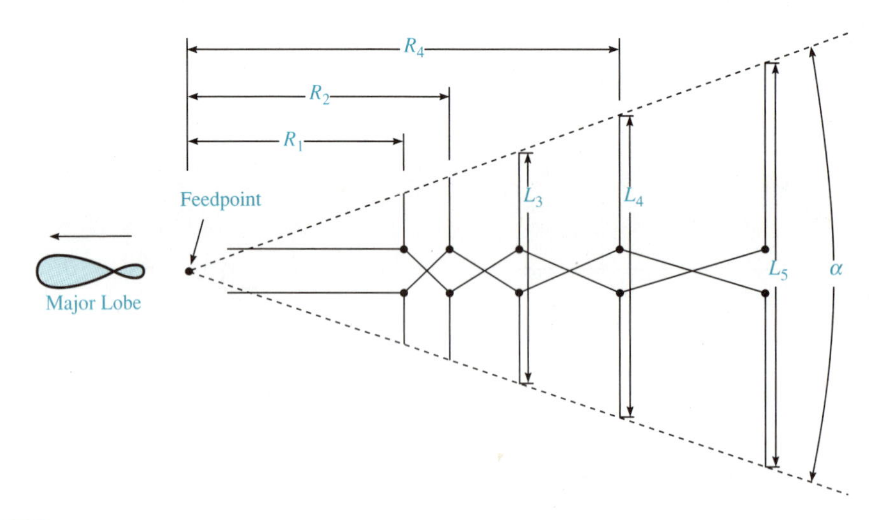

Figure 4.16 Log-periodic antenna.

basic structure of the log-periodic antenna is similar to the broadside array in that it is a series of dipole antennas connected in series. This, however, is where the comparison ends. Examination of the figure shows that the lengths and the spacings between the dipoles is not the same, but rather increase as we progress through the array. These lengths and spacings are related such that adjacent elements have a constant relation to each other, as follows:

$$\frac{R_2}{R_1} = \frac{R_3}{R_2} = \frac{R_4}{R_3} = \frac{1}{\tau} = \frac{L_2}{L_1} = \frac{L_3}{L_2} = \frac{L_4}{L_3} \tag{4.11}$$

$$\frac{1}{\tau} = \frac{R_{n+1}}{R_n} = \frac{L_{n+1}}{L_n}$$

Where: R = Dipole spacing

L = Dipole length

τ = Design ratio (< 1)

The ends of the dipoles lie along a straight line, and the angle where they meet is designated by α. For a typical antenna design, $\tau = 0.7$ and $\alpha = 30°$.

The antenna is called log-periodic because the antenna impedance varies repetitively when plotted as a function of frequency, and varies periodically when plotted against the log of the frequency. Thus, the term **log-periodic** results.

4.4.4 Parabolic Reflectors

Often the antennas used for transmitting and receiving in communications systems cannot provide adequate range for that system. At these times one means of increasing an antenna's capability is by reflecting the signal that it either transmits or receives. The device that is the most efficient in so reflecting the signal is the **parabolic reflector**.

The problem faced when constructing a reflective surface is getting the largest amount of transmitted energy to travel forward in a narrow band. This difficulty is similar to that encountered when building a searchlight reflector or a reflecting telescope. Whenever light or electromagnetic waves strike a reflecting surface, they are reflected in a way familiar to us all: *The angle of incidence equals the angle of reflection*. That is, if a wave strikes a curved surface at an angle made with an imaginary line perpendicular to that surface, a normal line, the wave reflects in the opposite direction at an angle exactly equal to that with which it hit. There is, however, a difference between spherical and parabolic surfaces, as shown in Figure 4.17.

In Figure 4.17(a) we see that when a spherical surface is used as a receiving antenna, the results are very random. When the energy strikes the surface at any point on the sphere, it is bounced around with no particular order to it. In contrast, Figure 4.17(b) shows a parabolic surface where a receiving antenna is placed at the focal point of the parabola. Here, you can see that all of the energy striking the parabolic surface is directed to the receiving antenna at the focal point—a very efficient arrangement.

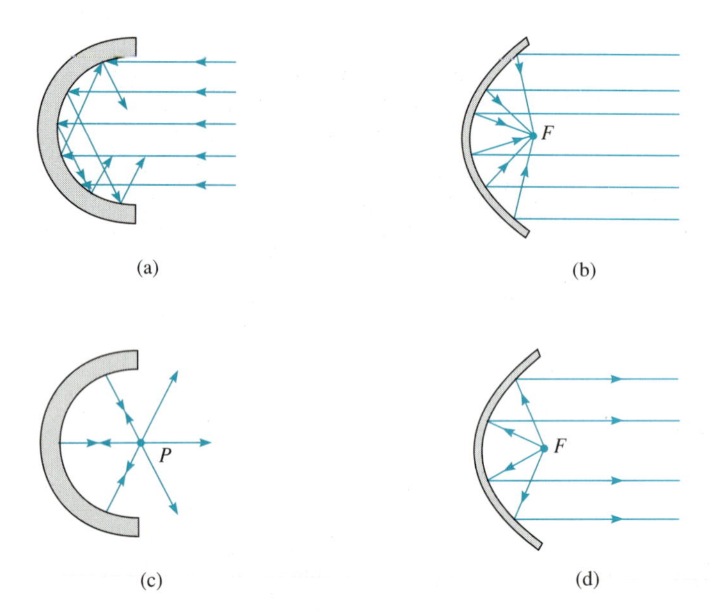

Figure 4.17 Spherical and cylindrical surfaces.

Figures 4.17(c) and (d) show the same spherical and parabolic surfaces for the transmitting case. In Figure 4.17 (c) there is a transmitting antenna placed at point P around the spherical surface. This results in a random form of transmission with regards to direction. As a matter of fact, there are some areas where the transmitted energy is actually cancelled out by reflections off the surface. Figure 4.17(d) shows a parabolic reflector with the transmitting antenna placed at the focal point once again. The energy transmitted from this antenna is all directed out in the desired direction. Thus, you can see that the parabolic reflector is much more than something that looks impressive on the top of a building; it is a very effective and efficient method of energy transmission for communications systems.

4.4.5 Helical Antennas

A *helix*, by definition, is a spiral. A **helical antenna**, therefore, features a wire wound in the shape of a screwhead, or spiral. Most often, this type of antenna is used at relatively high frequencies, in order for the dimensions to be large compared with wavelength.

Figure 4.18 shows a helical antenna. The helix, or spiral, is fed at one end and is usually connected to the center of a coaxial transmission line, whose outer conductor is attached to the ground plane. The basic geometry of the helix is described in terms of the diameter D and its turn spacing S. For an N-turn helix, the total length of the antenna is equal to NS, and its circumference is πD. The length of the wire per turn is then

$$L = \sqrt{S^2 + \pi D^2} \tag{4.12}$$

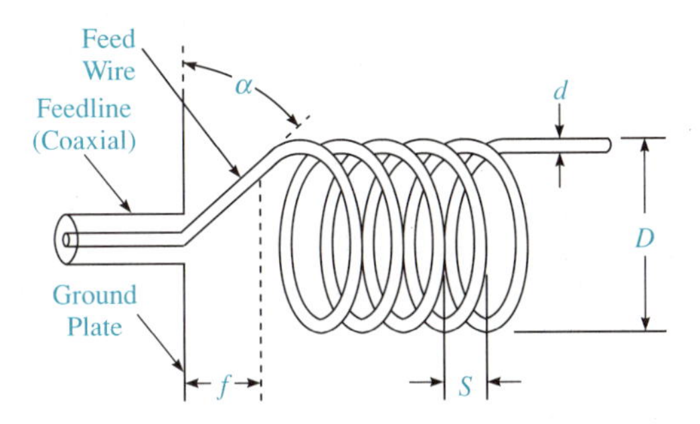

Figure 4.18 Helical antenna.

The pitch angle α is that angle made between a line tangent to the helix and the plane perpendicular to the helical axis (as shown in Fig. 4.18). The angle is equal to $\tan^{-1} \dfrac{S}{\pi D}$.

A typical helical antenna operating in the axial mode has a circumference of approximately one full wavelength and a spacing of approximately one-quarter wavelength. The pitch angle may range from 12 to 18°, with 14° being considered optimum. The gain and beam width depend on the number of turns and the spacing between them, that is, on the total length of the helix. The feed-point impedance is resistive and on the order of 100 Ω at the frequency where the circumference equals one wavelength.

In some applications it is necessary to note if the helix is wound with a right-hand or left-hand pitch, because the orientation determines whether there is right or left-hand circular polarization. Helical antennas have found considerable use in space telemetry at the ground stations. The circular polarization in the helix is useful in space transmission because the ionosphere causes the waves to be rotationally polarized.

4.5 Summary

This chapter introduced the theory of antennas. It emphasized that without an antenna, there will be no communications system. Antenna basics were presented to familiarize you with their terms and applications. Following this introduction, such terms as radiation pattern, directive and power gain, near and far fields, radiation resistance, efficiency, EIRP, polarization, and beam width/bandwidth were covered.

Specific types of antennas were then presented, such as dipoles, arrays, log-periodic, and helical. Parabolic reflectors, which are used in conjunction with many different types of antennas, also were discussed.

Questions

4.2 Antenna Basics

1. Why do we call the Hertz antenna a dipole?
2. What is a Marconi antenna?
3. Why is an antenna a reciprocal device?
4. Define **dipole**.
5. How is an open ended transmission line an antenna?
6. Explain how an antenna is an impedance matching device.

4.3 Antenna Terminology

7. Why is a radiation pattern a relative measurement?
8. Define **isotropic source**.
9. Define **near** and **far fields**.
10. Define **EIRP**.
11. Name three types of antenna polarization.
12. Define **directive gain**.
13. Why is an isotropic antenna used as a reference for measuring directive gain?
14. What is **directivity** of an antenna?
15. How does **power gain** differ from **directive gain**?
16. What is another term used for **near field**?
17. What is another term used for **far field**?
18. What is the **radiation resistance** of an antenna?
19. What is the expression for **antenna efficiency**?
20. What terms, beside power, are used to calculate antenna efficiency?
21. If the antenna directive gain of a system is increased, what effect does this have on EIRP?
22. Define **beam width**.
23. Define **bandwidth** of an antenna.
24. Why is the feed point input resistance of an antenna an important consideration?

4.4 Antenna Types

25. Describe an **antenna array**.
26. What can a dipole antenna be related to?
27. The half-wave dipole radiation pattern has a very distinguishable shape to it. What is it?
28. What is the purpose of the **director** in an antenna array?
29. What is the purpose of the **reflector** in an antenna array?
30. How is a **broadside array** constructed?
31. How is an **end-fire array** constructed?
32. Is the log-periodic antenna unidirectional or omnidirectional?
33. Describe the construction of a log-periodic antenna.
34. Why is a parabolic reflector much better to use than a spherical reflector?
35. Define **helix**.
36. What is the **pitch angle** in a helix antenna?
37. What determines whether a helix antenna has a right-hand or left-hand pitch?

Problems

4.3 Antenna Terminology

1. If the power measured at a given point on a test antenna is 5 mW, the power at a reference antenna is 0.5 mW, and the efficiency is 85%, what is the power gain of the antenna?
2. If the power radiated from an antenna is 5 W and the power dissipated is 1 mW, what is the efficiency?
3. If a transmitter antenna has a directive gain $A_t = 10$ and radiated power of 150 W, determine the EIRP, and the power density 15 km away.
4. If the radiated power from an antenna is 150 W and the dissipated power is 2.5 W, what is the antenna efficiency?
5. An EIRP of 65 dBm is required. If the radiated power is 100 W, what antenna directive gain is needed?
6. An antenna must have a minimum efficiency of 80%. If the radiated power is measured as 350 W, what is the maximum amount allowable for the dissipated power?
7. If the system in Problem 6 has a tolerance on the efficiency of $\pm 5\%$, what is the range of dissipation power assuming that the radiated power remains constant?
8. The system presented in Problem 3 encounters a problem where the directive gain of the antenna falls to only 5. What is the new EIRP and power density at 15 km?

4.4 Antenna Types

9. If the design ratio τ for a log-periodic antenna is changed from the ideal value of 0.7 to 0.5, what effect does this have on the dipole lengths and spacings of the antenna?
10. The pitch angle α of a helical angle is ideally 14°. If the diameter of the helix is held to a constant 2.2 cm, what spacing must be incorporated to result in an ideal pitch angle?
11. For the conditions in Problem 10, what is the length of wire per turn L?

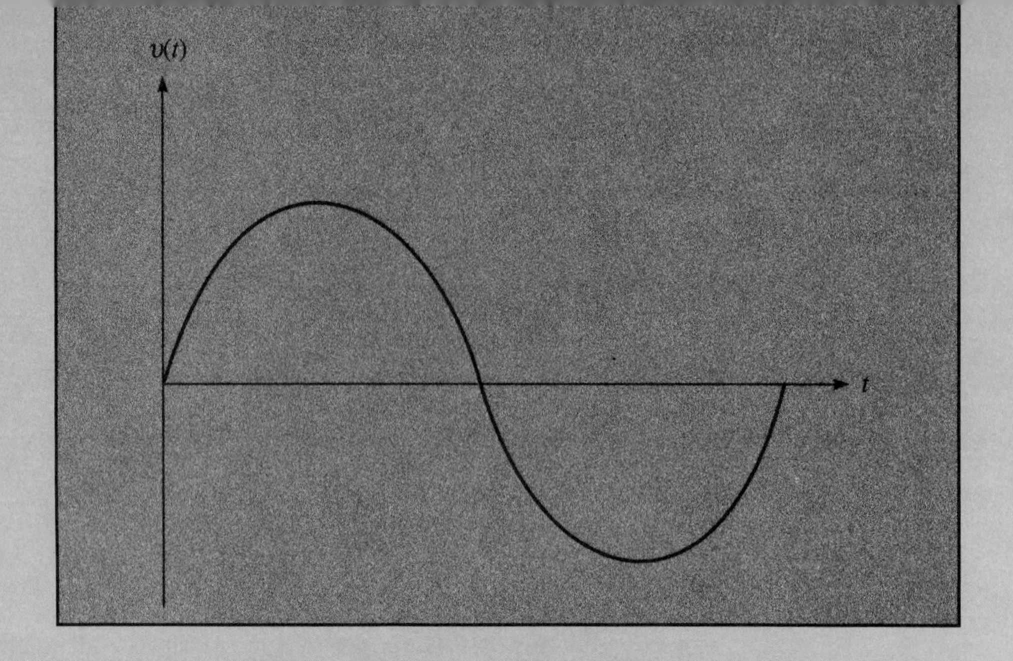

5

Outline

Objectives

- To introduce the spectrums of individual signals
- To distinguish the difference between time and frequency domain and where each is used in communications
- To introduce the Fourier series and the use of the series without the need for calculus
- To introduce distortion and the method to calculate its percentage in specific systems

Key Terms

spectra
frequency domain
symmetry
distortion
pulse width
repetition rate
time domain
Fourier series
rectangular pulse
harmonic
harmonic distortion

Signal Spectra

5.1 Introduction

Many electronic communications texts regard the topics of signal spectra and Fourier series as just an extra part of an existing chapter, which virtually downgrades these topics to secondary importance. Actually, the breakdown of signals into their spectral components is an important way to understand what happens to a communications signal. It would be very difficult indeed to teach students about communications and the signals that make up communications systems if it were not possible to tear each of the signals apart and analyze their content. It is important that all communications signals, no matter how complex, can be broken into elementary signals that can be analyzed by means of the Fourier series. Through this means it is possible, therefore, to analyze some very complex signals simply by knowing the basics of signal spectra and Fourier analysis, and the Fourier series.

5.2 Time Domain and Frequency Domain

In Chapter 1 we covered the differences between time domain and frequency domain. There we introduced you to the two types of displays and where they can each best be used. Now we will provide a quick review of both time domain and frequency domain so you may relate it to signal spectra and to the theory of the Fourier series.

You will recall that the concept of time domain can be represented by a single sine wave as shown in Figure 5.1. This is a signal that varies in amplitude along the y axis and in time along the x axis. Thus, it is said to be in the **time domain**. The mathematical expression for this signal is

$$v(t) = A \sin (2\pi f)t \tag{5.1}$$

or

$$v(t) = A \sin \omega t$$

Where

$v(t)$ = Voltage as a function of time

A = Peak amplitude of the signal

f = Frequency in hertz

t = Time in seconds

$\omega = 2\pi f$

(Throughout this text the term ω will be used instead of $2\pi f$.)

The instrument that is used to view the time-domain display is the oscilloscope. By placing the signal on the vertical deflection plates (vertical input), the scope provides the horizontal sweep in time that may be changed by the operator on the front panel of the scope. This adjustment, known as the **sweep time**, is usually expressed in seconds, milliseconds, or microseconds.

If we took this same signal and placed it in the frequency domain it may look like that shown in Figure 5.2. There would be an amplitude representation on the y

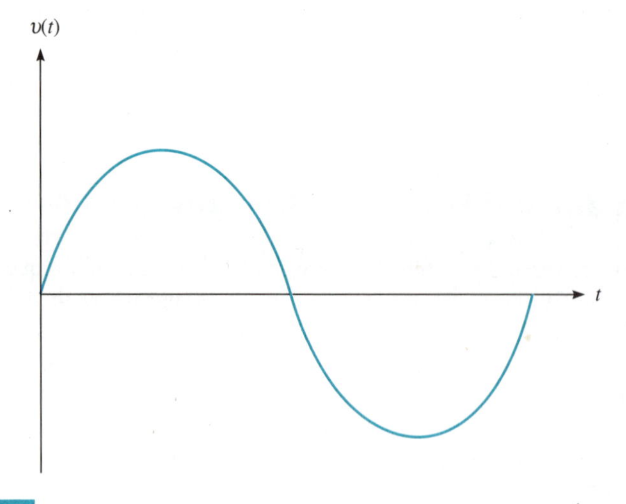

Figure 5.1 Time domain.

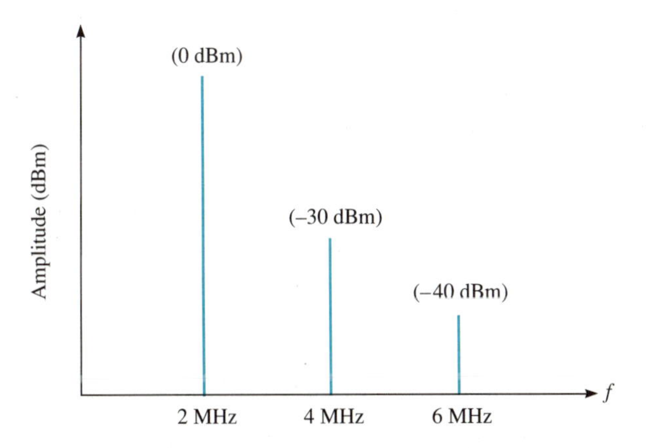

Figure 5.2 Frequency domain.

axis, just as in the time domain, but there would be a frequency representation on the x axis. The signal would consist of the fundamental (2 MHz, for example), a second harmonic (4 MHz), a third harmonic (6 MHz), and so on. This presentation shows the entire picture of the signal rather than the masked one shown in the time-domain presentation. One large advantage of the frequency domain, as was mentioned previously, is that you can see if there is any excess harmonic contribution to the signal. If, for example, the signal presentation in Figure 5.2, which shows that the fundamental was at a level of 0 dBm, and the second harmonic was at a level of -2 dBm. There is, therefore, an excessive amount of harmonic content in the signal that would probably make it undesirable for use in your system. However, the signal relationships shown in Figure 5.2, with a 0 dBm fundamental and a -30 dBm second harmonic, are acceptable.

With the spectrum analyzer, the instrument that displays the frequency domain, the signal is once again applied to the vertical deflection plates of the device, and the horizontal sweep is generated within the instrument. The operator can also control the increments that the horizontal presentation makes across the face of the cathode-ray tube (CRT). This time, however, the increments are not seconds, milliseconds, and microseconds, but increments of frequency such as 10 kHz/cm, 100 kHz/cm, or 20 MHz/cm. So the operator can tell the frequency of specific signals by obtaining a rough reading from the analyzer and then use the frequency-per-centimeter scale to make finer measurements. The amplitude on the spectrum analyzer is read using a decibels-above-1-mW dial, which tells the operator what the top of the display is (0 dBm, -10 dBm, etc.). Then the operator can use the fact that each division on the screen is 10 dB with small increments of 2 dB between to determine the actual absolute power level and frequency.

Example 5.1

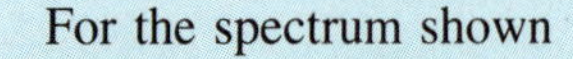

For the spectrum shown

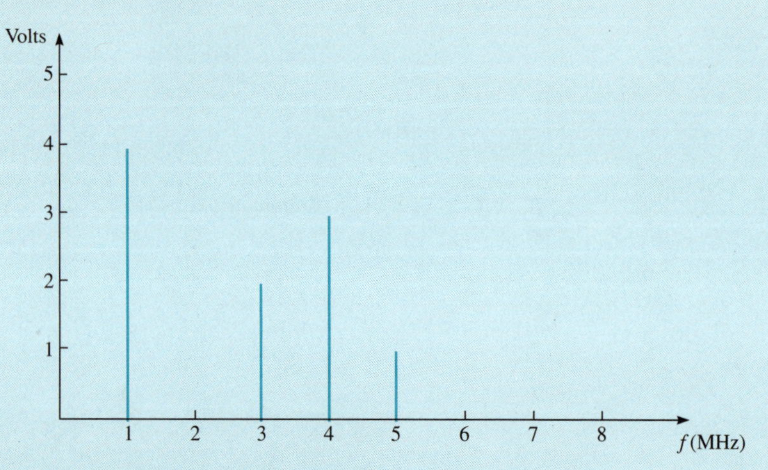

1. How many sine waves are shown?
2. Which has the greatest amplitude?
3. Is any of the signals a harmonic or any other?

Solution

1. There are four signals present: 1 MHz, 3 MHz, 4 MHz, and 5 MHz.
2. The one with the greatest amplitude is the signal at 1 MHz.
3. The signals 3, 4, and 5 MHz could all be harmonics of the 1-MHz signal.

Thus, you can once again see how advantageous it is to have signals in the frequency domain. However, you may not often have a spectrum analyzer available to view these signals, or there may only be a scope to display the signals you have. In this case you must have some method to change the time-domain display to a frequency-domain presentation. The method used to do this is a mathematical one, the Fourier series.

5.3 Fourier Series

As we mentioned, the Fourier series is the method used to change a time-domain signal to a signal in the frequency domain. We shall see later that this series will take the time representation of $v(t)$ and break it into a series of frequency components. We will also see various types of signals, which frequency components are present, and which are not.

The Fourier series was derived in 1826 by the French mathematician and physicist Jean Baptiste Joseph Fourier (1768–1830). Basically, the **Fourier series** is a mathematical technique used to break up a complicated expression into a series of simple ones that can be studied more easily than the original expression.

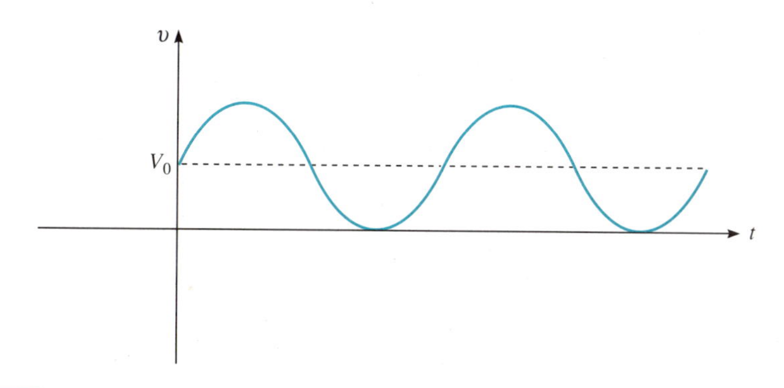

Figure 5.3 Sine wave with dc offset.

This sounds like a typical dictionary definition but if you look carefully it is simply saying that we can take complex waveforms (nonsinusoidal) and break them into their individual frequency components for simple analysis. For the sine wave originally presented in Figure 5.1, not too much analysis can occur. It simply is the relationship described in Equation 5.1; the only possible difference may be if the sine wave has a dc offset, as shown in Figure 5.3. In this case, not only must the ac portion of the signal be considered, but the dc-offset portion, as well. This would result in the following:

$$v(t) = V_0 + A \sin \omega t \tag{5.2}$$

Where: V_0 = dc-offset voltage of the signal

If we were to look at the entire frequency spectrum of a sine wave, we would find that there are harmonics present. Thus, neither Equations (5.1) or (5.2) are the complete pictures of a sine wave. The actual series for a sine wave is as follows:

$$v(t) = V_{dc} + V_1 \sin \omega t + V_2 \sin 2\omega t + \cdots + V_n \sin (n\omega t) \tag{5.3}$$

Where: n = Harmonic number

Equation (5.3) represents all of the harmonics possible with the sine wave. You can see that there is a spot for the dc component of the wave (If there is no dc offset, this term is zero), and an amplitude and frequency for each of the harmonics; 1, 2, 3, and so on. Equation (5.3) results in the presentation shown in Figure 5.4.

Thus, you can now see how we have taken a time-domain signal, such as the sine wave of Figure 5.1, which was $v(t) = A \sin \omega t$, and converted this time dependent signal into a frequency spectrum, as shown in Equation (5.3) and Figure 5.4. The process is no more complex than evaluating single sine-wave signals and adding them together.

The Fourier series is used most when the signals are nonsinusoidal in nature. The Fourier series is actually representing sinusoidal components in nonsinusoidal periodic waveforms. Such signals as square waves, triangular waves, ramps, and rectified sine waves are some of the waves that can be evaluated using the Fourier

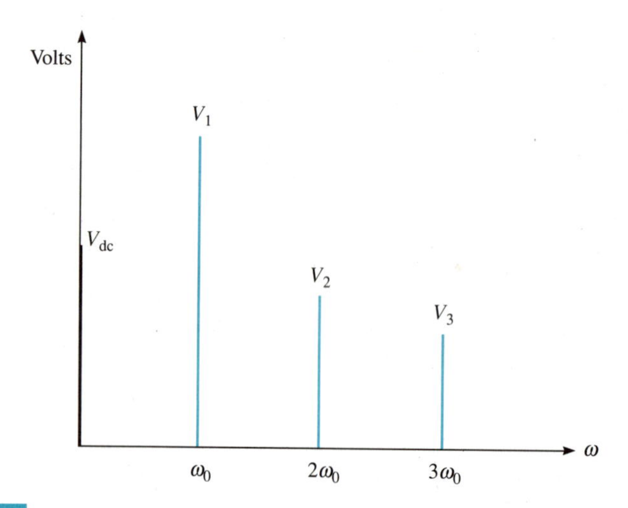

Figure 5.4 Sine wave representation.

series, but the signals must be periodic functions. This is a necessity for evaluating the signals.

The general form of the Fourier series is a series of terms such as

$$f(t) = A_0 + A_1 \cos \alpha + A_2 \cos 2\alpha + A_3 \cos 3\alpha + \cdots + A_n \cos N\alpha + B_1 \sin \beta + B_2 \sin 2\beta + B_3 \sin 3\beta + \cdots + B_n \sin N\beta \quad (5.4)$$

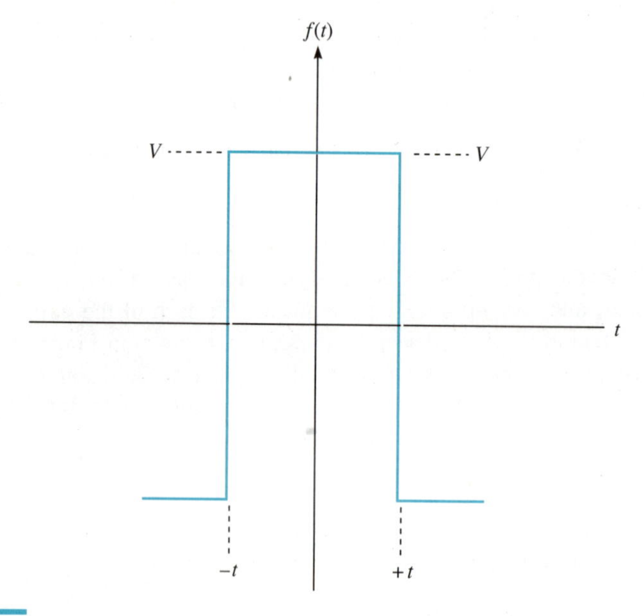

Figure 5.5 Even symmetry.

This equation shows that a Fourier series is made up of an average-value component (A_0) and a series of harmonically related (α, 2α, 3α, β, 2β, 3β, etc.) sine and cosine waves, with a **harmonic** being a multiple of the fundamental signal. From this equation it is possible to determine coefficients of any complex periodic signal.

To use the Fourier series for the signals listed, a certain symmetry must be associated with the signals. Symmetry may be *even, odd*, or *half-wave*. We will investigate all three types and relate them to commonly used signals.

The first type of symmetry to be investigated is **even symmetry**. As shown in Figure 5.5, the signal is symmetrical about thc amplitude (vertical) axis. This relationship is sometimes referred to as **mirror symmetry**. For all even functions, cosine terms are dominant because a cosine wave is itself an even function. This can be seen by drawing a cosine wave and checking periodic points. Since the cosine terms are the only terms used in Equation (5.4), the Fourier series consists of an average value and the cosine terms. For an even function

$$f(t) = f(-t) \tag{5.5}$$

In words, Equation (5.5) can be stated as the magnitude of the function at $+t$ is equal to the magnitude of the function at $-t$. You can see this is true by referring back to Figure 5.5.

The second type of symmetry we mentioned was **odd symmetry**. This type of symmetry, sometimes referred to as **skew symmetry**, refers to signals symmetrical around a line drawn midway between the vertical and horizontal axis for the signal (Fig. 5.6). For an odd, or sine, function, the sine function dominates and all of the A coefficients are equal to zero in Equation (5.4). Thus, the Fourier series for an odd function is an average term (dc) and its associated sine terms. For an odd function

$$f(t) = -f(-t) \tag{5.6}$$

Equation (5.6) says that the magnitude of the signal at $+t$ is equal to the negative of the magnitude of the signal at $-t$. This can be confirmed by referring to Figure 5.6.

The final type of symmetry is **half-wave symmetry**. A signal with half-wave symmetry is shown in Figure 5.7. The wave is defined as half-wave symmetrical if the first half of the cycle (from 0 to $T/2$) repeats for the second half of the cycle ($T/2$ to T), but is of the opposite sign. This can be verified in Figure 5.7. You can see that the wave from $t = 0$ to $t = T/2$ has an of amplitude $-$V, whereas from $t = T/2$ to $t = T$ the wave has an amplitude of $+$V. This verifies the wave as half-wave symmetrical since the first half cycle is the same as the second half cycle only with an opposite sign. For half-wave symmetry all of the even harmonics for both the sine and cosine functions in Equation (5.4) are equal to zero. Thus, for a half-wave function

$$f(t) = -f\left(\frac{T}{2} + t\right) \tag{5.7}$$

Equation (5.7) was verified using Figure 5.7 in our discussion above.

Before using the general Equation (5.4) and applying the concepts of symmetry, it is necessary to calculate the coefficients of the equation to determine what the values

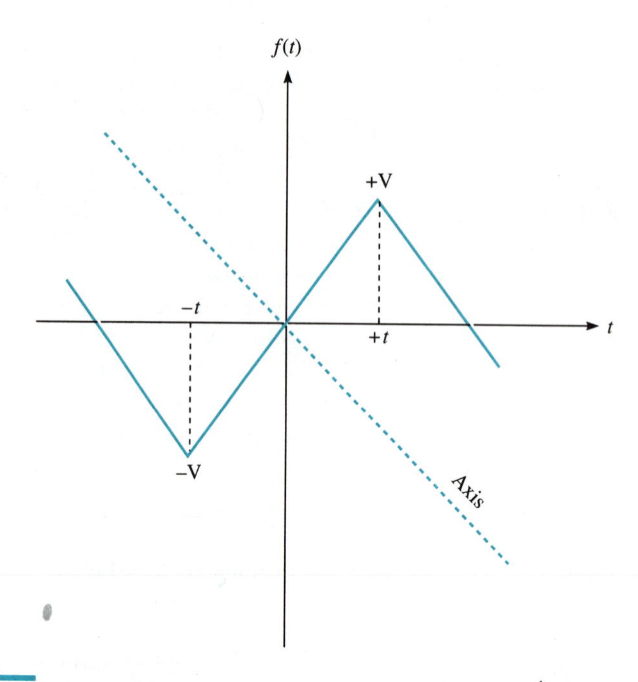

Figure 5.6 Odd symmetry.

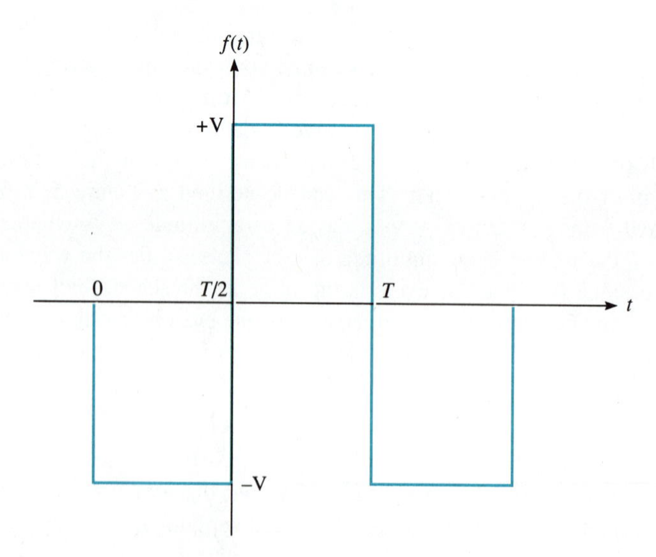

Figure 5.7 Half-wave symmetry.

of A_0, A_1 to A_n and B_1 to B_n are to be to satisfy the conditions for the signal. These are calculated as follows:

$$A_0 = \frac{1}{T}\int_0^T f(t)\,dt \tag{5.8}$$

$$A_n = \frac{2}{T}\int_0^T f(t)\cos Nt\,dt \tag{5.9}$$

$$B_n = \frac{2}{T}\int_0^T f(t)\sin Nt\,dt \tag{5.10}$$

The solutions of these equations involve integral calculus, which is beyond the scope of this text. We, therefore, supply coefficients for specific signals in Table 5.1, which

Table 5.1 Fourier Series

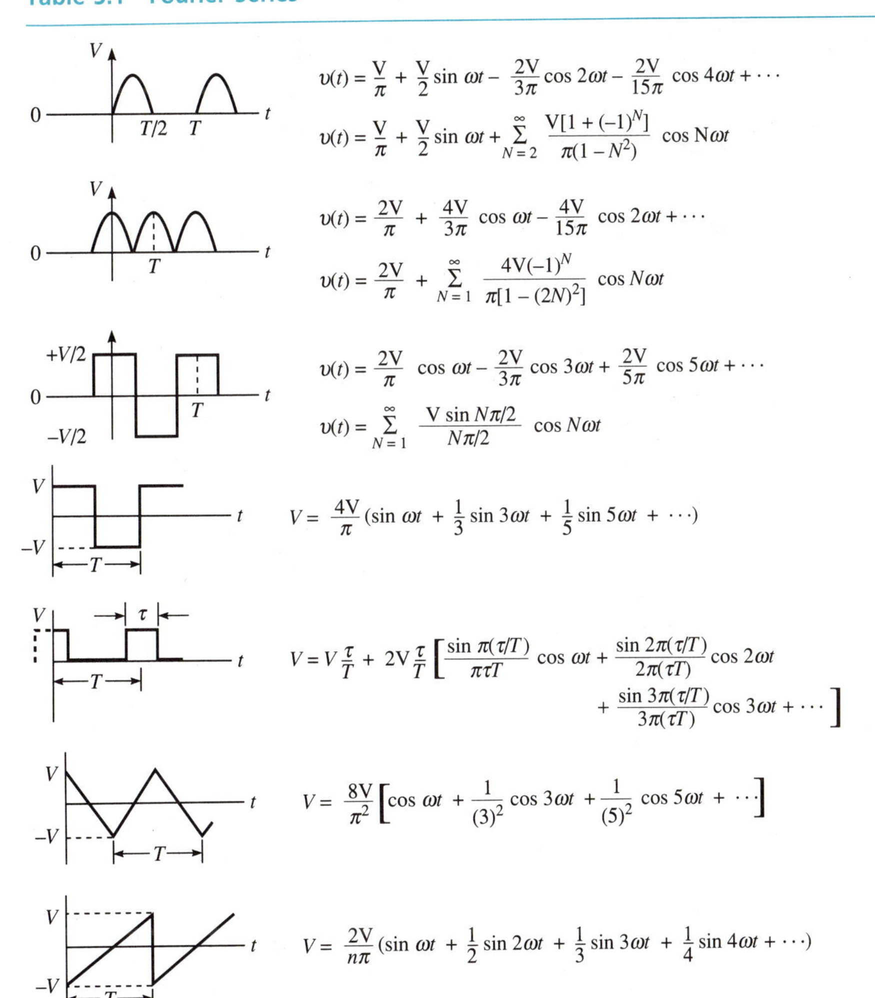

$$\upsilon(t) = \frac{V}{\pi} + \frac{V}{2}\sin\omega t - \frac{2V}{3\pi}\cos 2\omega t - \frac{2V}{15\pi}\cos 4\omega t + \cdots$$

$$\upsilon(t) = \frac{V}{\pi} + \frac{V}{2}\sin\omega t + \sum_{N=2}^{\infty}\frac{V[1+(-1)^N]}{\pi(1-N^2)}\cos N\omega t$$

$$\upsilon(t) = \frac{2V}{\pi} + \frac{4V}{3\pi}\cos\omega t - \frac{4V}{15\pi}\cos 2\omega t + \cdots$$

$$\upsilon(t) = \frac{2V}{\pi} + \sum_{N=1}^{\infty}\frac{4V(-1)^N}{\pi[1-(2N)^2]}\cos N\omega t$$

$$\upsilon(t) = \frac{2V}{\pi}\cos\omega t - \frac{2V}{3\pi}\cos 3\omega t + \frac{2V}{5\pi}\cos 5\omega t + \cdots$$

$$\upsilon(t) = \sum_{N=1}^{\infty}\frac{V\sin N\pi/2}{N\pi/2}\cos N\omega t$$

$$V = \frac{4V}{\pi}\left(\sin\omega t + \frac{1}{3}\sin 3\omega t + \frac{1}{5}\sin 5\omega t + \cdots\right)$$

$$V = V\frac{\tau}{T} + 2V\frac{\tau}{T}\left[\frac{\sin\pi(\tau/T)}{\pi\tau T}\cos\omega t + \frac{\sin 2\pi(\tau/T)}{2\pi(\tau T)}\cos 2\omega t + \frac{\sin 3\pi(\tau/T)}{3\pi(\tau T)}\cos 3\omega t + \cdots\right]$$

$$V = \frac{8V}{\pi^2}\left[\cos\omega t + \frac{1}{(3)^2}\cos 3\omega t + \frac{1}{(5)^2}\cos 5\omega t + \cdots\right]$$

$$V = \frac{2V}{n\pi}\left(\sin\omega t + \frac{1}{2}\sin 2\omega t + \frac{1}{3}\sin 3\omega t + \frac{1}{4}\sin 4\omega t + \cdots\right)$$

presents some widely used waveforms and the Fourier series used for each of them.

To illustrate how to use the information in Table 5.1, let us consider the following example.

Example 5.2

Given: The signal shown in the figure below

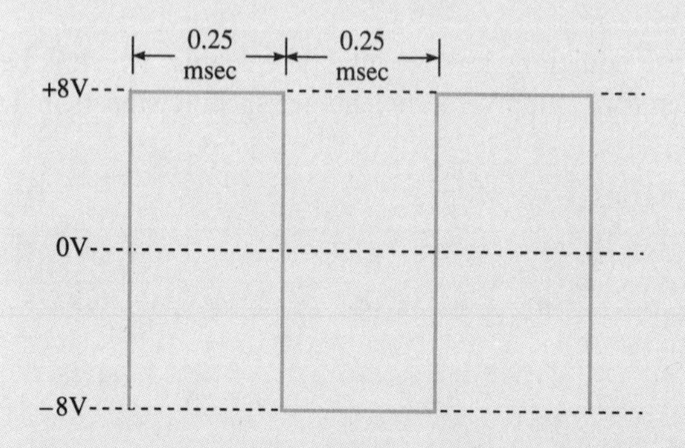

Find:
1. The coefficients of the first seven harmonics
2. Draw the spectrum of the waveform

Solution

1. Upon investigation we see that the waveform is a square wave with a +8V to −8V swing of the signal. This signal has an even excursion above and below zero, and thus the average dc component is 0 V. Also, if we compare this square wave with Table 5.1, we find that the Fourier series listed designates only odd components possible with this waveform. The Fourier series for this wave is

$$v(t) = \frac{2V}{\pi}\cos\omega t - \frac{2V}{3\pi}\cos 3\omega t + \frac{2V}{5\pi}\cos 5\omega t - \frac{2V}{7\pi}\cos 7\omega t + \cdots$$

For our example $V/2 = 8$ V, so $V = 16$ V. Since we are concerned with only the coefficients of the waveform, the calculations will be from the following:

$$Vn = \frac{2V}{N\pi}$$

Where: $V = 16$ V
N = Harmonic number

With this information we can devise the following chart:

N	Voltage (V)
0	0 (dc)
1	10.19
2	0
3	3.40
4	0
5	2.04
6	0
7	1.46

2. With the coefficients calculated, we now need to draw the frequency spectrum. First we need to find what frequency the waveform has for a fundamental. We know that the entire waveform requires 0.5 msec (millisecond) for one cycle T, so we take the reciprocal of T and find the frequency:

$$f = \frac{1}{T}$$

$$= \frac{1}{0.5 \text{ ms}}$$

$$= 2000 \text{ Hz}$$

$$= 2 \text{ kHz}$$

With this information we can find $3f$ (6 kHz), $5f$ (10 kHz), and $7f$ (14 kHz), and plot the spectrum, as shown in the figure below. The amplitudes are the amplitudes of the coefficients found in part (1) of this example.

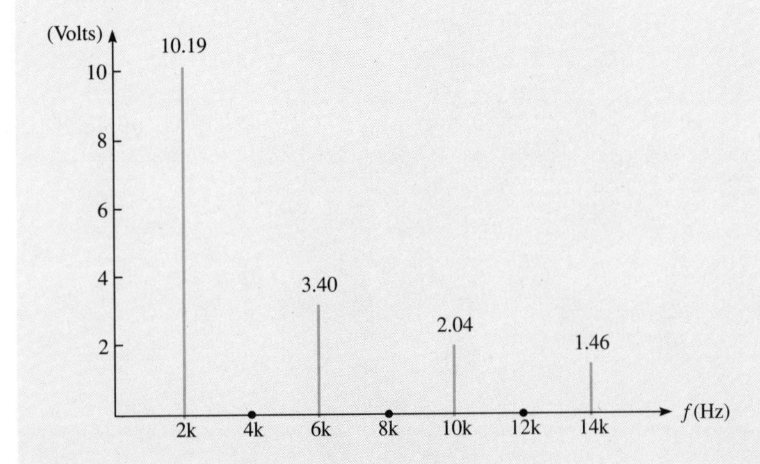

Example 5.3

Sketch the frequency spectrum of the first four harmonics of the waveform shown in the figure below.

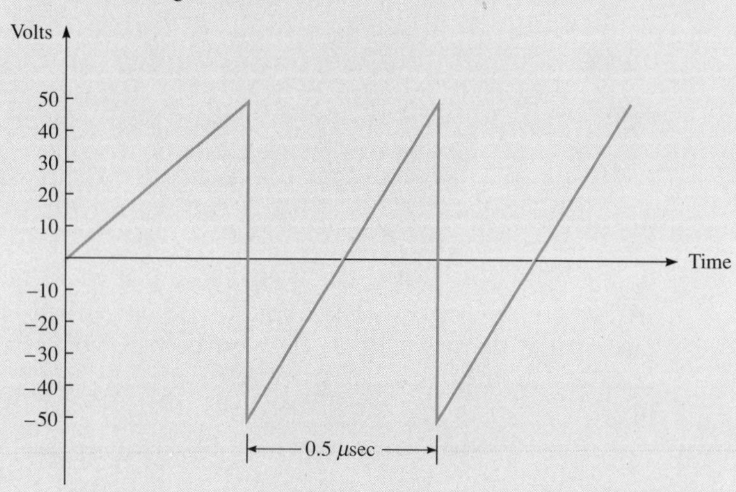

Solution

1. The first task is to determine the frequency. We see that the period of the waveform is 0.5 μs, and therefore

$$
\begin{aligned}
f &= \frac{1}{T} \\
&= \frac{1}{0.5 \times 10^{-6}} \\
&= 2 \text{ MHz}
\end{aligned}
$$

2. The harmonic levels must now be found. Table 5.1 shows that the sawtooth is made up of all harmonics and that the amplitudes are

$$V_a = \frac{2\text{V}}{n\pi}$$

Since $V = 50$ V, and we need f_1, f_2, f_3, and f_4, the amplitudes are

$$f_1 = \frac{100}{\pi} = 31.83 \text{ V}$$

$$f_2 = \frac{100}{2\pi} = 15.91 \text{ V}$$

$$f_3 = \frac{100}{3\pi} = 10.61 \text{ V}$$

$$f_4 = \frac{100}{4\pi} = 7.95 \text{ V}$$

The spectrum for the signal is shown in the figure below.

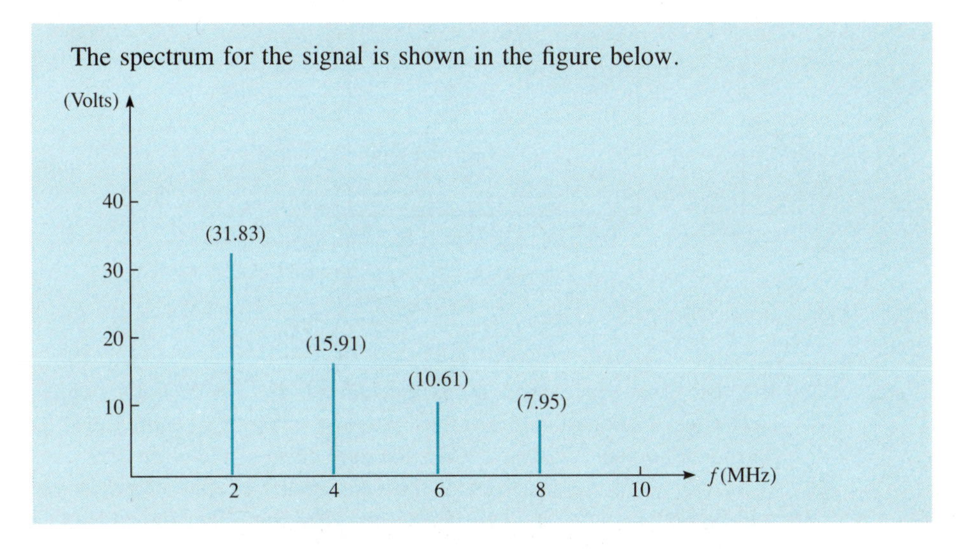

One of the more complex waveforms to analyze and one that finds much use in industry is the **rectangular pulse**. As you can see in Figure 5.8, this waveform differs from all the other waveforms we have discussed thus far. It has a burst of electromagnetic energy for a short period of time and then is completely off for an extended period. The period that the pulse is present is called the **pulse width** (τ). The time between pulses is termed the **repetition rate** (T). The important relationship between these two parameters is called the **duty cycle** of the waveform. In words it is the ratio of the pulse width to the repetition rate. Expressed mathematically it is

$$\text{Duty cycle} = \frac{\tau}{T} \tag{5.11}$$

If Equation (5.11) is multiplied by 100, the duty cycle is obtained as a percentage—a valuable parameter to know in many applications as it may determine how much power is to be dissipated by components within the system. For example, if we had a continuous-wave (CW) signal that had a 10 W level of power, all of the components within a particular system would have to be able to dissipate at least 10 W. If that same system had a pulsed input that had a 10% duty cycle (10 ms pulse with a repetition rate of 100 ms, for example), the signal would only be on for 10% of the time and thus would not need to dissipate any large amount of power because there would not be a large amount of power present. Most of the components would only need to be 1 W components to have the system operate safely. Even though the pulse may still be a 10 W pulse. Pulsed signals are used for many radar applications.

The **pulsed signal**, like all the other signals we have discussed, consists of a series of harmonically related signals. Table 5.1 shows the pulsed waveform and its Fourier series.

$$V = \frac{V\tau}{T} + \frac{2V\tau}{T}\left[\frac{\sin x}{x}(\cos \omega t) + \frac{\sin 2x}{2x}(\cos \omega t) + \frac{\sin 3x}{3x}(\cos \omega t) + \cdots + \frac{\sin Nx}{Nx}(\cos \omega t)\right] \tag{5.12}$$

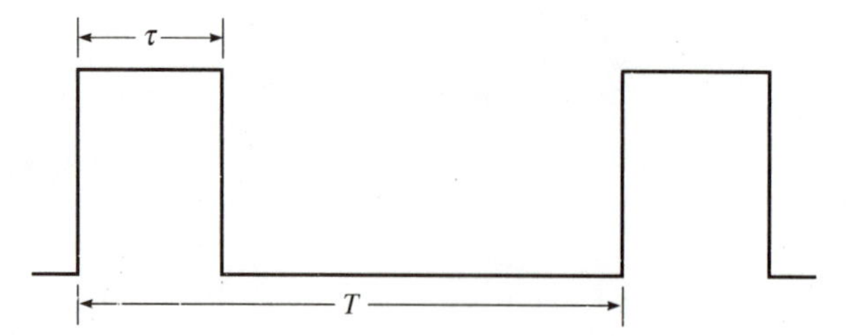

Figure 5.8 Rectangular pulse.

We recognize this expression as the familiar sin x/x function, which is a decaying sine wave. In Equation (5.12) the value of x is the expression $N\pi(\tau)/T$), which is the harmonic number N times π times the duty cycle of the pulse. This decaying function is exactly what the spectrum of the pulse will look like. As a matter of fact, there will be both positive and negative terms when the calculations for the pulse series are made. Obviously, the spectrum analyzer will only display positive signals, so the negative values are turned around, meaning that the absolute values of the spectrum parts are displayed. This should be kept in mind when doing problems with pulsed waveforms. When asked to draw the spectrum, you should plot all the positive values.

Let us now take a look at the example pulse that we referred to above, the 10% duty-cycle pulse. Figure 5.9 shows the pulse and the frequency spectrum associated with it, and we can see that the spectrum exhibits the decaying sine-wave character-

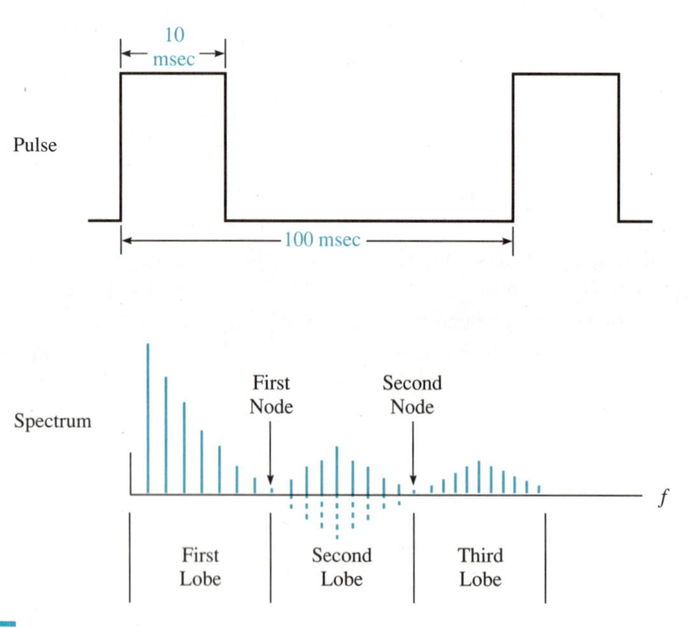

Figure 5.9 10% duty cycle pulse.

istic. This is illustrated by the dashed lines in the second set of frequency lines. The actual lines that would be displayed on a spectrum analyzer are the solid lines, but the sin x/x function is clearly seen if the second set of lines are drawn in the negative direction as the dashed lines are. You can also see how the spectra is divided into **lobes**, areas that define the spectra of the signal. The first lobe contains all harmonics from zero to the first null of the spectrum. The second and third nulls are those that have all of harmonics between the first and second nulls and between the second and third nulls, respectively.

The most interesting spectral lines for pulsed signals occur when the pulse width and the repetition rate are changed. In other words, the duty cycle of the pulsed waveform is changed. The effects that are to be discussed are when the duty cycle is decreased. You will recall that the duty cycle is the pulse width τ divided by the repetition rate T. To decrease the duty cycle we must, therefore, either decrease the pulse width keeping the repetition rate constant or increase the repetition rate keeping the pulse width constant.

If we hold the repetition rate T constant and then decrease the pulse width, we find some interesting occurrences taking place. Beginning with a duty cycle of 20%, we obtain the spectrum shown in Figure 5.10(a). You can see that this spectrum is one that exhibits the typical decaying sine-wave characteristic and has a series of nulls and lobes as described previously. If we now reduce the pulse width until we have a duty cycle of 10% (Fig. 5.10b), we see that the amplitude of the spectrum is beginning to decrease. The lobes are also becoming broader as the null points move farther and farther out, because, as the pulse width decreases there is more bandwidth required

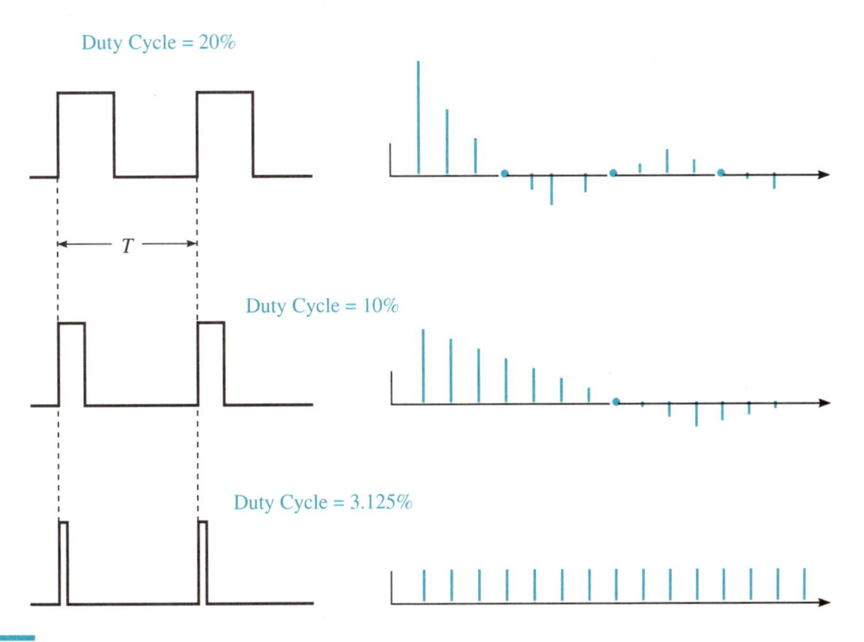

Figure 5.10 Decreasing pulse width.

for the waveform. The null points must therefore move farther out on the frequency spectrum to accommodate the required bandwidth.

We now get to a critical point when reducing the pulse width of a pulsed signal. As shown in Figure 5.10(c), this point lies where the duty cycle is reduced to 3.125%, or the pulse width equals $T/32$. At this point the pulsed signal is considered to be an impulse that theoretically requires an infinite bandwidth. This, of course, is not the case in practice, but it does explain why the spectrum is now a uniform amplitude for all frequencies.

In the second method for decreasing the duty cycle, we increase the repetition rate T and hold the pulse width constant. Figure 5.11(a) shows the spectrum for a duty cycle of 20%. The spectrum, as expected, is the same as before. If we now increase T until the 10% duty cycle is obtained (The rate is now 10 times that of the width of the pulse), we see that the spectrum is much denser, because as we increase the rate we are decreasing the frequency ($f = 1/T$), as in Figure 5.11(b). More and more harmonics can now be in the first lobe since these harmonics are getting closer to the fundamental frequency, which is decreasing. As T approaches infinity, as in Figure 5.11(c), it is no longer possible to distinguish between the frequency components, and there is a continuous curve with the sin x/x shape to it. It is interesting to note that when we decreased the pulse width to decrease the duty cycle, the null points

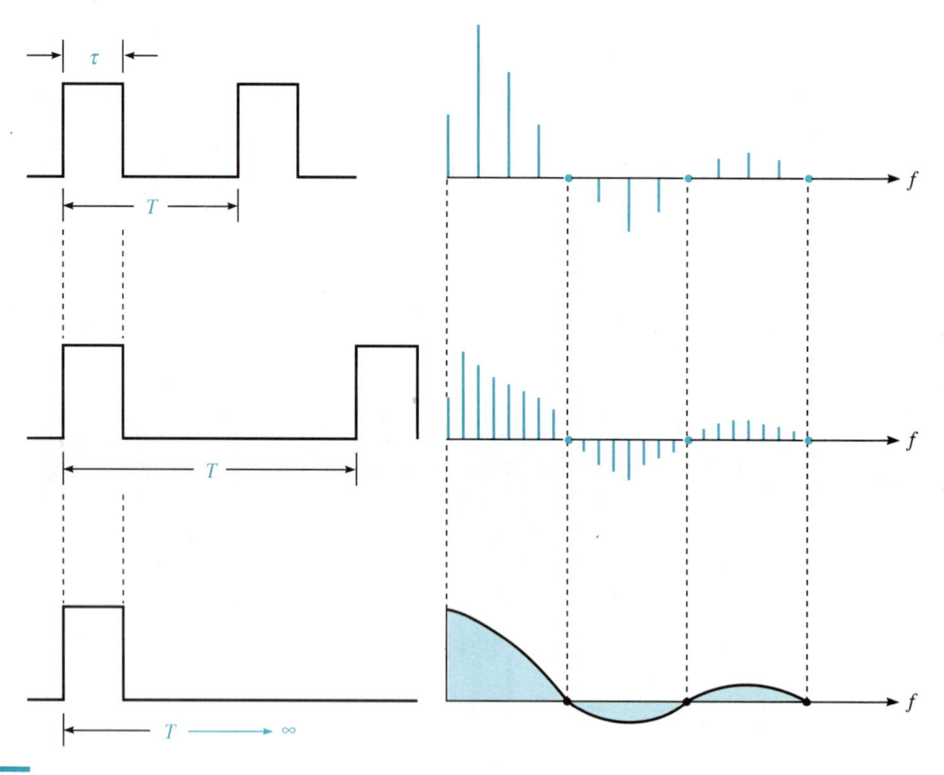

Figure 5.11 Increasing repetition rate.

shifted because more bandwidth was needed. You will note that when the repetition rate was increased to produce the same duty cycle, the null points did not change but stayed exactly at the same points. This occurs because, as we change the repetition rate, we do not require more bandwidth. We are decreasing the frequency of the signal, but are not changing the null points.

Now that we've discussed the concept of signal spectra and shown some examples of such spectra, we present a circumstance where this theory can be put to use. Such a circumstance occurs when a signal is put through a filter to obtain specific frequency components for a required application. To illustrate this, let us consider an example.

Example 5.4 We use the same square-wave signal shown in Example 5.1. The square wave is centered at 0 V, has a +8V to −8V excursion, and a total period of 0.5 ms. This signal is recreated in Figure 5.12(a), and its associated spectrum is shown in Figure 5.12(b).

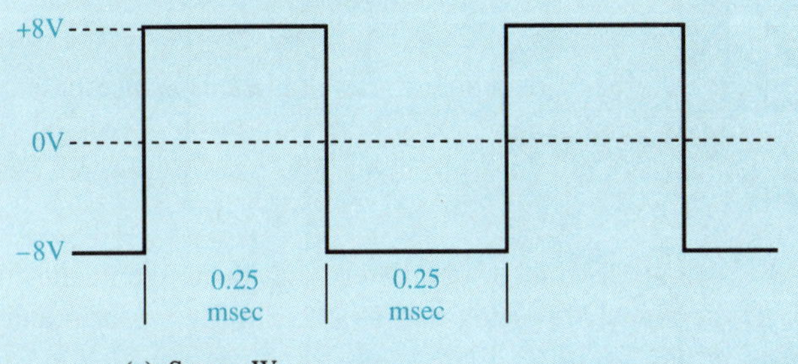

(a) Square Wave

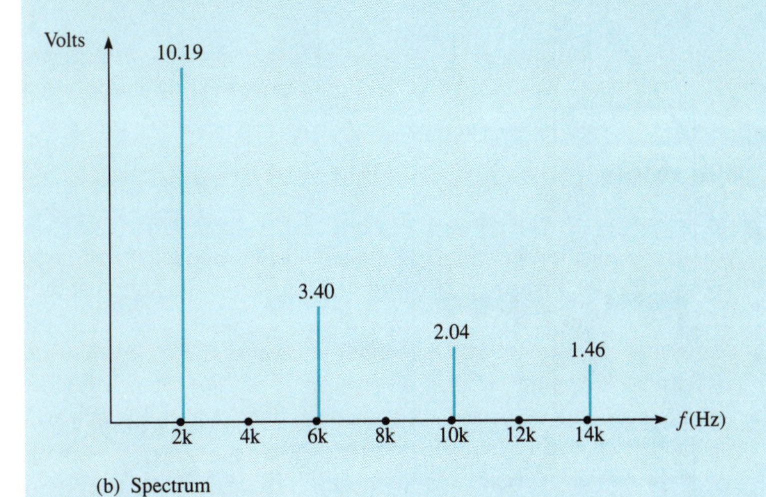

(b) Spectrum

Figure 5.12 Square wave.

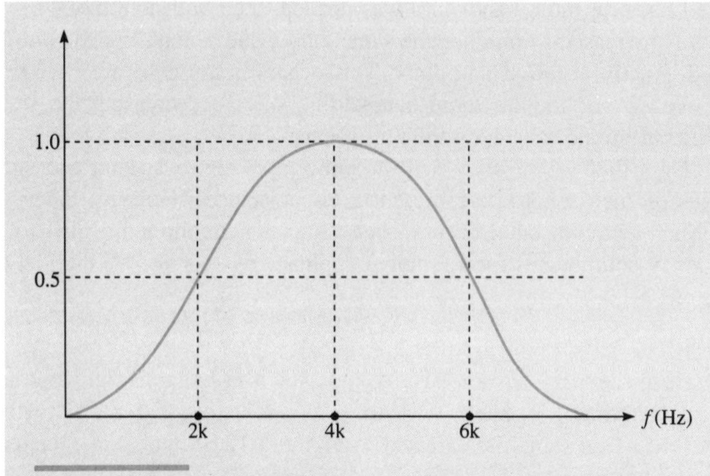

Figure 5.13 Bandpass filter.

The filter to which this waveform is applied is a bandpass filter shown in Figure 5.13. Its band center is at 4 kHz and drops to half power at both 2 kHz and 6 kHz. We now need to determine what the output spectrum of the filter is.

Solution

To solve this example we must first determine, once again, the fundamental frequency of the signal, by using the period of the waveform and taking its reciprocal. That is,

$$\begin{aligned} f &= \frac{1}{T} \\ &= \frac{1}{0.5 \text{ ms}} \\ &= 2 \text{ kHz} \end{aligned}$$

Another means of determining the fundamental frequency is to examine the spectral plot of Figure 5.12(b) and note that the main signal at an amplitude of 10.19 V is at 2 kHz.

Now that we've determined the fundamental frequency, we can proceed to the rest of the spectrum. First, we must consider which frequency components will be passed completely by the filter and which will be attenuated. The only frequencies that will be passed completely by this filter occur in the range from approximately 3 kHz to 5 kHz. As we can see from Figure 5.12(b), the signal we are using has no components at all in this frequency range, and we, therefore, do not need to be concerned with them.

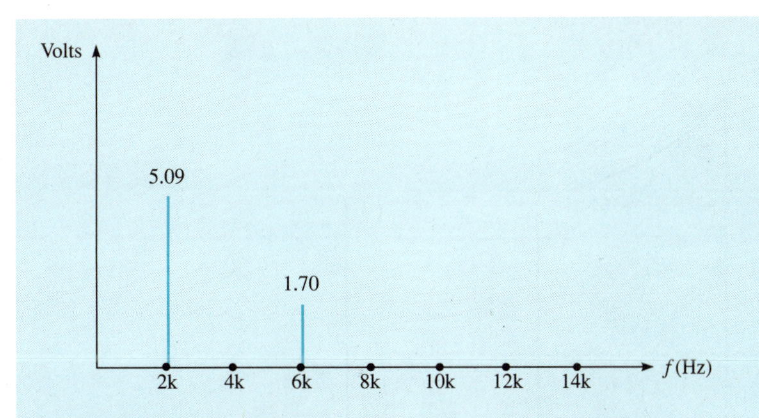

Figure 5.14 Resulting spectrum.

Two components in our signal are at 2 kHz and 6 kHz, both of which will be attenuated by one-half in the filter, as the response in Figure 5.13 shows. Thus, the 2 kHz spectrum component will exit the filter at an amplitude of 5.09 V, and the 6 kHz component will be at an amplitude of 1.70 V. These are the only components that will be present at the filter output, because the other components are either completely attenuated or they do not exist in our signal. The final output spectrum will be as shown in Figure 5.14. This figure shows a 2 kHz component at 5.09 V and a 6 kHz component at an amplitude of 1.70 V as predicted.

5.4 Distortion

Distortion of a signal means changing its usual or normal shape. The distortion of the signals we have been discussing usually comes in two forms: **phase distortion** and **harmonic distortion**. This distinction means that the input signal to a device may be changed, or modified, by changing its phase, or by adding harmonics to the original signal.

Phase distortion of a signal often occurs with square-wave signals, causing the signal to distort as shown in Figure 5.15. The distortion is usually caused by a reactive component within an amplifier or active circuit that causes a sag or tilt in the signal where it should be flat. Careful design will eliminate such distortion.

Harmonic distortion usually occurs when an active circuit, which is designed to be a linear device, is driven into a nonlinear region. When this occurs, the fundamental signal does not increase in amplitude but the harmonics rise according to how far the device is driven nonlinearly. To calculate the amount of distortion a single harmonic contributes, use the relationship shown in the following example.

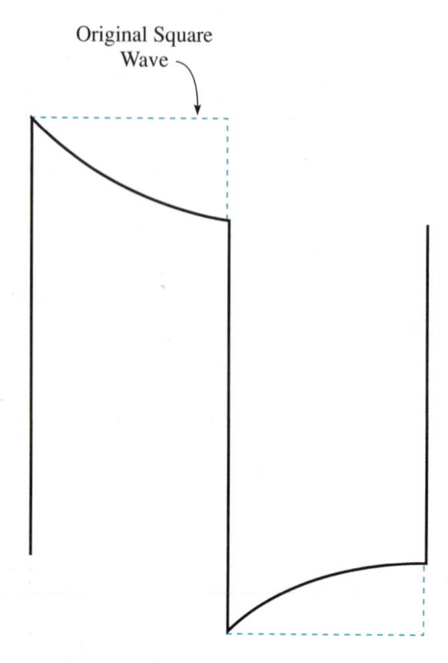

Figure 5.15 Phase distortion.

Example 5.5 In this example, the fundamental will be at 15 V and the third harmonic at 1.5 V. The third harmonic distortion will be designated as D_3.

$$D_3 = \frac{\text{Third harmonic}}{\text{Fundamental}} = \frac{1.5\text{ V}}{15\text{ V}} = 0.10, \text{ or } 10\% \tag{5.13}$$

(Whichever distortion you require, simply use the value of voltage for that harmonic and compare it with the fundamental)

If the **total harmonic distortion** is to be calculated, we simply square all of the distortions, add them together, and take the square root. That is,

$$\text{THD} = \sqrt{D_2 + D_3 + \cdots + D_n} \tag{5.14}$$

Where: THD = Total harmonic distortion

D_2 = Second harmonic distortion

D_3 = Third harmonic distortion

D_n = Nth harmonic distortion

To illustrate this idea consider the following example.

Example 5.6

Given: $D_2 = 0.37$
$D_3 = 0.20$
$D_4 = 0.09$

Find: Total harmonic distortion

Solution

$$\text{THD} = \sqrt{D_2^2 + D_3^2 + D_4^2}$$
$$= \sqrt{(0.37)^2 + (0.20)^2 + (0.09)^2}$$
$$= \sqrt{(0.14) + (0.04) + (0.01)}$$
$$= \sqrt{(0.19)}$$
$$\text{THD} = 0.44, \text{ or } 44\% \quad \text{(total distortion)}$$

Thus, you can see that each of the individual harmonics contributes to the total harmonic distortion of the signal. The original second harmonic was 37%, but when all of the harmonics are added, the total distortion is 44%.

5.5 Summary

It is important to understand the spectrum of the signals that will be used for communications systems. This chapter has provided an introduction to signal spectra by defining and illustrating the differences and uses of both time domain and frequency domain. The Fourier series was then presented to completely characterize the signals.

Finally, the concept of distortion was introduced in Section 5.4. This is a very valuable section because the essential purpose of a communications system is to apply an *intelligence* to a carrier, transmit it, and then demodulate it to obtain the exact signal that was transmitted with minimal or no distortion.

Questions

5.2 Time Domain and Frequency Domain

1. Why is the frequency domain used for the Fourier series?
2. What is the mathematical expression for a single sine wave in the time domain?
3. What is one advantage of the frequency domain over the time domain?
4. What instrument is used to display the frequency domain?
5. What instrument is used to display a time domain signal?
6. If it is desired to know what level each of the harmonics of a signal are, what should be used to measure them?
7. What are two parameters that can be obtained from a frequency domain display?

5.3 Fourier Series

8. Define **Fourier series**.
9. Name three types of signals that can be evaluated using the Fourier series.
10. Define **harmonic**.
11. Describe **even** and **odd** symmetry.
12. How does a pulsed signal differ from other signals?
13. Define **duty cycle**.
14. What does the spectrum of a pulse resemble?
15. What effect does decreasing the pulse width of a signal while keeping the repetition rate constant have on the spectrum?
16. What is the cirtical point for reducing the pulse width?
17. Are Fourier series operations used more for sinusoidal or non-sinusoidal applications? Why?
18. Describe **half-wave symmetry**.
19. How can a filter improve harmonic performance?

5.4 Distortion

20. Define **distortion**.
21. What causes **phase distortion**?
22. Define **harmonic distortion**.
23. What is **total harmonic distortion**?
24. When an amplifier is driven into a nonlinear area the second, third, and all harmonics are increased while the fundamental amplitude remains constant. How does this affect total harmonic distortion?

Problems

5.3 Fourier Series

1. For the waveform in Figure 5.16,
   a. Find the dc value
   b. Find the first five components of the signal
   c. Draw the spectrum

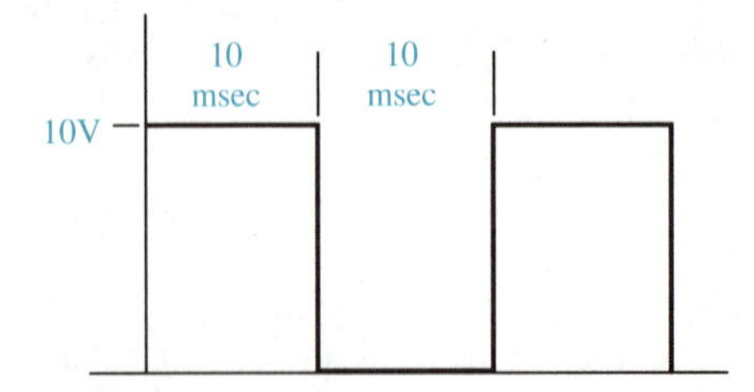

Figure 5.16

2. For the waveform shown in Figure 5.17,
   a. Find the dc value
   b. Find the first seven components
   c. Draw the spectrum

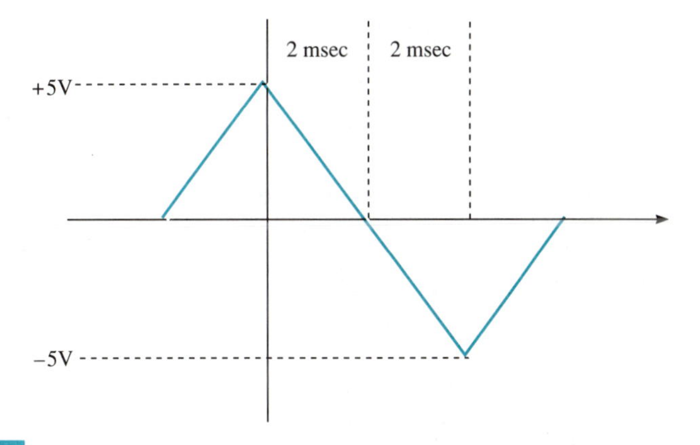

Figure 5.17

3. Given the waveform in Figure 5.18,
   a. Find the dc component
   b. Determine the peak amplitude of the first eight components
   c. Plot the sin x/x function
   d. Draw the spectrum

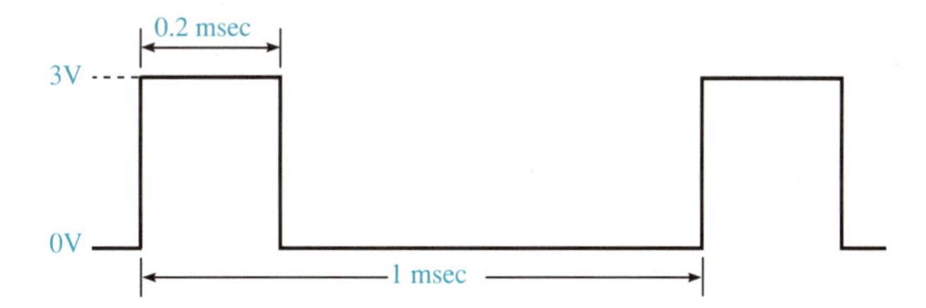

Figure 5.18

4. The signal shown in Figure 5.19(a) is applied to a filter with the response shown in Figure 5.19(b). Draw the spectrum at the output of the filter.

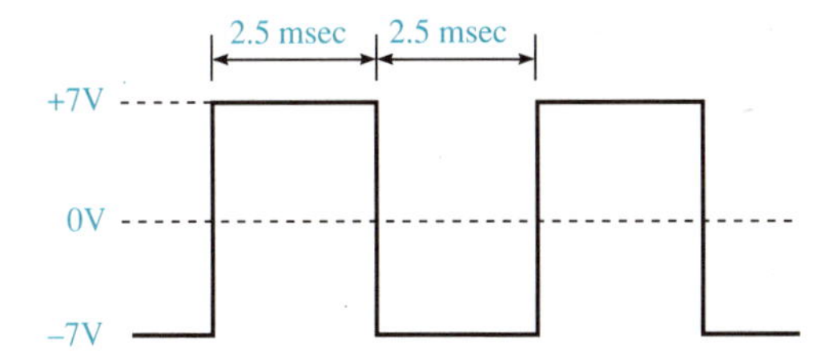

Figure 5.19(a)

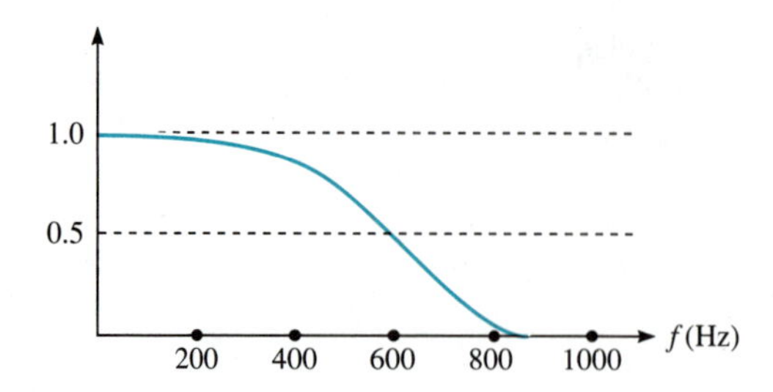

Figure 5.19(b)

5. For the signal in Figure 5.20, determine the first six components and draw the spectrum.

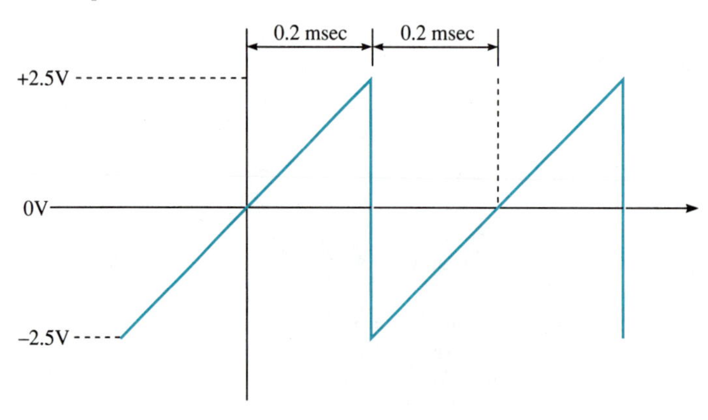

Figure 5.20

5.4 Distortion

6. Determine the percentage of second-order, third-order, and total harmonic distortion for the spectrum shown in Figure 5.21.

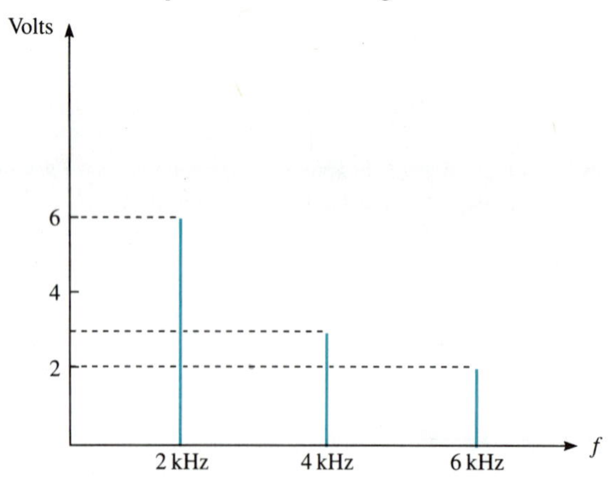

Figure 5.21

7. If the fundamental component of a wave has an amplitude of 8 V, and the maximum second-order distortion that is allowed is 7%, what is the maximum voltage level for the second harmonic?
8. Using the results of Problem 7, what must the maximum voltage level of the third harmonic be to maintain a maximum total harmonic distortion of 8%?
9. The fundamental of a signal has been measured as 6 V and the second harmonic is 2 V. If it is known that the total harmonic distortion for the system is 37%, what is the amplitude of the third harmonic of the signal?
10. In Problem 9, what is the effect on the total harmonic distortion if the second harmonic is suddenly increased to 3.5 V?

6

Outline

Objectives

- To introduce the concepts of noise
- To discuss how noise affects communications systems
- To describe both correlated and uncorrelated noise with examples of each
- To introduce noise parameters such as noise figure, noise factor, and noise temperature

Key Terms

noise	burst noise	external noise
harmonic distortion	noise current	thermal noise
total harmonic distortion	noise factor	flicker noise
uncorrelated noise	noise figure	noise power density
internal noise	correlated noise	noise bandwidth
shot noise	intermodulation distortion	signal-to-noise ratio
		noise temperature

Noise

6.1 Introduction

The concept of noise is very difficult for many people to grasp. When this topic is raised, students usually dash to specific texts to try and glean some quick information rather than gradually developing a basic understanding that will last and be applicable for many cases. Let us now take the time to understand the basic ideas behind noise and to relate them to communications systems.

To begin our discussions, we provide a simple definition of the term noise that relates specifically to communications systems.

Noise

> Noise is any unwanted electrical signal within a communications system that interferes with the sound or image being communicated.

This definition characterizes noise as something that is unwanted in a communications system. This is something that you can relate to when you think of the times that the television signal from your favorite station had some interference on it that caused ''snow'' to appear; instead of what you want to see. This is a noise that interferes with the desired signal.

Noise within a system is very difficult, if not impossible, to characterize because of its random nature, that is, its lack of periodicity. It does not repeat itself like the many types of signals we have discussed thus far. The only exception is the very common 60-Hz ''hum'' that occurs within a power-supply system when inadequate filtering is included in the design or a filter capacitor goes bad. This form of noise is easily characterized because it is a 60-Hz sine wave that is periodic. This, however, is the exception when discussing noise.

The completely random nature of noise is what makes it difficult to characterize. But what does *random* mean compared with the typical signals we have discussed thus far? Figure 6.1 (a) depicts a sine wave, which we have discussed previously in this text. You can see that this sine wave is well behaved in that it starts at zero, proceeds up to a maximum, through zero, to a negative maximum and back to its original starting point. The period and frequency are easy to determine; They are T and $1/T$, respectively. Also, the amplitude of the wave is designated as A.

Part (b) of Figure 6.1 is an entirely different story. We see that it has one thing in common with the signal in Figure 6.1(a)—it begins at zero. This is where the similarities begin and end, however. Nothing repeats during the entire duration of the signal (it is random), and we can, therefore, not determine a period T or frequency $1/T$. Also, there is no definite amplitude associated with the signal since it varies drastically from a low to a high value. This signal is truly random and is in a true sense a noise signal. Consider our definition of noise and look at Figure 6.1. You will see that the signal shown indeed interferes with reception in our communications system, and is therefore an unwanted signal.

In order to picture the idea of random noise and to understand that you cannot determine its frequency, consider the visual presentation of noise on the spectrum-

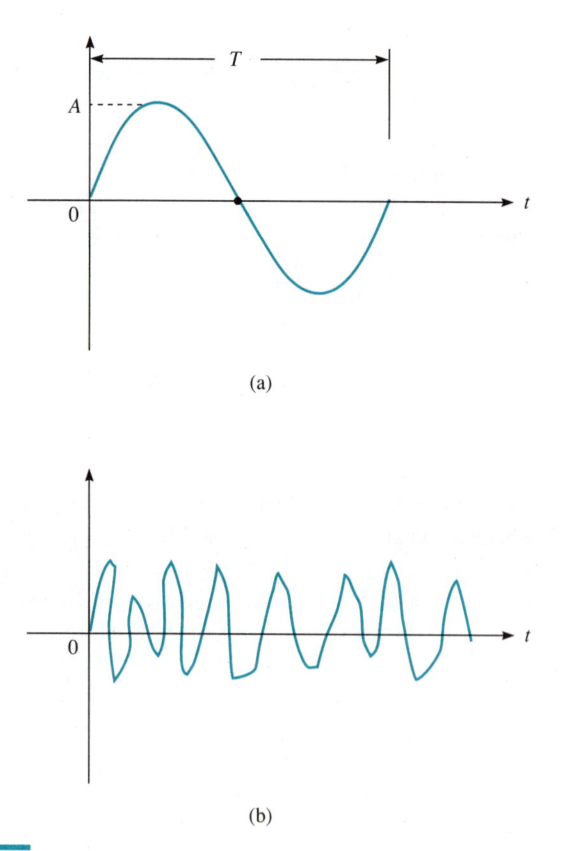

Figure 6.1 Periodic and random signals.

analyzer display. If you notice that when you first turn on the analyzer and until you turn it off, there is a ''grass'' level at the bottom of the display. This is the internal noise level of the spectrum analyzer. When you examine this level, you see that it appears throughout the entire frequency spectrum of the analyzer. There is no distinction as far as frequency is concerned.

Often in communications systems a feedback system is incorporated to eliminate an unwanted signal. Such a system is shown in Figure 6.2. You can see that the signal source output is fed into a sampling circuit. The output of the sampler is then sent to a phase-shift circuit, which shifts the phase of only the unwanted signal by 180°, cancelling the unwanted signal and providing an output of only the desired signals. In the case shown in Figure 6.2, we want the fundamental frequency f_0, and want to eliminate the third harmonic $3f_0$, which is very high at the signal output, as shown in Figure 6.2(b). Figure 6.2(c) shows the final output with the feedback circuitry inserted.

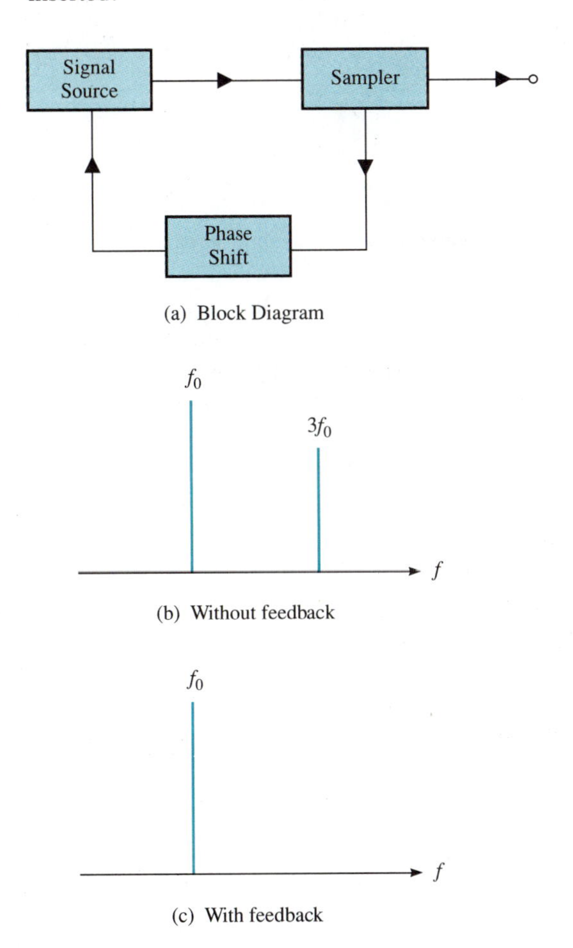

Figure 6.2 Feedback system.

The concept of feedback works well for signals such as the example shown above. It is relatively easy to determine the frequency of the unwanted signal and design the circuitry to shift that frequency by 180° to eliminate it. Noise cannot be handled in the same way, however, because it is random and an exact frequency cannot be determined for the phase-shift circuit design. Thus, feedback is not the answer for eliminating, or minimizing, noise in a communications system.

To determine the best methods for handling noise within a system, it is necessary to understand the different types of noise that may be present within that system. Noise can be broken down into two basic types: correlated and uncorrelated. Consider that the term *correlate* means "closely related." In **correlated noise** the desired signal and the noise component are closely related. So, in order to have correlated noise we must first have a signal present; there will not be noise if there is no signal. Thus, the correlation between signal and noise. On the other hand, **uncorrelated noise** occurs when there is no signal; it is a noise generated within the individual components of the system. Let us now investigate each type of noise to gain an understanding of noise, and to determine how to either eliminate or minimize each type.

6.2 Correlated Noise

Correlated noise, that noise generated by an input signal, is of two types: harmonic distortion and intermodulation distortion. You will note that we referred to these as noise, but labeled them as types of distortion. In Chapter 5 we defined distortion as a change in the usual, or normal, shape of the original signal. This is what a noise component of a signal does to that signal—changes its shape or characteristics. Thus, we can consider a distortion of the signal a noise caused by the input signal and the circuitry the signal is entering.

In Chapter 5 we also discusssed harmonic distortion and made a series of calculations for specific harmonics and the total harmonic distortion for a signal. As you will recall, the ideal signal is one that contains only the fundamental, or desired signal. If, for example, we wanted a signal at 5 MHz at a level of 6 V, these would be the only components of the spectrum we would like to see. Although there may be a small contribution from the second or third harmonic, their amplitudes should be very low. If, however, the amplitudes of the second and third harmonics are at such a level that they interfere with the fundamental, they can cause considerable problems with the operation of the circuits the signal must feed. If, for example, our 5-MHz fundamental has a second harmonic (10 MHz) at a level of only 4 V, and a third harmonic (15 MHz) at a level of 2 V, we see from Chapter 5 (Equations 5.13 and 5.14) that we have a second harmonic distortion of 67%, a third harmonic distortion of 33%, and a **total harmonic distortion** of 75%. This is an intolerable amount of distortion of the fundamental frequency and would render the signal totally useless.

The large amount of distortion in our example is generally due to a nonlinearity within some active device through which the signal is being transmitted. That is, the transistor being used for an amplifier is not properly biased, causing the input level

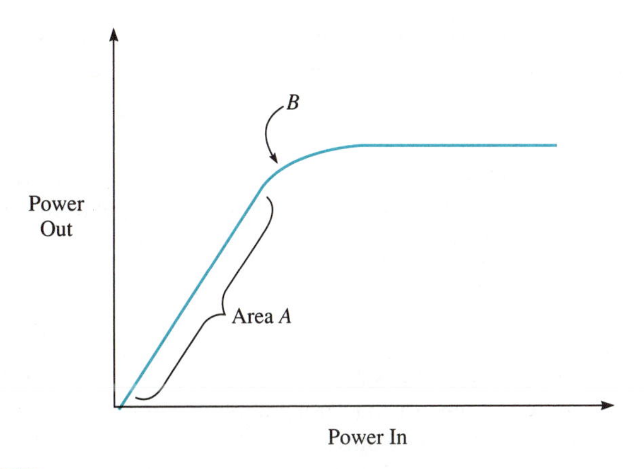

Figure 6.3 Amplifier response curve.

into the amplifier to drive the transistor into a compression region. You will recall from previous electronic-circuit courses that a properly biased class-A amplifier will have a linear output over a specified input power range. This linear range is designated as area *A* in Figure 6.3. You can see that for an increment of power into the amplifier there is an increment out that is proportional to the gain of the amplifier. The gain of the amplifier determines the slope of the response curve through this area. In this linear region there is a good relationship between the fundamental frequency and its associated harmonics.

If we continue to increase the input power to the amplifier, we will reach a point where the output power no longer increases but remains constant. At this point the fundamental signal level will cease to increase and the power being applied to the amplifier will cause the harmonics of the signal to increase. This point is called the **compression point** of the amplifier and is shown in Figure 6.3 at point *B*. The amplifier is now operating in a nonlinear region as can be seen in the figure. Any increase in input power will no longer increase the output power; only the harmonics increase and will eventually cause severe problems. This represents the noise (correlated) we call **harmonic distortion**.

The second type of correlated noise is **intermodulation distortion**. This type of distortion is caused by two or more frequencies being applied to the input of a nonlinear amplifier (high-power amplifier, for example), or an amplifier that has been driven into a nonlinear area. Intermodulation products are formed when the individual signals and their harmonics, generated by the nonlinear action, are combined to form other signals. These signals can cause severe problems within a communications system designed to operate at a certain frequency and only that frequency. Most of the time the system can live with the harmonics generated by a single frequency because it was designed to operate at that frequency and the harmonics have been compensated for. However, if a second or third signal is applied to the input and generates additional harmonics, there is no provision made to handle these additional signals. A common area in which this phenomenon occurs is the environment around an airport—one of the busiest areas for electronic signals. The volume of incoming signals and the need for clear reception require specific precautions be taken when designing a communications system to operate in the area of an airport.

To further understand intermodulation consider the following example. If we have a primary signal F_1 and an additional signal F_2, which is an interfering signal at virtually the same level, what spectrum will you see on a spectrum analyzer?

The spectrum will consist of:

$$F_1, F_2, F_1 + F_2, F_1 - F_2, 2F_1 + F_2, F_1 + 2F_2, 2F_2 - F_1, \ldots 3F_1 + F_2, F_1 + 3F_2, 3F_1 - F_2, \ldots$$

In other words, there will be a multitude of combinations of the two signals. When there is a "2" in the combinations, or intermodulation products, we refer to the noise as a **second-order intermodulation distortion.** When there is a "3", the noise is called a **third-order intermodulation distortion**, and so on for the higher order products. These are convenient ways of characterizing and identifying the type, or order, of intermodulation distortion (correlated noise).

Example 6.1 details the discussion above.

Example 6.1

Given: Signal, F_1 = 1 MHz
Signal, F_2 = 3 MHz
(at the input of a nonlinear device)

Find: The spectrum of these two signals

Solution

From our discussions we see that there will be a multitude of signals present at the output of the nonlinear device:

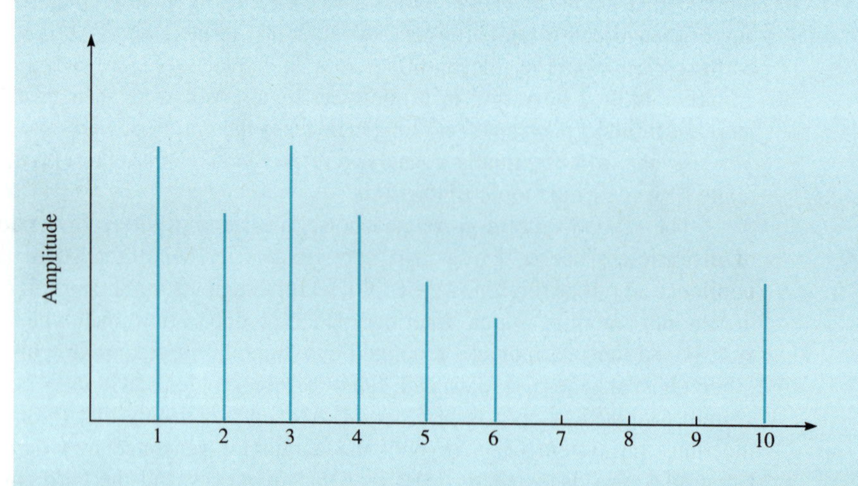

Figure 6.4 Output spectrum.

1. There are the signals themselves, 1 MHz and 3 MHz.
2. There are the sum and difference frequencies, which are 4 MHz and 2 MHz.
3. There are the second-order components (we will list only the ones we discussed in the text), which are 5 MHz ($2F_1 + F_2$), 7 MHz ($F_1 + 2F_2$), using only the positive values for now.
4. There are the third-order components, which are 6 MHz ($3F_1 + F_2$), and 10 MHz ($F_1 + 3F_2$).
5. There also is a 0 Hz component, which is $3F_1 - F_2$.

The spectrum for these signals is shown in Figure 6.4.

(There will also be harmonics of each of these signals, which are not shown in Fig. 6.4.)

6.3 Uncorrelated Noise

As previously stated, **uncorrelated noise** is noise present even when there is no signal. It can be generated within the individual components of the system or be a noise from a completely independent outside source. Uncorrelated noise is therefore divided into two separate categories: external and internal.

6.3.1 External Noise

External noise is of three types: atmospheric, extraterrestrial, and man-made. All of these are sources which are present in varying degrees in the air and can possibly interfere with the operation of a circuit or system. However, because the greatest source of noise in communications systems is from internally generated noise (which we will cover in the next section), external sources will be mentioned only briefly. Nevertheless, it is important to know the external sources so you can recognize them and appropriately shield circuits and systems so these external sources will have a minimal effect on operation.

The first type of external noise is **atmospheric noise**, commonly referred to as **static electricity**. It occurs in the form of impulses of energy that cover a wide band of frequencies. The noise propagates through the atmosphere similar to electromagnetic waves and is thus effected by atmospheric conditions just as a desired signal would be. You will recognize atmospheric noise in a receiver as a cracking sound with no signal present or the noise you hear between stations on a radio receiver.

The second type of external noise, **extraterrestrial noise**, is generated outside the earth's atmosphere. It is divided into two categories: solar and cosmic. **Solar noise**, as the name implies, is generated directly from the sun's heat. The most common and recognizable type of solar noise is sun-spot activity, which occurs in

11-year cycles. Solar noise can therefore be planned and accounted for by communications people. **Cosmic noise**, sometimes called **black body noise**, occurs throughout the galaxy and occurs because of the temperature associated with distant stars. This temperature causes stars to contribute noise similar to that described for the sun. This type of noise is not cyclic but is determined by the positions of the stars and their temperatures relative to that of the earth.

The third type of external noise is the one we have the most control over—**man-made noise**. This noise is generated by such things as automotive ignition systems, commutators in electric motors, fluorescent lights, and high-power switching equipment. Anything that causes a spark to be generated is potentially a source of man-made noise. This type of noise can be substantially reduced by incorporating proper shielding methods.

6.3.2 Internal Noise

Internal noise, just as the name implies, is noise generated *within* a device or component. This is the type of noise a circuit designer must be very aware of in order to produce as low a noise device as possible. Low-noise components are vital to the success of any communications system, particularly the receivers of such systems. In many cases the level of the received signal is very low, but if the signal is not higher than any internally generated noise it will never be detected. Thus, an understanding and description of each type of internal noise, its cause, and how to keep it to a minimum is essential for good low-noise, high-sensitivity communication systems.

The following four common types of internal noise can be encountered within a communications system: thermal noise, shot noise, flicker noise, and burst noise.

The first type, thermal noise, most often comes to mind when internal-noise sources are mentioned. This is also the most common type to be used for analysis of noise within a system. Thermal noise basically occurs when current flows through some resistive element within a device or component, such as a transistor, a diode, or simply a carbon resistor. It has been said many times that the most efficient and effective noise source you can make is a simple carbon resistor with a current flowing through it.

To better understand the statements above, consider the actions of a current as it flows through a resistance. (We use *resistance* rather than *resistor* because we want to make this a general explanation rather than pinning it down to a specific component. Thus, by using resistance we can be referring to resistors or junctions of active devices.) With a current flowing in a resistance the electrons are free to move about. As the temperature of the resistance is increased, either by external means or by an increase in the current flow through the device, the electrons receive more energy and move around the material more so that they have many more interactions with molecules and other electrons within the material. This can be related to the system used for lottery drawings in some states. You have seen how a group of ping pong balls with numbers are placed in a clear plastic dome. When it's time for the drawing, a stream of air is turned on to make the balls move about inside the dome. As the air is increased, the balls move about more rapidly and many collisions occur. If we call the dome our resistive material, the ping pong balls our electrons, and the stream of air our temperature, then we have a direct analysis. The random movement of the ping pong balls in the dome produces a random drawing of numbers for the lottery. The random movement of electrons within the resistive device produces a current. The current flowing through the resistive element produces a voltage drop across it. This random type of voltage or current produced by an elevation of temperature is called **thermal noise**. As the name implies, this type of noise is directly proportional to temperature.

Thermal noise is completely random and, as such, never repeats itself. This produces a frequency spectrum that has a power level that is flat with frequency. That is, the noise spectrum has a constant level over a very wide frequency range, which can be verified by turning on a spectrum analyzer with no signal input. The "grass" level we referred to before has a constant level over the entire screen. If you change frequencies, the level remains constant, because of the properties of the random thermal noise. Since virtually all frequencies are present in the thermal noise, it is sometimes referred to as **white noise**. The analogy here is to white light, which contains all colors. Similarly, white noise contains all frequencies.

An important parameter of thermal noise, or any noise, is the average power of the noise. This is important since measurement of the noise characteristics of a system or device is a comparision of the noise that is applied to the input to that which is present at the output. Thus, an accurate reading of the input noise level is necessary.

The average noise power delivered to a system at some temperature T is described as

$$N_{\text{th}} = kTB \tag{6.1}$$

Where: N_{th} = Thermal-noise average power (in watts)
k = Boltzmann's constant (1.38×10^{-23} W/kHz)
T = Temperature in K (Kelvin = °C + 273)
B = Bandwidth in which the measurement is made in hertz

Example 6.2

Given: A system operating at 31°C in a bandwidth of 200 Hz.

Find: The thermal-noise power

Solution

$$N_{\text{th}} = kTB$$
$$= 1.38 \times 10^{-23}(31^\circ + 273)(200)$$
$$= 8.39 \times 10^{-19}\text{ W}$$

It should be pointed out that the average noise power is independent of the resistance of the device being tested. This can be seen in Equation (6.1), which also reinforces our statement that the noise power is directly proportional to the temperature of the device. The other variable, the bandwidth, is like a faucet that can be opened or closed with great accuracy. If we do not want a lot of water, we simply do not turn the

Figure 6.5 Noise source.

faucet on or turn it on very slightly. (A system with a narrow bandwidth allows a limited number of frequencies to pass and thus a low level of noise.) If we require a great deal of water, we turn the faucet to its maximum on position. (A system with a wide bandwidth allows a great many frequencies to pass and also a large level of noise.) So, we see how the bandwidth also directly controls the noise power if all other parameters are held constant. If we consider the average noise power for a system in a 1-Hz bandwidth, we will find that it is designated as N_0 and expressed mathematically as

$$N_0 = kT \tag{6.2}$$

The term N_0 is called the **noise power density**.

A noise source that characterizes thermal noise is shown in Figure 6.5, which consists of a source E_n and a resistance R. If this source were connected to a circuit with an impedance of R equal to the resistance of the source, the source R and the load R would both have a voltage $V_n/2$ across them. By using Equation (6.1) and the relationship $P = E/R$, we can discover the expression for the noise voltage V_n as

$$V_n = \sqrt{4RkTB} \tag{6.3}$$

Example 6.3

Given: A device is operating at a temperature of 22°C with a bandwidth of 4 kHz.

Determine:
1. Noise power density
2. Total noise power
3. Noise voltage for an internal resistance of 75 Ω and a 75 Ω load.

Solution

1. The noise power density from Equation (6.2) is

$$N_0 = kT = 1.38 \times 10^{-23}(22 + 273)$$
$$= 4.07 \times 10^{-21} \text{ W/Hz}$$

2. The total noise power is given by Equation (6.1) as

$$N = kTB = N_0 \times B = 4.07 \times 10^{-21}(4 \times 10^3)$$
$$= 1.63 \times 10^{-17} \text{ W}$$

3. Noise voltage is given in Equation (6.3) as

$$V_n = \sqrt{4RkTb} = \sqrt{4RN} = \sqrt{4(75)(1.63 \times 10^{-17})}$$
$$= 0.0677\ \mu\text{V (microvolts)}$$

The second type of internal noise, **shot noise**, originally came from vacuum-tube applications. To understand this analogy, consider the construction of a vacuum tube. A filament on the inside of the tube is heated by running current through it. A cathode, which is a metal cylinder, is wrapped around the filament. There may then be a wire mesh grid structure if the tube is a triode (three-element device), or just another metal cylinder that is the plate of the tube. When the filament is heated, electrons are "boiled" off the cathode and proceed through the vacuum of the tube toward the plate. The plate has a positive voltage (usually around +150 to 200 V) which attracts and accelerates the electron toward the plate structure. When the electrons strike the plate, a noise is created similar to dropping a handful of BB's (or shot) onto a metal plate, thus the name *shot noise*. Figure 6.6 is a two-dimensional picture of a tube with two elements, showing how the electrons strike the plate and cause the shot noise.

This same type of noise is also produced within a semiconductor when an electron with a charge q consisting of an average current I_{dc} crosses a potential barrier. When these electrons are gathered into the collector region, a noise is generated that is a shot noise. This noise occurs because there is no repetitive, direct path the electrons can take as they progress through the semiconductor and cross the potential barrier. Mathematically, shot noise is expressed as

$$I_n = \sqrt{2qI_{dc}B} \tag{6.4}$$

Where: I_n = Root-mean-square average noise current (in amperes)
q = Charge of an electron, 1.6×10^{-19} C
I_{dc} = dc bias (in amperes)
B = Bandwidth (in hertz)

Like the thermal noise previously covered, shot noise is flat with frequency. Equation (6.3) is valid for frequencies comparable to the inverse of the transmit time

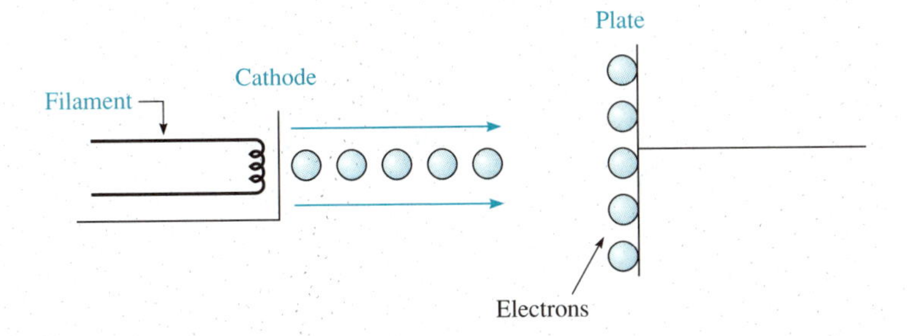

Figure 6.6 Vacuum tube analogy for shot noise.

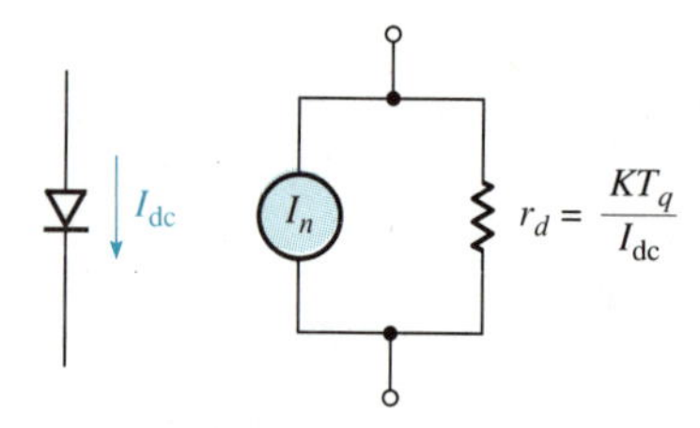

Figure 6.7 Equivalent noise generator setup.

of the electrons across a junction area. This frequency, $f = 1/T$, is well into the gigahertz range. (T is the transit time of the electron to cross the junction within the semiconductor.) An equivalent noise generator for a single-diode junction is shown in Figure 6.7. We see that the diode with a current flowing through it I_{dc} results in a current source with a current of I_n and a source resistance equal to kTq/I_{dc}, with the parameters defined above in Equation (6.3). To understand how this concept applies to a junction, consider the following example.

Example 6.4 If $I_{dc} = 2.1$ mA (milliamperes), and the noise of the diode is measured in a bandwidth of 7 MHz, determine:

1. Noise current
2. Equivalent noise voltage

Solution

1. Noise current is given by Equation (6.4).

$$I_n = \sqrt{2qI_{dc}B}$$

$$= \sqrt{2(1.6 \times 10^{-19})(2.1 \times 10^{-3})(7 \times 10^{6})}$$

$$= 68 \text{ nA (nanoamps)}$$

2. From Figure 6.7 we see that the resistance of the junction r_d is

$$r_d = \frac{kTq}{I_{dc}}$$

$$= \frac{26 \text{ mA}}{I_{dc}}$$

$$= 12.38\ \Omega$$

(The quantity, kTq can be between 25 and 40 mV (millivolts) at room temperature, depending on the junction doping. Generally, 26 mV will do for the majority of cases.)

$$E_n = I_n(r_d)$$

$$= 68 \times 10^{-9}(12.38)$$

$$= 0.84\ \mu\text{V}$$

The third type of internal noise is **flicker noise**, which is associated with defects in the surface structure of semiconductors. The noise power contained in this type is dependent on bias current through a particular device. The frequency response of flicker noise is much different from that of thermal or shot noise; whereas thermal and shot noise have a very flat response over a broad range of frequencies, flicker-noise power decreases with frequency. This power is inversely proportional with frequency and exhibits this $1/f$ characteristic for low frequencies. Above a few kilohertz the noise power is relatively low, but is basically flat over a wide range of frequencies.

Figure 6.8 shows frequency versus noise power for flicker noise. The $1/f$ section is from f_1 to f_2, which is basically 0 Hz to some low kilohertz value. The system then takes on the form of white noise from that point on and has a constant flat response through the rest of the frequency spectrum.

Flicker noise is sometimes referred to as **pink noise** because most of the noise power is concentrated at the low-frequency end of the spectrum, as distinguished from the white-noise spectrum, which is higher frequency noise.

The $1/f$ noise (flicker) is produced in our old friend the carbon resistor when it carries dc current. Thus, if you are operating at the low end of the frequency spectrum with carbon resistors, you may want to replace them with metal-film resistors to eliminate this source of noise.

The final type of internal noise is called **burst noise**, another type of low-frequency noise. It has been shown that when this noise occurs, there is a sudden and random shift in bias-current levels. These are very short duration changes that just as suddenly and randomly return to their original state. The result is an impulse type of noise spike in the system, the result of which is a "pop," or burst, within the system. If there is sufficient amplification within the system and a speaker is attached, the pop can be heard. This pop in the system led to the name "popcorn noise."

Burst noise, like flicker noise, is very device-sensitive. This characteristic makes any mathematical model of the noise virtually impossible, and the phenomenon is

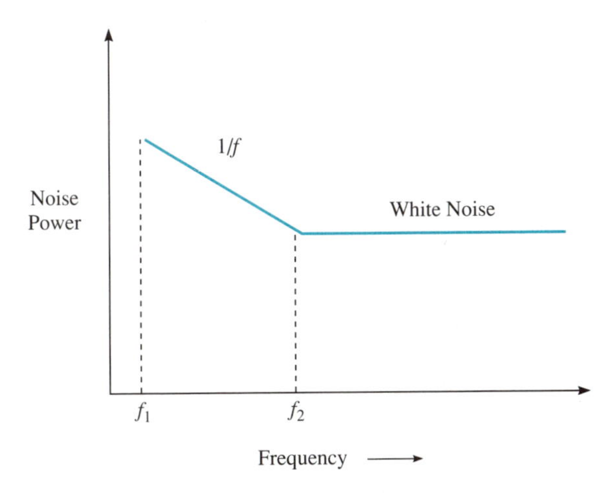

Figure 6.8 Noise power vs. frequency for flicker noise.

therefore very difficult to predict. However, it is known that the noise increases with bias level and is of a low frequency, inversely proportional to the square of frequency ($1/f^2$).

6.4 Noise Bandwidth

In most calculations the noise bandwidth is assumed to be an ideal rectangular shape. This assumption allows us to analyze what is happening and to ignore any effects of such components as capacitors and inductors in the circuits. Unfortunately, this does not reflect the real world. In real-life situations there are capacitors and inductors in every communications circuit and, as such, they will alter this ideal shape of the noise bandwidth.

An example of the statements above is shown in Figure 6.9. Figure 6.9(c) shows a low-pass filter, which is placed at the output of a noise source. Figure 6.9(a) displays an ideal situation where the capacitor in the filter is not considered and a nice rectan-

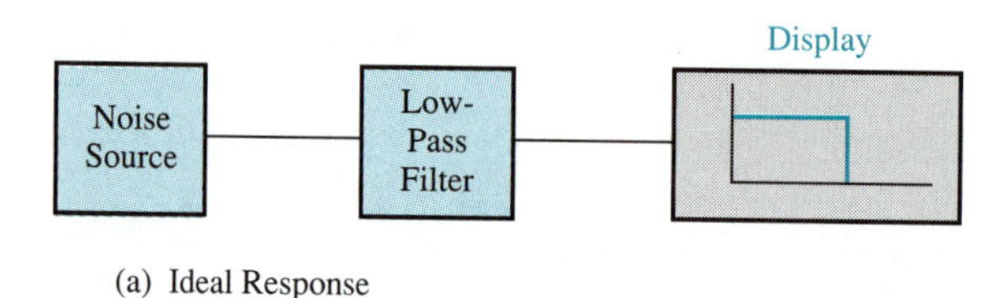

(a) Ideal Response

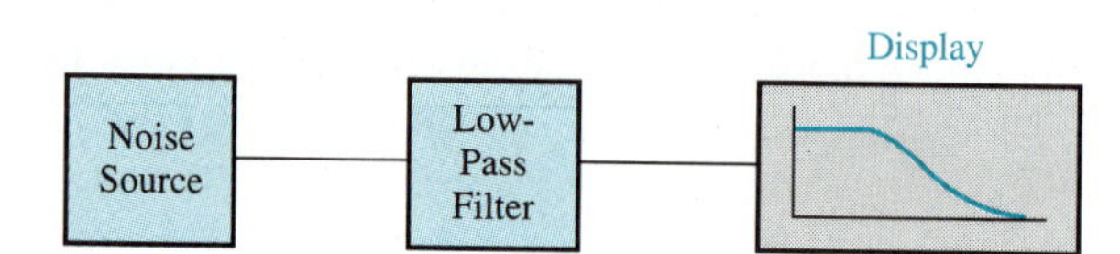

(b) Real World Response

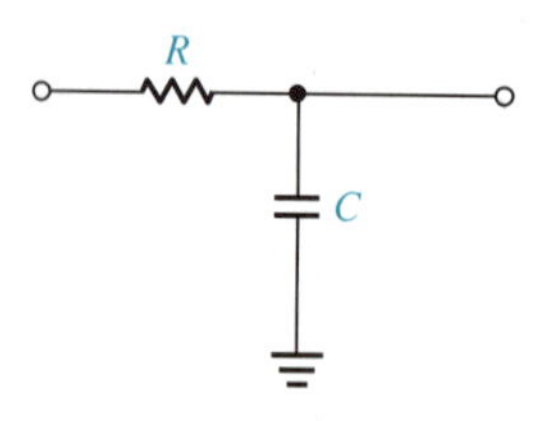

(c) Low Pass Filter

Figure 6.9 Effects of capacitance on noise.

gular response is present. Figure 6.9(b) is the real-world representation of the low-pass filter on the noise source. We see that the noise follows the actual response of the filter, rolling off at a specific cutoff frequency because the capacitor is now exhibiting a reactance that gets smaller and smaller as the frequency increases. This causes more of the signal to be shunted to ground and less to be sent to the display. Thus, there is a definite bandwidth associated with the noise at this point.

One point should be made now about the response of a noise spectrum: The noise level never reaches zero. If we look at the **filter transfer function** (the expression that describes what the filter is doing as far as voltage in and out as a function of frequency in this case), we would see that the response should go to zero as the frequency approaches infinity. This, however, is not the case, because even if the capacitor becomes a "perfect short" and the resistor exhibits no internal capacitance, there would still be noise generated within the spectrum analyzer (display). This low value of noise level is called the **noise floor** and, as you recall from previous discussions, will equal kTB. This term was called N_{th} in our previous discussion (Equation 6.1). We now will refer to it as the noise floor of a particular device and it will be the limiting factor in noise calculations and measurements.

The question arises as to what bandwidth should be used in calculating the output noise of a specific device (the filter in our case). We see from Figure 6.9 that there is more noise in the circuit beyond f_c, the cutoff frequency. This particular case can be expressed as a rectangular response curve with a width equal to the equivalent noise bandwidth, which is

$$B_{\text{eq}} = \frac{\pi}{2} f_c \tag{6.5}$$

(f_c, the cutoff frequency, is the point where the response of the filter falls off by 3 dB from the passband loss)

Equation (6.5) indicates that for our low-pass filter, the total noise power adds up to 57% ($\pi/2 = 1.57$) more than an ideal flat response up to the cutoff frequency f_c. This is a good approximation and usually comes out very close to the noise power that is actually measured.

For more complex shapes a mathematical model must be obtained for each individual case. With the shape determined, the noise spectral-density function is now integrated over the entire frequency-response shape. If the shape is not flat ($1/f$ noise, for example), the shape must be considered and the integration done accordingly. Because this text does not cover integration procedures, we will not elaborate further on the mathematical procedures. Suffice it to say that complex wave shapes require complex mathematical analysis.

6.5 Noise Parameters

To work with noise and its characteristics, it is necessary to understand the parameters used to describe it. This section will cover signal-to-noise ratio, noise factor, noise figure, and noise temperature. Each of these terms can be used to describe noise in a system. Some are more convenient to use for low frequencies, others for higher frequencies. Some are used more when speaking of low-noise devices; others have more meaning when describing general noise characteristics.

The most basic, and probably the most easily understood, term associated with noise is the **signal-to-noise ratio**. The idea of a signal entering or leaving a device is relatively easy to understand. There may be a sine, square, triangular, or any other type of wave entering a device as the desired signal, designated S_i. The desired signal coming out of that same device, hopefully of the same shape, is designated S_o. These are the desired *signals* associated with a device.

Along with the desired signal at the input to a device is a certain amount of noise that may have been created in any number of ways. It may be an external noise or any type of internal device generated in previous stages and fed onto the device we are analyzing. Regardless of the source, there will be a certain level of noise, designated N_i, present at the input. Similarly, the input noise is amplified or attenuated by our device, and any internal noise the device itself generates is added to that noise. This noise at the output of the device is termed N_o. These terms are the *noise* components of the signal-to-noise ratio.

Having defined the two types of components involved, it is now time to define signal-to-noise ratio as a whole. Since the term is a ratio, we will be comparing two terms with the same units. These units may be volts, milliwatts, watts, or any other unit that describes a signal or a noise component.

The first term to look at is the input signal-to-noise ratio (SNR). This term is a comparison of the signal present at the input of a device to the noise present at the same point, that is,

$$\text{Input SNR} = \frac{S_i}{N_i} \tag{6.6}$$

Where: S_i = Signal input level
N_i = Noise input level

Consider the example of Equation (6.6) where an input signal has an amplitude of 2 V and the noise associated with this signal is measured at 0.15 V. The input SNR will then be 2/0.15, or 13.333.

The output signal-to-noise ratio (SNR) is a comparison of the signal present at the output of a device and the corresponding noise level at the same point. Expressed mathematically, it is

$$\text{Output SNR} = \frac{S_o}{N_o} \tag{6.7}$$

Where: S_o = Signal output level
N_o = Noise output level

An example of Equation (6.7) may be a signal level of 6 V measured at the output of an amplifier with a corresponding noise level of 0.5 V. For this example the output SNR would be 6/0.5, or 12.

The parameters discussed above (input versus output SNR) are very useful in calculating the actual noise characteristics of a specific circuit. The ideal limit of a system's sensitivity is set by the noise present at the input. In practice, however, the sensitivity is often limited by noise generated within the system or component itself. The number that indicates how closely the ideal is approached is called the **noise factor**, F. This term is defined as the ratio of the input SNR to the output SNR. Mathematically, it is

$$F = \frac{\text{Input SNR}}{\text{Output SNR}} \tag{6.8}$$

or

$$F = \frac{S_i/N_i}{S_o/N_o}$$

We can illustrate the concept of noise factor by using the examples presented above for input SNR and output SNR. Recall that we calculated the input SNR as 13.33 and the output SNR as 12. Using these numbers, we find that the noise factor for our example will be 1.11, a good noise factor because it is very close to a 1.0 value. The 1.0 value is an ideal number that means that the system or component has contributed no additional noise to the system. To illustrate, consider the example used to explain Equation 6.6: the input SNR had a signal level of 2 V and a noise level of 0.15 V. As we calculated before, the input SNR was 13.33. The output signal level was then reported to be 6 V, which means that the device had a gain of 3. If the circuit is ideal and does not contribute any noise, the input noise should also only be increased by a factor of 3. This would mean that the output noise level should be 0.45 V. If we assume this to be the case, the calculations would be done in the following manner:

$$\text{Input SNR} = \frac{2}{0.15} = 13.33$$

$$\text{Output SNR} = \frac{6}{0.45} = 13.33$$

$$\begin{aligned} F &= \frac{\text{Input SNR}}{\text{Output SNR}} \\ &= \frac{13.33}{13.33} \\ &= 1.0 \end{aligned}$$

For many calculations the term noise factor can be used. There are, however, many cases where another term is needed to express noise. Such a term is noise figure. **Noise figure** is found by taking the log of the noise factor and multiplying it by 10. That is,

$$\text{Noise figure (in decibels)} = 10 \log F \tag{6.9}$$

To illustrate this idea, consider the two cases (Equations 6.6 and 6.8) for which we calculated a noise factor of 1.11 and 1.0, respectively, for the first case

$$\begin{aligned} \text{N.F.} &= 10 \log F \\ &= 10 \log (1.11) \\ &= 10(0.0453) \\ &= 0.453 \text{ dB} \end{aligned}$$

For the second case ($F = 1.0$) the calculations are

$$\begin{aligned} \text{N.F.} &= 10 \log F \\ &= 10 \log (1.0) \\ &= 10(0) \\ &= 0 \text{ dB} \end{aligned}$$

Thus, we see that in the ideal case of a device or system that contributes no additional noise, there is a noise figure of 0 dB. This is why we have said that the first example was well designed because its noise figure was only 0.45 dB, which is very low.

The following examples will illustrate how the noise figure is calculated and what constitutes low noise and high noise.

Example 6.5 Low Noise

Given:

Input signal = 5 V
Input noise = 0.5 V
Output signal = 10 V
Output noise = 1.2 V

Find:

1. Noise factor
2. Noise figure in decibels

Solution

1. $$\text{Input SNR} = \frac{S_i}{N_i} = \frac{5}{0.5} = 10$$

$$\text{Output SNR} = \frac{S_o}{N_o} = \frac{10}{1.2} = 8.33$$

$$\text{Noise factor} = F = \frac{\text{Input SNR}}{\text{Output SNR}}$$

$$= \frac{10}{8.33}$$

$$= 1.20$$

2. $$\text{Noise figure} = 10 \log F$$

$$= 10 \log 1.2$$

$$= 10(0.079)$$

$$= 0.79 \text{ dB}$$

Example 6.6

High Noise

Given: Input signal = 5 V
Input noise = 0.5 V
Output signal = 10 V
Output noise = 4.5 V

Find: **1.** Noise factor
2. Noise figure in decibels

Solution

1. $$\text{Input SNR} = \frac{S_i}{N_i}$$

$$= \frac{5}{0.5}$$

$$= 10$$

$$\text{Output SNR} = \frac{S_o}{N_o}$$

$$= \frac{10}{4.5}$$

$$= 2.22$$

$$\text{Noise factor} = F = \frac{\text{Input SNR}}{\text{Output SNR}}$$

$$= \frac{10}{2.22}$$

$$= 4.5$$

2. Noise figure $= 10 \log F$

$$= 10 \log 4.5$$

$$= 10(0.653)$$

$$= 6.53 \text{ dB}$$

When two or more amplifiers or devices are cascaded together, as in a communications transmitter or receiver, the total noise figure of the system is a function of each of the individual noise figures of that system. The input noise figure of the system will, however, generally depend only on what the noise figure of the first stage is. This is guaranteed if the gain of this first stage is reasonably high. The mathematical relationship for total noise factor which is converted to noise figure is

$$NF = F_1 + \frac{F_2 - 1}{G_1} + \frac{F_3 - 1}{G_1 G_2} + \frac{F_4 - 1}{G_1 G_2 G_3} + \cdots \quad (6.10)$$

Where: NF = Total noise factor
F_1 = Noise factor of the first stage
F_2 = Noise factor of second stage
F_3 = Noise factor of third stage
F_4 = Noise factor of fourth stage
G_1 = Gain of first stage
G_2 = Gain of second stage
G_3 = Gain of third stage

All of the noise factors shown in Equation (6.10) are as described earlier. Also, the gains are not in decibels but are also ratios. Generally the noise figures and gains are given in decibels and you are required to convert them to either noise-factor or gain ratios, place them in Equation (6.10) to calculate total noise factor, and then change the final answer to decibels.

Example 6.7 The circuit shown in Figure 6.10 is a three-stage amplifier with the following characteristics:

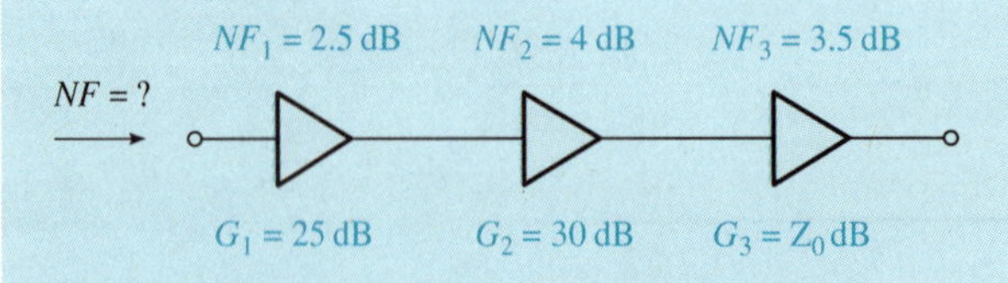

Figure 6.10 Noise figure example.

$$NF_1 = 2.5 \text{ dB}, G_1 = 25 \text{ dB}$$

$$NF_2 = 4 \text{ dB}, G_2 = 30 \text{ dB}$$

$$NF_3 = 3.5 \text{ dB}, G_3 = 20 \text{ dB}$$

(Gains are power gains, dB = 10 log P)

Solution

The first task is to convert all of the noise figures and gains into ratios. Recall that $NF = 10 \log F$, and for the gains Gain (in decibels) = 10 log P. With these equations, the following ratios are obtained:

$$F_1 = 1.77$$

$$F_2 = 2.50$$

$$F_3 = 2.25$$

$$G_1 = 325$$

$$G_2 = 1000$$

$$G_3 = 100$$

Using Equation (6.10), we obtain

$$NF = 1.77 + \frac{2.5 - 1}{325} + \frac{2.25 - 1}{(325)(1000)}$$

$$= 1.77 + \frac{1.5}{325} + \frac{1.25}{325{,}000}$$

$$= 1.77 + 0.0046 + 0.0000038$$

$$= 1.7746$$

$$NF \text{ (dB)} = 10 \log 1.7746$$

$$= 10(0.2491)$$

$$= 2.491 \text{ dB}$$

As we mentioned, the noise figure is usually stated in decibels. There are times, however, when noise is expressed as a **noise temperature**. The formula to calculate noise temperature is

$$NF = 10 \log \left(\frac{1 + T_n}{290} \right)$$

Where: NF = Noise figure in dB
T_n = Noise temperature in K

Converting all of the noise figure calculations in this chapter to noise temperature, we obtain

Example	Noise Figure (dB)	Noise Temp (K)
5.4	0.79	58
5.5	6.53	1015
5.6	2.49	224

Thus, we see that noise is a very important part of any communications system, and there are many different ways to express noise. You should be very careful of the units used for a particular system and make the calculations accordingly.

6.6 Summary

This chapter introduced the concepts and definitions of noise. It showed how noise affects the operation of communications systems and how to calculate specific noise parameters for a particular system.

The chapter dealt with both correlated and uncorrelated noise, dividing uncorrelated noise into external and internal noise.

Parameters such as the noise figure, noise factor, and noise temperature were then investigated to complete the characterization of noise in a communications system.

Questions

6.1 Introduction

1. Define **noise**.
2. Name one type of noise that is not random.
3. Why is it impossible to use a feedback system to remove noise?
4. Why is noise so difficult to characterize?

6.2 Correlated Noise

5. What is **correlated noise**?
6. Name two types of correlated noise.
7. Define **compression point**.
8. What frequencies are contained in second-order intermodulation products?
9. What is one cause of harmonic distortion?
10. How is intermodulation distortion meausured?
11. Distinguish between second order and third order intermodulation distortion.

6.3 Uncorrelated Noise

12. Name two types of external noise.
13. Name two types of internal noise.
14. Why is thermal noise also called **white noise**?

15. Why is flicker noise called $1/f$ noise?
16. Why is thermal noise the most common type?
17. What is **uncorrelated noise**?
18. How can external noise be controlled?
19. How can internal noise be controlled?
20. Define **shot noise**.
21. Define **burst noise**.
22. Distinguish between **noise power** and **noise power density**.

6.4 Noise Bandwidth

23. Define **noise floor**.
24. What is a **noise bandwidth**?

6.5 Noise Parameters

25. What is **input SNR**?
26. Define **noise factor**.
27. Define **noise figure**.
28. What is output SNR?
29. Define **noise temperature**.
30. Explain the conditions necessary to have the first stage noise figure in a multistage system by basically the noise figure of the entire system.

Problems

6.2 Correlated Noise

1. We have signals $f_1 = 5$ MHz and $f_2 = 7$ MHz applied to the input of a nonlinear device. What is the output spectrum of the device? Include the fundamentals and the second and third intermodulation distortion only.
2. If two signals are applied to the input of a nonlinear device: $f_1 = 2$ MHz and $f_2 = 2.6$ MHz, what are the third-order intermodulation products at the output?

6.3 Uncorrelated Noise

3. What is the noise power density of a device at 33°C?
4. A device is operating at 18°C and has a bandwidth of 10 kHz. What is the noise power density and the total noise power?
5. A system operates at a temperature of 40°C. The maximum noise power allowed is 9×10^{-17} W. What is the maximum bandwidth the system can have?
6. A device is operating at 25°C with a bandwidth of 15 kHz. Find the noise power density, total noise power, and the noise voltage for an internal resistance of 50 Ω and a load of 50 Ω.
7. A dc current in a diode is found to be 3.6 mA. Determine the noise current and the equivalent noise voltage.

6.4 Noise Bandwidth

8. For a cutoff frequency of 150 MHz, what is the equivalent noise bandwidth?
9. The cutoff frequency in Problem 8 is increased to 200 MHz. What is the percentage of increase in the equivalent noise bandwidth?

6.5 Noise Parameters

10. An input signal is at 18 V and has a noise level of 0.22 V. What is the SNR?

11. An input signal is at 7 V with an associated noise of 0.5 V. The output is 21 V with a noise level of 1.6 V. What is the noise factor and the noise figure?

12. A two-stage amplifier has the first stage with a gain of 20 dB and a noise figure of 1.5 dB. The second stage has a gain of 26 dB and a noise figure of 3.2 dB. Find the resultant input noise figure of the amplifier.

13. A system has an input SNR of 18 and an output SNR of 16. What is the noise factor, noise figure, and noise temperature of this system?

7

Outline

Objectives

- To introduce the components that make up communications systems
- Present the theory and uses of oscillators (R–C, L–C, and crystal), rf (radio frequency) amplifiers, mixers, and filters
- Discuss and relate each component to a particular portion of a communications system

Key Terms

oscillator
Barkhausen criteria
Wien bridge
resonance
piezoelectric
amplifier
interelement capacitance
local oscillator
conversion loss
filter
low-pass filter
bandpass filter
feedback
regenerative feedback
lead–lag network
crystal
temperature coefficient
stray capacitance
mixer
nonlinear
isolation
noise figure
high-pass filter
ultimate attenuation

Communications Components

7.1 Introduction

Thus far in this text, we have discussed the basics of communications systems; introduced the idea of propagation, and looked at the antennas and transmission lines used to transfer energy into a propagating medium; investigated signal spectra; and looked at the properties and effects of noise on a communications system. With this detailed background, we are now ready to look at the heart and soul of every communications system—its components.

The components used in a communications system, or any system, are similar to the pieces used in a chess game. If you are familiar with the game of chess you know that each piece can only move a very specific way; some move diagonally, some forward, and some may only move one space at a time. Knowing how each piece is allowed to move enables you to plan your moves and to anticipate your opponent's moves in advance. Without this knowledge, you could not win many chess matches. Similarly, without knowledge of the components used for communications systems, you would not be able to design or to evaluate a system with any success.

We will cover four categories of components used in communications systems: oscillators, which generate the signals; rf amplifiers, which increase the amplitude of the signals; mixers, which combine signals to form new ones; and filters, which can eliminate unwanted signals and pass the desired ones.

7.2 Oscillators

In most cases the term *oscillation* induces us to think of an undesired condition, as in the case where we need a pure amplified signal with no interference from other signals. There are times, however, when we purposely design a circuit to oscillate at a certain frequency or range of frequencies. The largest majority of signal sources originate with some form of oscillator. To understand the concept of an oscillator, let us begin with a general definition of the term itself.

Oscillator

> A device, employing positive feedback, which, by using a dc input and frequency-determining components, produces an output that fluctuates between two states, (+) and (−). This fluctuation may be around a zero point, or a positive or negative dc level.

This general definition takes in a large scope of oscillators, and some of the terms in this definition need to be expanded on. To ensure that you have sufficient background on the topic of oscillators, we will begin with the basics.

For most oscillator applications, it is necessary that an oscillator begin oscillating and sustain that oscillation until the operator is finished with that particular signal and turns it off. Some applications require only one shot of oscillations when they are turned on, but because these are in the minority, we will not discuss them here. We will, however, mention the "one-shot" applications later in this chapter. In general, we will discuss oscillators that are turned on and are required to sustain the required oscillations.

In order for an oscillator to produce self-sustaining oscillations, certain criteria must be met by the circuit. The circuit must have

1. An input power source
2. Gain
3. Frequency-determining components
4. Positive feedback

These four criteria must *all* be met for the circuit to sustain its oscillations. This situation is similar to the triangle needed to have a fire: heat, fuel, and oxygen. If any one of the three components are missing, there will be no fire. Similarly, if any one of the four criteria are missing, the oscillator will not sustain its oscillations. Remember this when you are troubleshooting an oscillator that is not operating. These criteria represent a list of absolute necessities for a functioning oscillator.

Now we need to explain each item in more detail.

An **input power source** is necessary because the oscillator requires power to operate. As we will see when we get into specific types of oscillator circuits, some type of device (solid state or vacuum tube) is required for the oscillator to operate. This device needs a certain number of volts to perform the functions for which it was designed. The dc power used to provide the necessary voltage is also the input used to produce an output from the completed oscillator. In one sense, the oscillator is actually a dc-to-rf converter because the only input to the device is a dc voltage, and the output of the device is an rf signal (This signal may also be an audio signal if that is what is desired).

Gain is needed in the oscillator circuit (A_v) to overcome any internal losses within the circuit. This gain, as we will see later on, need only be large enough to produce a product with feedback (B) that is equal to 1: that is, $A_vB = 1$. This usually means that the gain of the active device need not be very great.

Example 7.1

The forward system gain (A_v) for an oscillator circuit has been measured as 15. Find the feedback factor necessary to begin oscillations in the circuit.

Solution

For the system to operate properly, the product of A_v and B must be unity. Thus,

$$A_vB = 1$$

$$15B = 1$$

$$B = \frac{1}{15}$$

The frequency of oscillations at the output of the device is determined by **frequency-determining components**. These components may be resistor–capacitor combinations (R–C), inductor–capacitor combinations (L–C), resistor–inductor–capacitor combinations (R–L–C), cavity resonators, tuning forks, or a number of other combinations that result in a resonant circuit that produces only the desired frequency at the output of the oscillator.

The final criterion, **feedback**, brings us back to our original definition of an oscillator. For a feedback scheme to be effective in sustaining oscillations within a device, it must conform to certain criteria itself. This criteria, termed the **Barkhausen criteria**, states (1) that the signal must be exactly in phase with the original input signal at the loop closure point, and (2) the steady-state gain around the feedback loop must be exactly equal to unity. To fully understand these requirements, let us refer to Figure 7.1 for a more complete explanation.

As we can see in Figure 7.1, the term at the input, designated V_{in}, is the input power parameter listed in the original self-sustaining criteria. This input is sent to the amplifier portion of the loop (A_v), where a portion is fed back. The feedback network (B) causes any necessary phase inversion and makes up any gain that may be needed to have the product of A_v and B equal unity. The feedback voltage V_{fb}, is now added to V_{in}, and the signal is reinforced. This operation is expressed mathematically as

$$\frac{V_{out}}{V_{in}} = \frac{A_v}{1 - A_vB} = A_{fb} \tag{7.1}$$

Where: A_v = Forward system gain (noninverting)

B = The fractional part of the output that is fed back (feedback factor)

A_{fb} = Overall system gain with feedback

Equation (7.1), with a noninverting A_v, is a representation of positive, or **regenerative, feedback**. This type of feedback is absolutely necessary for operation

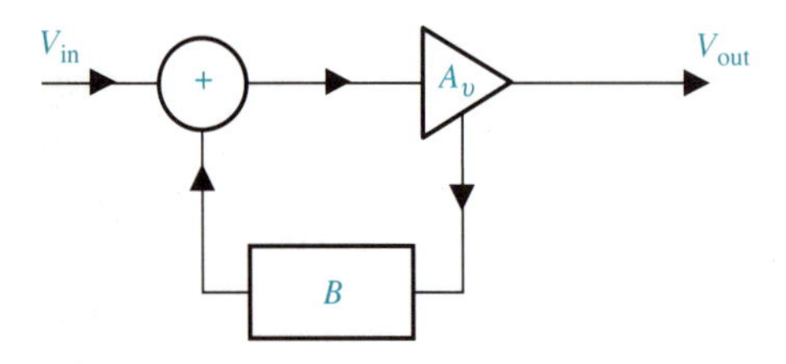

Figure 7.1 Feedback system for an oscillator circuit.

of an oscillator because the signal that is fed back must reinforce the original signal to keep the circuit in operation.

One point may be disturbing to you who have looked at Equation (7.1) and said that if we fulfill the Barkhausen criteria for sustained oscillations we could be in trouble. Recall that the criteria says that the feedback must be exactly in phase with the original input signal at the loop closure point. This can take place by having a noninverting amplifier and no phase shift in the feedback factor. This should cause no problem. The apparent problem arises when we fulfill the second part of the criteria that the overall steady-state gain around the feedback loop must equal unity ($A_vB = 1$). If we plug this part into Equation (7.1), we have $A_v/1 - 1$, or $A_v/0$, meaning that the overall system gain of the circuit goes to infinity. Under normal conditions or when negative feedback is incorporated, this would be disastrous. If we allowed the gain of a normal amplifier to go to infinity, we would have a circuit that would put out many frequencies different from those desired. In other words, it would be oscillating. But this is exactly what we want our circuit to do—oscillate. Thus, a mathematical gain of infinity is a desirable and necessary condition.

We have mentioned many times that there is a specific criterion for sustaining oscillations in a circuit. The question that has not been addressed as yet is what does it take to get the circuit to oscillate in the first place? Figure 7.2 shows an oscillator circuit consisting of the gain module (A_v) and the feedback factor (B) along with a plot of the output voltage (V_{out}) as a function of time. You can see that a portion of the output voltage has to build up in amplitude from time $t = 0$ to some further specified time t. To get the circuit to start from zero and build the signal up to its

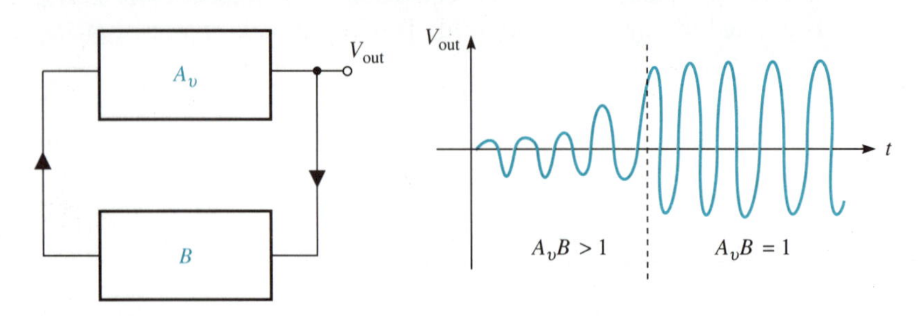

Figure 7.2 Conditions for beginning oscillations.

proper amplitude, a condition is required where A_vB is greater than unity. That is, $A_vB > 1$. With this condition present, the circuit has a high enough gain to cause it to amplify the signal and reinforce it to produce oscillations. Once the proper amplitude is reached, circuit elements are either switched in or out to sustain the oscillations ($A_vB = 1$). We have therefore begun the oscillations and guaranteed that they will be sustained.

Example 7.2 Using the feedback factor from Example 7.1, $B = 1/15$, determine the overall system gain with feedback (A_{fb}), if the forward gain (A_v) is increased to 45 to ensure that the circuit sustained its initial oscillations.

Solution Using Equation (7.1), we obtain

$$A_{fb} = \frac{A_v}{1 - A_vB}$$

$$= \frac{45}{|1 - (45 \times 1/15)|}$$

$$= \frac{45}{|1 - 3|} = \frac{45}{2}$$

$$= 22.5$$

Now that we have the basic idea of how a circuit can be used as an oscillator, we will look at some specific types: R–C, L–C, and crystal oscillators. All of these circuits are used in communications systems for various applications.

7.2.1 R–C Oscillators

R–C oscillators are circuits that use resistors and capacitors in different combinations to provide the necessary phase shift in the positive feedback loop to produce and sustain oscillations. The two types of configurations we will look at are the R–C phase-shift oscillator and the Wien bridge oscillator.

The **phase-shift oscillator** uses a series of R–C networks to produce the required phase shift necessary to produce oscillations. A typical circuit using an operational amplifier as the gain module is shown in Figure 7.3, in which you can see three R–C networks in the feedback path from the output to the input. These networks (C_1 R_1, C_2 R_2, and C_3 R_3) may each have a maximum phase shift of 90°. The circuit will oscillate at the frequency where the total phase shift through the three R–C networks is equal to 180°. This accounts for one-half of the total phase shift required to produce and sustain oscillations; the other half is achieved within the operational amplifier itself. As you will note, the input is at the negative terminal, which is the inverting input of the operational amplifier, creating a signal at the output that is 180° out of

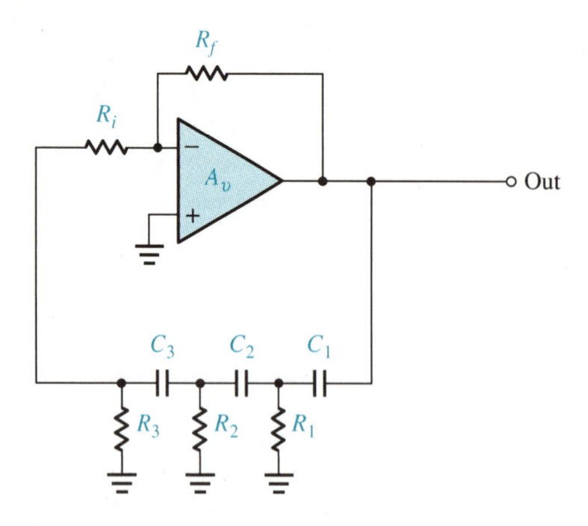

Figure 7.3 R–C phase shift oscillator.

phase with that of the input. Thus, the combination of the operational amplifier (180°) and the R–C networks (180°) results in a positive feedback at the frequency of oscillations.

By going through some complex calculations, we find that the feedback factor (attenuation) in the three R–C networks will exhibit a B of $1/29$. This means that the operational amplifier must have a gain of at least 29 in order to begin the oscillations in the ciruit ($A_vB = 1$, $A_v \times 1/29 = 1$, therefore $A_v = 29$). This gain is set by the values of R_i and R_f at the input of the operational amplifier and in the feedback path, respectively.

The frequency of operation for the R–C phase-shift oscillator is

$$F_r = \frac{1}{2\pi\sqrt{6}RC} \tag{7.2}$$

This equation may be further broken to

$$F_r = \frac{0.0649}{RC}$$

Example 7.3

Given: $R_i = 6.2\ \text{k}\Omega$
$C = 0.002\ \mu\text{F}$
$R = 15\ \text{k}\Omega$

Determine:
1. The value of R_f required for the circuit to oscillate
2. Frequency of oscillation

Solution

1. Since the closed-loop gain of the system must equal 29 from the discussions above, R_f/R_i must equal 29.

$$R_f = 29R_i$$
$$= 29(6200)$$
$$= 179.8 \text{ k}\Omega$$

2. Frequency of oscillation is given by Equation (7.2).

$$F_r = \frac{1}{2\pi\sqrt{6}RC}$$
$$= \frac{1}{2\pi(2.45)(15000)(0.002 \times 10^{-6})}$$
$$= \frac{1}{1.885 \times 10^{-4}}$$
$$= 5.305 \text{ kHz}$$

The second type of R–C oscillator is the **Wien bridge oscillator**. This circuit also uses R–C networks, but in a much different way from the R–C phase-shift circuit. You may recall the term Wien bridge from one of your previous classes as a ''capacitance comparison'' bridge, or simply a Wien capacitance bridge. This type of bridge, shown in Figure 7.4, has two resistive branches (R_1 and R_2) and two reactance branches (R_3, C_3 and R_4, C_4). The mathematical relationship of these branches is

$$\frac{C_3}{C_4} = \frac{R_2}{R_1} = \frac{R_4}{R_3} \tag{7.3}$$

and

$$\frac{C_4}{C_3} = \frac{1}{\omega R_3 R_4} \tag{7.4}$$

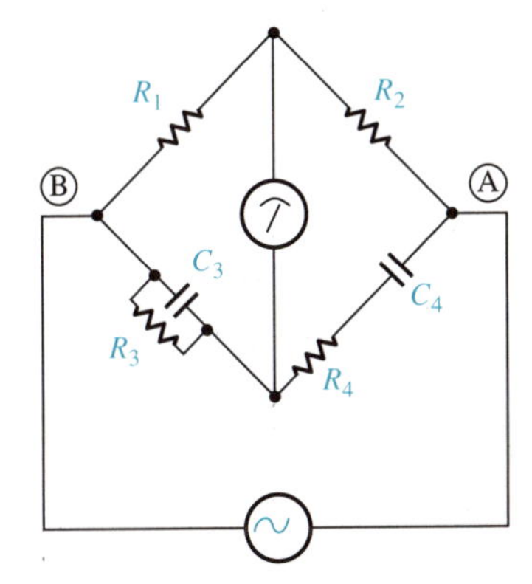

Figure 7.4 Wien capacitor bridge.

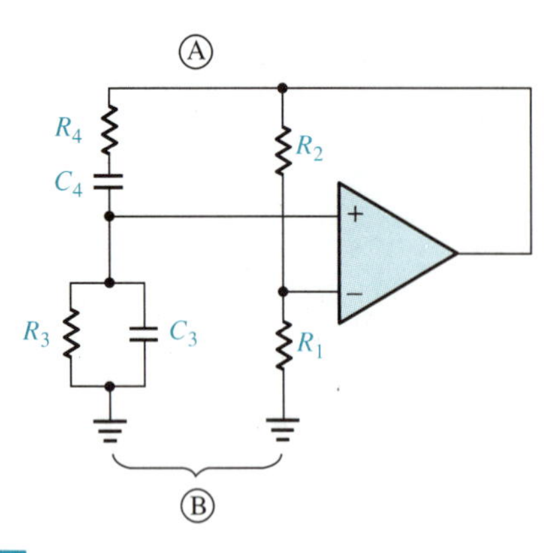

Figure 7.5 Wien bridge oscillator.

The equation of interest to us when discussing the Wien bridge oscillator is Equation (7.4). One term in this equation is absolutely necessary in designing an oscillator: That term is ω the frequency term, which tells us what the frequency of the bridge, and thus the oscillator, will be. By adjusting the values of either the capacitors or resistors in the appropriate leg, we can change the frequency of operation.

Now that we have a general idea of what the Wien bridge does, let us see how it fits into our scheme for a communications oscillator. The Wien bridge oscillator is shown in Figure 7.5. Comparing this circuit with the bridge in Figure 7.4, we see that the signal to the bridge in Figure 7.4 is applied between points A and B. If we connect point B to ground and make it a common point, we have the same situation in Figure 7.5; that is, the signal, which is the feedback signal from the operational amplifier, is applied between points A and B also (Point A is the junction of R_2 and R_4, C_4 and point B is the junction of R_1 and R_3, C_3).

The R–C networks perform a dual function within the Wien bridge circuit. The combination of R_1 and R_2 forms a voltage divider allowing the operational amplifier to perform as the amplifying element needed to have $A_vB = 1$ and providing a critical portion of the bridge circuit. The combination of R_3, C_3 and R_4, C_4 also have dual functions. First, they are like the R_1 and R_2 combination in that they are a critical part of the bridge circuit. Second, they form what is called a **lead–lag circuit**, which determines the frequency of operation of the oscillator.

The **lead** portion of the network consists of the combination of R_3, C_4, and the **lag** portion of the combination consists of R_4, C_3. To understand what this lead–lag circuit is and how it operates, let us separate it from the complete oscillator and look at it closely. We can see in Figure 7.6 that an equivalent circuit for the network is a reactive voltage divider, where the R_4, C_4 combination is Z_1 and the R_3, C_3 combination is Z_2. When a resonant frequency is reached with this circuit ($R = X_c$), there is a $-45°$ phase shift across the Z_1 combination and a $+45°$ phase shift across the Z_2 combination. As a result the total phase shift across the entire network is 0°. At frequencies below resonance, the lead network has control due to the high reactance of C_4. The circuit and response curve is shown in Figure 7.7(a). You can see that the

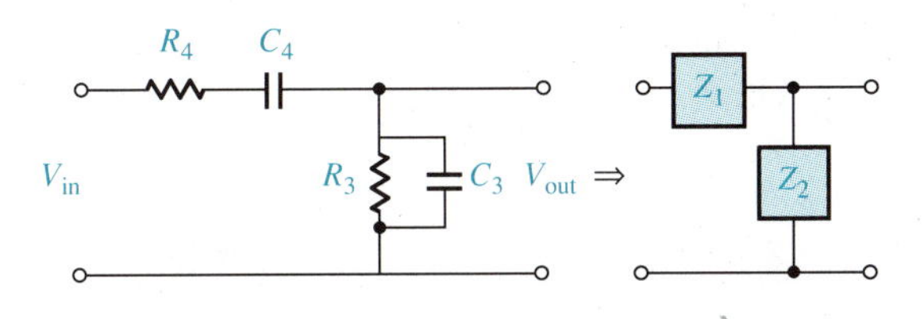

Figure 7.6 Lead-lag circuit.

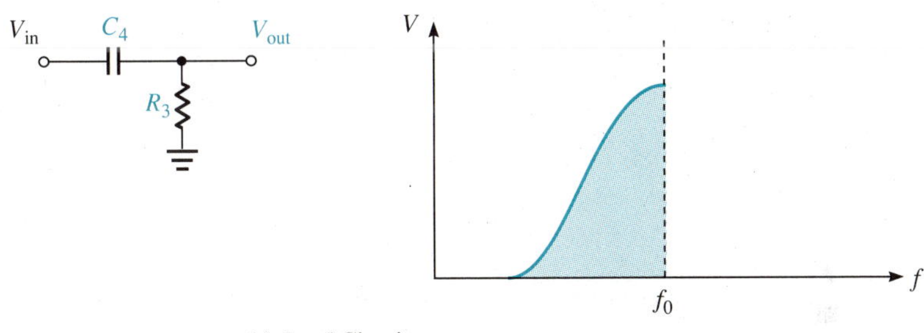

(a) Lead Circuit

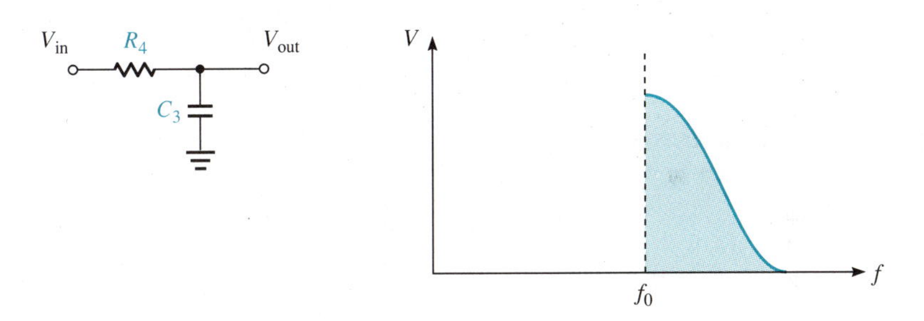

(b) Lag Circuit

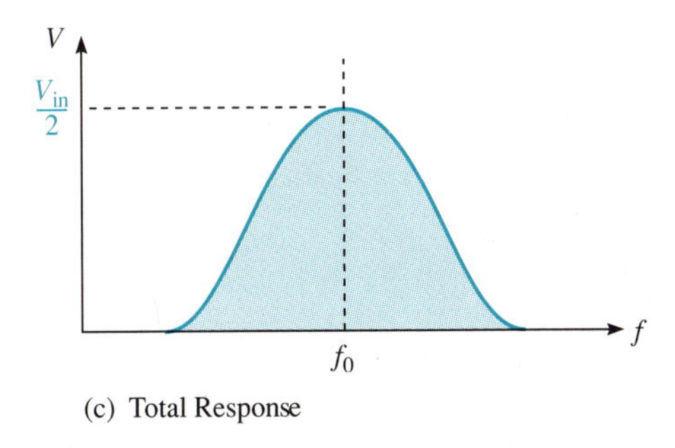

(c) Total Response

Figure 7.7 Lead-lag response curves.

capacitor is in series with the signal, and when it exhibits a large reactance it will not allow the signal to pass except as the resonant frequency is approached. Similarly, for frequencies higher than f_0, the lag circuit takes control because of the reactance of capacitor C_3. As you can see in Figure 7.7(b), this capacitor is in shunt with the signal. As the frequencies get farther and farther from f_0, the reactance decreases and shunts the signal to ground.

The result of the actions discussed above is that we obtain the bell-shaped curve shown in Figure 7.7(c). This curve indicates that the circuit combination will produce an output at the resonant frequency f_0 and will attenuate all of the frequencies around it in accordance with the response curve. By using basic ac theory calculations for two impedances, as shown in Figure 7.6, and substituting in values for Xc and R, we can show that the V_{out}/V_{in} ratio will result in V_{out} being equal to $1/3\ V_{in}$ when we are at f_0. This amplitude is also shown in Figure 7.6.

To determine the resonant frequency f_0 of the circuit, we can use the following expression:

$$f_0 = \frac{1}{2\pi RC} \tag{7.5}$$

Where: $R = R_3 = R_4$
$C = C_3 = C_4$

Now that we have a basic understanding of how the individual pieces of the oscillator function, we can look at the basic operation of the entire circuit. When the bridge is balanced (f_0), the difference voltage generated between the (+) inverting and (−) noninverting terminals of the operational amplifier will be 0. The voltage divider, which consists of resistors R_1 and R_2, provides negative feedback at this point, which offsets the positive feedback established by the lead–lag network. The resistance ratio of this voltage divider is set at 2:1, which makes the gain of the operational amplifier 3. This fulfills the requirements set forth previously that the combination of gain (A_v) and feedback factor (B) must equal unity to sustain oscillations. Recall that the V_{out} of the lead–lag network was found to be $1/3\ V_{in}$ at resonance. A gain of 3 will, therefore, multiplied with this 1/3, produce the needed unity to sustain the oscillations at f_0. When the circuit is at resonance, only the frequency f_0 can pass through the lead–lag circuit, achieve the 1/3 ratio, be fed back in phase, and achieve a closed-loop overall gain of unity to cause the circuit to sustain oscillations.

Thus, there are basically two types of R–C oscillators that find applications in communications circuits. The application will usually determine which, if any, should be used.

7.2.2 L–C Oscillators

An **L–C oscillator** has a combination of inductance (L) and capacitance (C) in the feedback network of the circuit. You will recall from your basic ac-circuits course, how a combination of inductance and capacitance can cause a resonant circuit to be

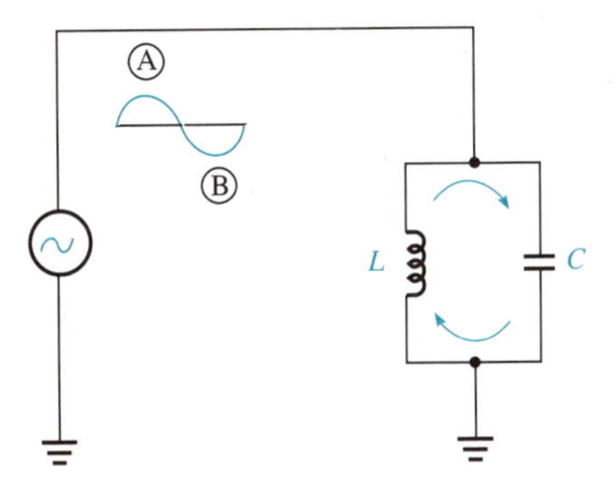

Figure 7.8 Tank circuit.

formed. This resonant circuit is sometimes called a **tank circuit**. The configuration used for such a circuit is shown in Figure 7.8. This parallel resonant circuit operates on the principle of a storing and transferring of energy. During the first half of a cycle, portion *A* of the figure, the capacitor absorbs at the same rate as it is being released by the inductor. This statement assumes that the capacitor is ideal (no losses). This is the capacitor charging up to its full capability. During the second half of the cycle, portion B of the figure, the inductor absorbs energy at the same rate as that which the capacitor is releasing it. This is the discharge of the capacitor on the reverse cycle. This operation continues and results in a ''flywheel'' effect, where the energy rocks back and forth charging and discharging the capacitor. The frequency of this action is determined by the values of the inductor and capacitor involved. This is the resonant frequency f_0, and is found mathematically by

$$f_0 = \frac{1}{2\pi\sqrt{LC}} \tag{7.6}$$

Now that we know what makes an L–C oscillator operate, we can look at some specific types of L–C oscillators. Bear in mind that the same criteria used for R–C oscillators apply to L–C oscillators; that is, input power, gain, frequency-determining devices, and feedback. The Barkhausen criteria must also be met: the feedback signal must be in phase at the loop closure point, and $A_vB = 1$.

The first L–C oscillator we will investigate is the **Colpitts oscillator**, shown in Figure 7.9. The circuit uses a single-transistor amplifier stage to achieve the required gain for oscillations.

Let us be sure that the configuration of this circuit satisfies all of the criteria set down thus far. Recall that the circuit must have input power, gain, frequency-determining devices, and feedback. Figure 7.9 shows us the following:

Input power:

V_{cc} to the transistor

Gain:

The transistor amplifier stage shown above the dashed line in the figure

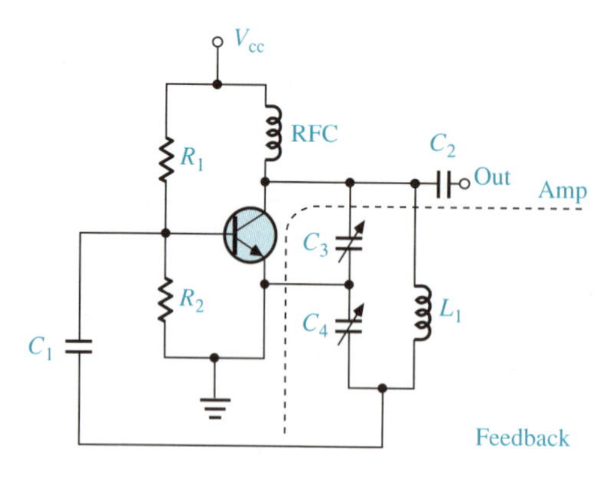

Figure 7.9 Colpitts oscillator.

Frequency-determining devices:
: The frequency of operation is determined by the combination of L_1, C_3, and C_4

Feedback:
: The feedback portion of the circuit is the components shown below the dashed line in Figure 7.9

Thus, you can see that the circuit contains all of the elements necessary to provide oscillations. The next question is whether the circuit can sustain the oscillations. Recall that the criteria were to have the feedback signal in phase and to have $A_vB = 1$. The signal is in phase back at the input of the transistor because the transistor itself shifts it by 180°, and the feedback network shown provides the other 180° of phase shift. The condition where the gain and feedback factor equal unity is taken care of by making the gain sufficient to cause this condition to occur. The feedback factor B for the Colpitts oscillator is determined as C_3/C_4. Thus, the gain of the transistor stage must be equal to C_4/C_3. This will result in $A_vB = 1$, which will sustain the oscillations of the circuit.

The frequency of the Colpitts oscillator is determined by

$$f_0 = \frac{1}{2\pi\sqrt{LC_{\text{total}}}} \tag{7.7}$$

Where: $C_{\text{total}} = \dfrac{C_3C_4}{C_3 + C_4}$

Equation (7.7) is valid only for circuits whose Q is greater than 10. For circuits with $Q < 10$, an additional term must be used, as shown in Equation (7.8). To have an accurate reading of frequency, Equations (7.7) and (7.8) must be multiplied together.

$$\sqrt{\frac{Q}{Q + 1}} \tag{7.8}$$

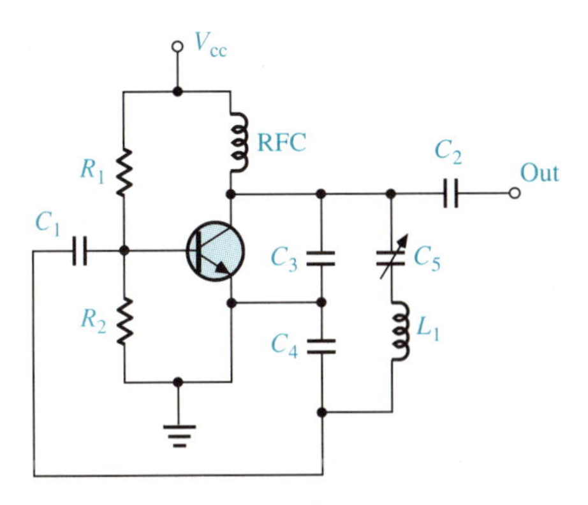

Figure 7.10 Clapp oscillator.

A second type of L–C oscillator, similar to the Colpitts, is the **Clapp oscillator**, a circuit configuration for which is shown in Figure 7.10. You can see how similar this circuit is to the Colpitts oscillator; the only variation is the insertion of an additional capacitor C_5. This capacitor is generally smaller than the previous capacitors C_3 and C_4, and as such can be used almost exclusively to change the resonant frequency of the oscillator circuit.

The total capacitance for the circuit is now the sum of the three capacitors in series. That is, it is the reciprocal of $1/C_1 + 1/C_2 + 1/C_3$. This value of capacitance can now be used to determine the resonant frequency of the circuit by using the same Equation (7.7) if the $Q > 10$ and using the combination of Equations (7.7) and (7.8) if $Q < 10$. Capacitor C_5 is not only used for tuning the circuit, it also provides a more accurate and stable frequency of operation since it is not affected by junction capacitances of transistors or stray capacitances of the circuit, which appear in parallel with C_3 and C_4. They do not appear in parallel with C_5, and thus do not affect the tuning of the circuit.

The final L–C oscillator circuit we will investigate is the **Hartley oscillator**. The circuit in Figure 7.11 shows that this circuit is actually the same circuit as the

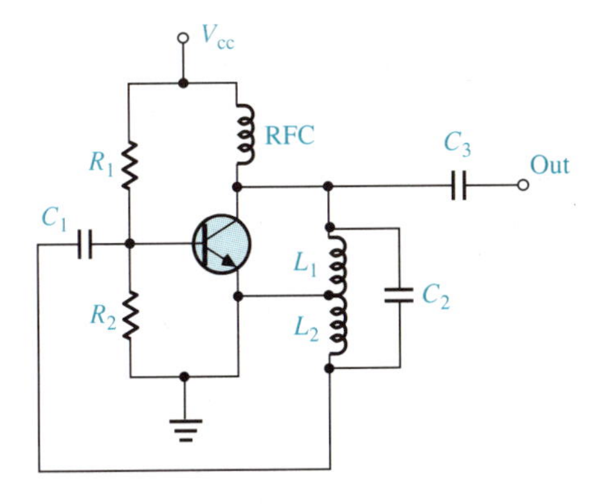

Figure 7.11 Hartley oscillator.

Colpitts, except that all of the capacitors are replaced by an inductor, and the inductor is replaced by a capacitor. The feedback is now provided by means of a tapped inductor instead of two capacitors designed to give the correct feedback factor. Now the ratio of turns in the inductor must provide this same type of feedback factor for the Hartley circuit. This feedback ratio B is now designated as L_1/L_2. With this feedback factor calculated, it now becomes apparent that the gain of the amplifier stage must be L_2/L_1 in order to sustain oscillations.

The frequency-determining components in the Hartley oscillator are L_t and C_2. L_t is equal to $L_1 + L_2$. With this relationship in mind, the resonant frequency of the Hartley oscillator is found as

$$f_0 = \frac{1}{2\pi\sqrt{L_t C}} \tag{7.9}$$

Where: $L_t = L_1 + L_2$

Equation (7.9) is also only valid for circuits where the $Q > 10$. For those circuits where $Q < 10$, Equations (7.8) and (7.9) must be multiplied together to find the true resonant frequency.

One component shown in Figures 7.9, 7.10, and 7.11 has not yet been mentioned—the **radio-frequency choke (RFC)**. Its purpose is to present a very high impedance to any ac or rf signals so that they are not present on the dc power lines. These components are placed in the collector portion of the amplifier circuit. This area has the amplifier ac or rf signal present and could cause some severe distortion of the signal as well as causing a ripple on the dc lines if the dc and ac or rf are not kept separate. Thus, the RFC is placed in the line to provide this very necessary isolation.

7.2.3 Crystal Oscillators

A necessary requirement for all communications systems is that the oscillators used to set up reference frequencies throughout that system be stable. That is, when an oscillator is set on a frequency, it will remain on that frequency and not drift around and cause problems with the entire system. Often a designer will decide to use a crystal oscillator to achieve this purpose. To understand how a crystal can perform such a task, we will look into the construction and operation of the crystal.

Quartz is the material used for crystals in oscillators.

Quartz

A mineral (silicon dioxide) occurring in hexagonal crystals in nature and having piezoelectric properties that are highly useful in radio and carrier communications. The crystals from which slabs are cut for oscillators are transparent and almost colorless.

The key term in the definition above is **piezoelectric**. When a mechanical force is applied to the quartz crystal it causes it to vibrate, and a voltage is developed at the frequency of this vibration. Similarly, if a voltage of a specific frequency is applied to the crystal, a mechanical vibration will result at that frequency. This is the **piezoelectric effect**. Maximum vibrations occur at the crystal's natural resonant frequency, which is determined by the physical dimensions of the crystal and how it was cut.

As shown in Figure 7.12, the quartz crystal is naturally hexagonal in shape. There are three axes associated with this crytalline configuration: optical, electrical, and mechanical. The axis that joins the end faces of the structure is called the **optical**, or z, **axis**. Any electrical stress placed on this axis does not produce the piezoelectric effect. The axis passing through the opposite corners of the hexagon is the **electrical**, or $\boldsymbol{x}$, **axis**. And finally, the axis that is perpendicular to the faces of the crystal is the **mechanical**, or $\boldsymbol{y}$, **axis**.

The most useful cut for a crystal is the Y cut, in which the faces are perpendicular to the y axis. However, crystals are not usually cut along the corresponding axes, but

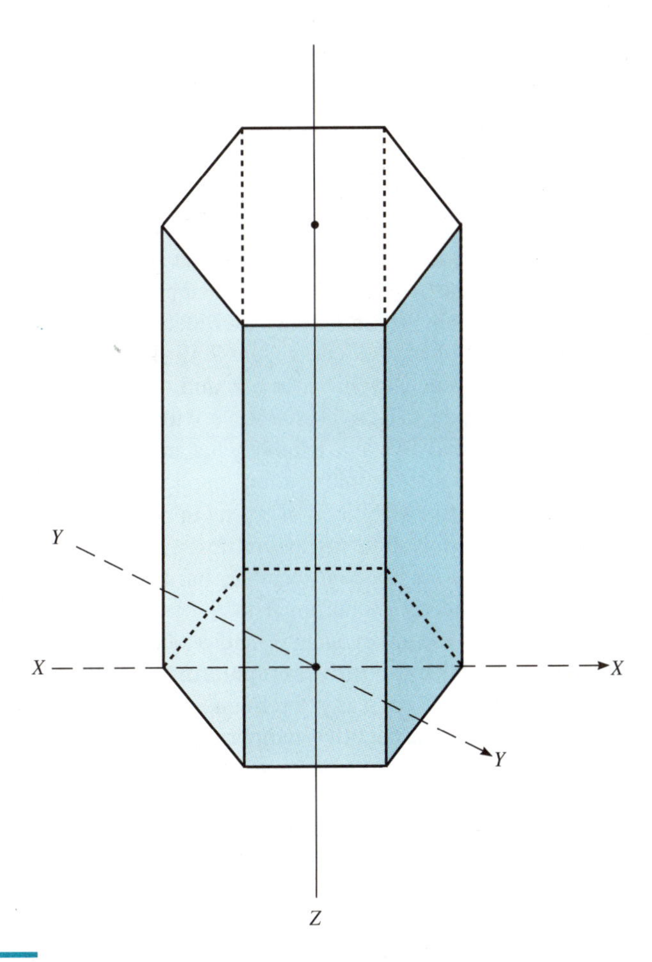

Figure 7.12 Quartz crystal.

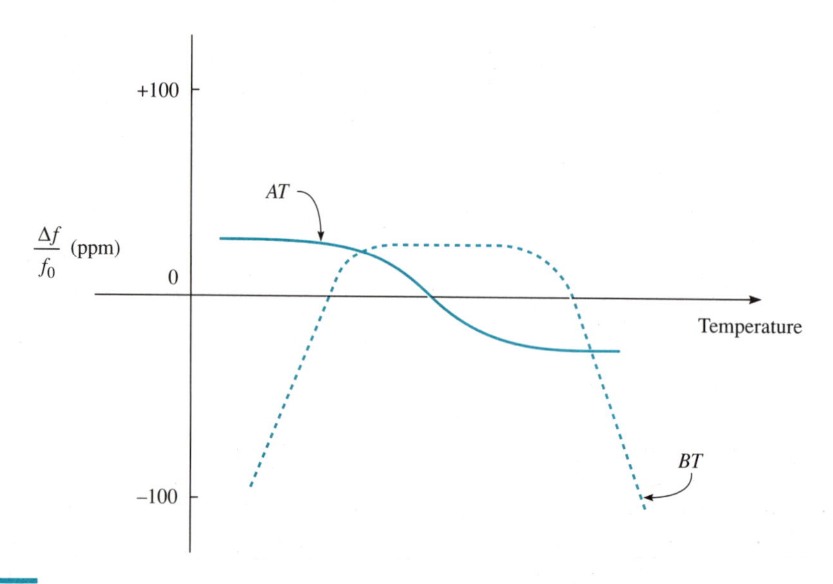

Figure 7.13 Temperature stability of crystal cuts.

are generally cut at an appropriate angle to the axis itself. As an example, if a Y cut is cut at an angle of 35° 20′ from vertical, it is called an AT cut, which is the most widely used form of crystal cut for crystal oscillator applications. If the cut is 49° 15′, the results will be a BT cut. The BT is not as widely used because it does not have the temperature stability of the AT cut. Figure 7.13 illustrates this fact further, presenting the variations in frequency in parts per million (ppm) as a function of temperature. You can see that the BT-cut crystal has a wide variation over a range of temperatures, whereas the AT-cut crystal is relatively flat and has no drastic variations. Thus, the AT is preferred for most communications applications.

Let us define the temperature stability of a crystal in more detail. When a crystal is subjected to a change in temperature, its natural resonant frequency will change. The change may be small or it may be considerable, but the frequency will change. If the frequency change is directly proportional to the temperature (an increase in temperature produces an increase in frequency, and a decrease in temperature decreases the frequency), it is called a **positive temperature coefficient**. If the reverse is true (an increase in temperature decreases the frequency and a decrease in temperature increases frequency) this is a **negative temperature coefficient**. It is vital that you understand this concept so that if you are ever called upon to produce a crystal oscillator that is subjected to temperature extremes, you can design compensation circuitry that will perform the proper operations to keep the frequency stable over the entire temperature range.

Temperature stability is expressed in hertz change per megahertz per degree Celsius (Hz/MHz/°C). To find out just how much of a frequency change you will have for a given condition, use the following mathematical relationship:

$$\Delta F = k(F \times \Delta C) \tag{7.10}$$

Where: ΔF = Change in frequency (in hertz)

k = Temperature coefficient (in hertz per megahertz per degree Celsius)

F = Natural resonant frequency (in megahertz)

ΔC = Change in temperature (in degrees Celsius)

To see how this works, consider Example 7.4.

Example 7.4

Given: A 15-MHz crystal has a positive temperature coefficient of 4 Hz/MHz/°C. Determine the crystal frequency if the temperature increases by 35°C.

Solution

If we use Equation (7.10), we find that

$$F = 4(15 \times 35)$$

$$= 4(525)$$

$$= 2100 \text{ Hz}$$

Since the natural frequency of the crystal is 15 MHz and $F = 2100$ Hz = 0.0021 MHz, the final frequency is

$$F = 15.0021 \text{ MHz}$$

Example 7.5

A 5-MHz crystal has a negative temperature coefficient of 3.5 Hz/MHz/°C. Determine the crystal frequency if the temperature increases by 42°.

Solution

Using Equation (7.10), we obtain

$$\Delta F = k(F \times \Delta C)$$

$$= 3.5(5 \times 42)$$

$$= 3.5(210)$$

$$= 735 \text{ Hz}$$

Since the crystal has a negative temperature coefficient, an increase in temperature causes a decrease in frequency. Thus, the new frequency is

$$F = 5 \text{ MHz} - 735 \text{ Hz}$$

$$= 4.999265 \text{ MHz}$$

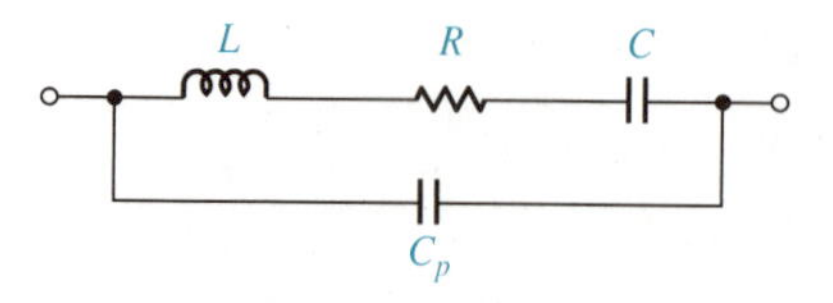

Figure 7.14 Equivalent circuit of a crystal.

Now that we have a basic idea of crystals, let us see how the component can be used in a communications circuit. To do this, let us present an equivalent circuit for a crystal (See Figure 7.14). The value C_p is the package capacitance of the crystal and usually ranges from 7 to 10 pF. C_s is the series capacitance, which is usually approximately 0.05 pF. The L component is the crystal inductance, which is a very large value for quartz crystals, usually in the 10's of henries of inductance. The last term is R, which is the internal resistance of the crystal. This is a very small value and is many times ignored altogether.

The equivalent circuit of a crystal is very important since it provides an explanation of why there are two resonances in a crystal. There are series and parallel resonances, the use of which determines how the crystal is used in the oscillator circuit. The series resonant frequency of the crystal is given by

$$F_s = \frac{1}{2\pi\sqrt{LC_s}} \tag{7.11}$$

and the parallel resonant frequency is given by

$$F_p = \frac{1}{2\pi\sqrt{LC_t}} \tag{7.12}$$

Where: $C_t = \dfrac{C_s C_p}{C_s + C_p}$

You can now see how important the equivalent circuit is for a designer who has to know both the series and parallel resonant frequencies of the crystal. To further understand how the series and parallel resonant frequencies are put to work, we will look at circuits where both the series and parallel frequencies are used.

Figure 7.15 shows a crystal oscillator circuit that uses a crystal operating at its series resonant frequency. This circuit is similar to the previously covered Colpitts

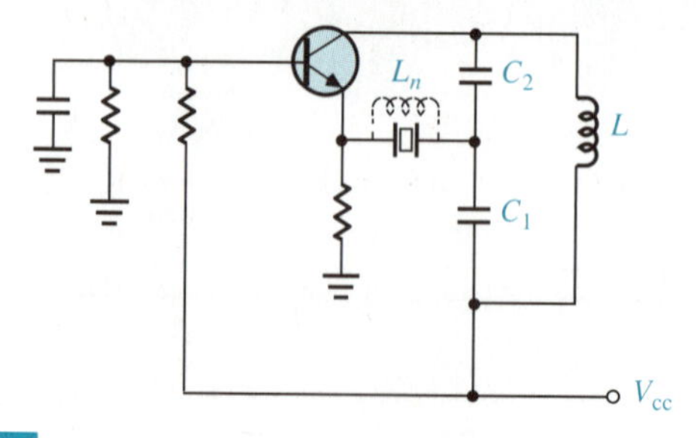

Figure 7.15 Crystal oscillator—series crystal.

oscillator, because the basic circuit is a Colpitts. The parallel combination of L and C_t form the frequency-determining portion of the circuit. C_t is the parallel combination of C_1 and C_2 ($C_t = C_1C_2/C_1 + C_2$). This combination should be tuned at or near the series resonant circuit of the crystal. The positive feedback for the circuit is through the crystal operating in its series resonant mode F_s. Notice at this point the symbol used for the crystal. It appears to be a capacitor with something in the center of the plates. That something is a special dielectric, which is a specific cut of quartz crystal that exhibits piezoelectric characteristics. The idea of a capacitor being its symbol is rather appropriate since, as we have seen from the equivalent circuit in Figure 7.14, the capacitive components of the crystal have a large effect on its frequency characteristics.

One component present in this circuit must be explained, the inductor L_n. This component neutralizes the C_p of the crystal. Recall from our discussions of the equivalent circuit and its parameters, that the value of C_p could be anywhere from 7 to 10 pF, a value of capacitance that could cause a difference in frequency and actually keep the circuit from locking to the crystal. Thus, by neutralizing it with the appropriate inductor, we can negate its effects on the circuit and enable it to lock to the crystal. To calculate what this inductor must be, we use the following equation:

$$L_n = \frac{1}{(2\pi F_0)^2 C_p} \tag{7.13}$$

Where: L_n = Neutralizing inductor (in henries)
F_0 = Series resonant frequency (in hertz)
C_p = Package capacitance of crystal (in farads)

A second circuit that uses a crystal for frequency stability is shown in Figure 7.16. This circuit, as you can see, has the crystal in a different location in the circuit than that of Figure 7.15. This circuit uses the crystal's parallel resonant frequency F_p rather that the F_s from the previous circuit. The feedback path for this circuit is through the combination of C_1 and C_2. This circuit is an R–C oscillator. You should recognize this from previous circuits and should not automatically look for inductors and capacitors to resonate. It is often the R–C combination that is the frequency-determining section, as is true for this case. The coil shown in this circuit is simply an rf choke, which provides a dc bias path through C_2.

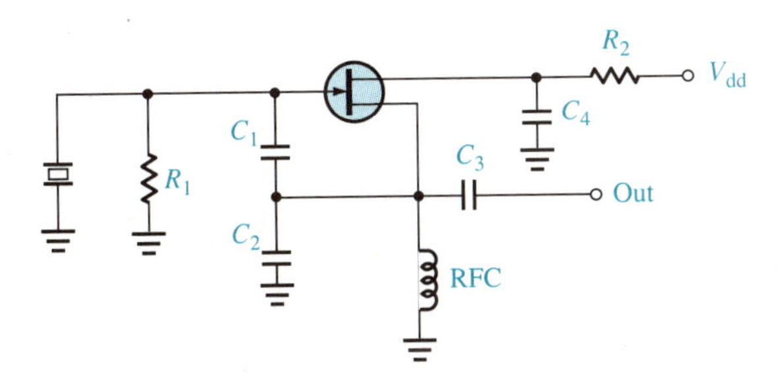

Figure 7.16 Crystal oscillator—parallel crystal.

Before we leave the topic of crystal oscillators, one point must be made very clear: the crystal is used in an oscillator circuit only for frequency stability, meaning that the crystal only *locks* the frequency on, it does not determine the frequency of the oscillator. You can vary the L or C or R of the circuit after the crystal has locked the frequency on, and the output frequency of the oscillator will not change. This is an important point since many people expect the crystal to perform a function it is not designed to do. So remember that *a crystal only holds the frequency of an oscillator, it does not determine it.*

We have seen how some of the typical circuits used for communications systems are developed and what the criteria are for starting up oscillations and sustaining them. We have also seen how to improve the stability of an oscillator by using a crystal. We now have the heart and soul of a communications system, the circuits that are the source of our signals, *oscillators*.

7.3 Rf Amplifiers

The rf amplifier is singled out here because many introductory courses that the student has had to this point have covered such areas as audio amplifiers, dc amplifiers, and the like. What is usually missing is an overview of what happens when the signals you are trying to amplify get up into the megahertz and hundreds of megahertz range. Suddenly the ideas and concepts that you learned previously are no longer acceptable. As a matter of fact, if you continue to practice these concepts with rf amplifiers, it is guaranteed that you will never get an rf amplifier to work. Thus, it is essential that the student understand that *higher frequencies require different procedures and concepts*, which is the objective of this section.

Two questions should be answered before getting any further into rf amplifiers and their applications in communications systems:

1. Where are rf amplifiers used in communications systems?
2. Why are they used?

To answer the first question let us consider a communications receiver and the signals present at the input to this receiver. Figure 7.17 shows a block diagram for a typical superhetrodyne receiver that may be used for communications applications. The signal coming in from the antenna is usually a very small signal, often in the -50 to -60 dB range or lower. If we are referring to satellite signals, we are talking very low level signals. In our diagram these signals are mixed with a **local oscillator** (LO), amplified and filtered at some lower frequency, and then processed as necessary. This sequence of events takes place if the signals are large enough to be mixed in the input mixer. We will often find that these signals are *not* large enough to overcome the noise of the receiver and are lost forever to this noise.

To make sure that the signals we want to receive are received, an rf amplifier is placed between the antenna and the mixer as shown in Figure 7.18. The amplifier now displays enough gain to exceed the noise level within the receiver, and the

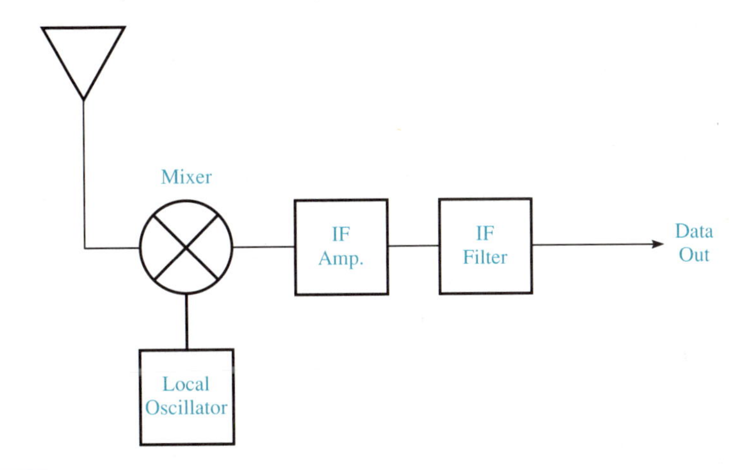

Figure 7.17 Receiver block diagram with no rf amplifier.

receiver operation described above is achieved. Thus, the rf amplifiers can be used at the front end of a receiver to ensure that there is proper system gain to allow the signal to be processed through the receiver and have the intelligence removed as intended.

The second question was partially answered in the explanation above: the rf amplifier is used to provide gain to the system to allow the system to perform. The second part of the answer is that we use rf amplifiers to provide a low noise figure at the input to the receiver so that the entire receiver will exhibit good noise properties.

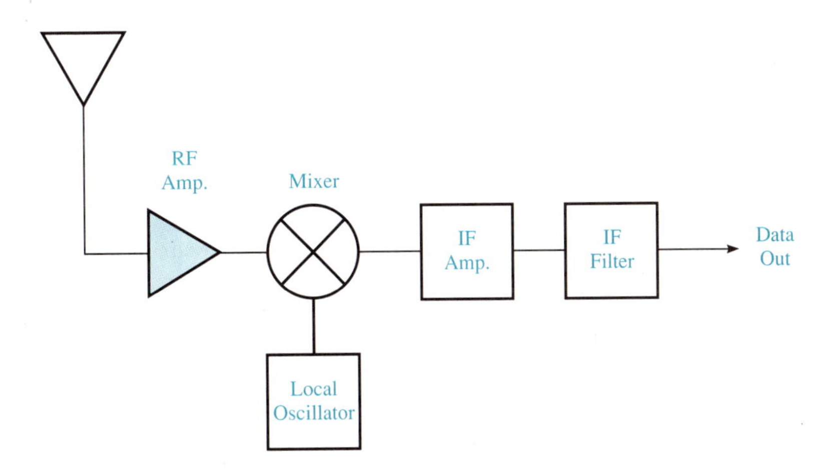

Figure 7.18 Receiver with rf amp.

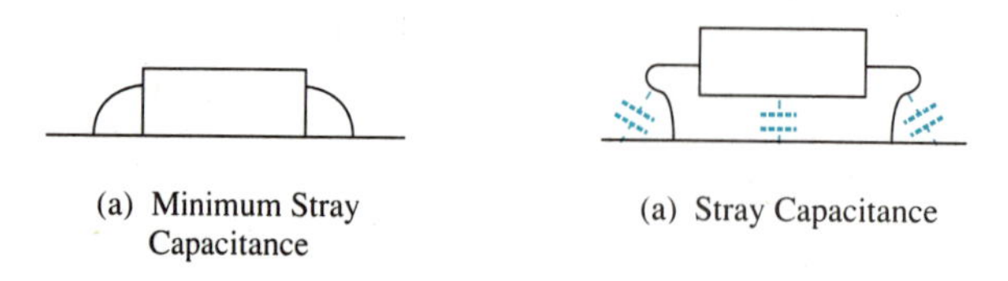

Figure 7.19 Stray capacitance.

This rf amplifier application is called a **low-noise amplifier** (LNA). You will recall from our discussions of noise and noise figures in Chapter 6 that when we have a group of cascaded components in a system, such as a communications receiver, the system (or input) noise figure is the one that controls the noise figure of the entire system. With this fact in mind, it should not be difficult to see that if we have an amplifier with a low noise figure right at the receiver antenna, we have a low-noise receiver. This explains why they are used.

Now that we have answered the two questions, we need to see how rf amplifiers are different from the amplifiers we studied in previous circuits courses. As we do this you should keep in mind that we are now dealing with higher frequencies, and we must therefore be concerned with dimensions and distances we did not worry about before. In previous circuits courses where low-frequency circuits were breadboarded and then built, there was no concern given to how long a lead was, where a component was placed with respect to another component, or if the transistor was placed in a socket or placed flat on the circuit board. We didn't concern ourselves because these things did not influence whether the circuit operated or not. Now, however, these factors do matter and will determine quickly whether the circuit works or not. All of the above factors must now be considered when designing and building rf amplifiers.

An additional factor that comes into play with rf circuits is capacitance. There are two types of capacitance that you should be aware of in rf circuits that were not a factor in low-frequency circuits: stray capacitance and interelement capacitance. **Stray capacitance** occurs when you do not build the rf circuit carefully. As we have mentioned above, there are very strict rules to follow when placing components in rf circuits. If these rules are not followed, capacitances can be set up that will cause degradation in circuit performance. The leads of components are the prime candidates for creating stray capacitance. If the component leads are not kept as short as possible, there will be a capacitance set up between the lead and ground and between the component and ground. Figure 7.19 shows examples of good and bad construction. You can see that if the component is placed flat on the circuit board and the leads as kept as short as possible, we will have a minimum or even zero stray capacitance, as shown in Figure 7.19(a). Figure 7.19(b) is the other extreme. The component is high above the circuit board and the leads are much longer than necessary. As a matter of fact, there are even extra bends in them, making things even worse. This situation results in considerable stray capacitance being set up between the component and ground as well as between the leads and ground. These capacitances are shown in Figure 7.19(b) as the dashed lines.

Example 7.6

We observe that the gain of the rf amplifier in our communications receiver is very erratic in gain. It should have 20 dB of gain but may vary by ±5 to 10 dB from this figure. What is the problem?

Solution

First, check to see if all of the dc voltages are as they should be. If the dc voltage fluctuates, then the gain of the amplifier will vary drastically.

If the dc voltages are tested and found to be proper, the next step is to apply the dc voltage, turn on the amplifier, and remove the input signal to the amplifier. If the amplifier still has a signal coming out of the output, the problem is solved—the amplifier is oscillating. The solution is to replace the transistor(s) within the amplifier and then find out what caused the oscillation. In many cases, the component arrangement is to blame. A simple shortening of leads or rearranging of components will allow the amplifier to work properly again.

Interelement capacitance is the capacitance between the collector and base elements, the base and emitter elements, or the emitter and collector elements of a transistor. There is always a certain amount of interelement capacitance in a device no matter the frequency at which we are operating. However, under low-frequency conditions this capacitance is virtually invisible because the reactance of the capacitance is so high that it has absolutely no effect on the signals being amplified. At high frequencies, however, it is a different matter. This capacitance now becomes very small and there is a distinct alternative path for the signals to take (other than the path we desire them to take). This alternative path, if it is from the collector (or drain) of a transistor to the base (or gate) can now make up a very low impedance feedback path that may shift the phase enough to cause severe oscillations within the circuit. You will recall from our discussions on the requirements for oscillations that the feedback signal must be in phase at the loop closure point. The loop closure point for interelement capacitance purposes is the base (or gate) of the device. If you couple this condition with the gain of the transistor, you can very definitely have a condition that will cause your rf amplifier circuit to oscillate, either at the amplified frequency or some other random frequency. To be sure that you do not encounter this undesirable condition, you should check the data sheet for devices with very low values of interelement capacitances. This is best accomplished by choosing devices specifically designed to operate at radio frequencies.

To understand some of the differences between the rf and low-frequency amplifiers, refer to Figure 7.20, a typical circuit used for low-frequency operations. This is an ac-coupled common-emitter amplifier with a frequency response up into the range of a few kilohertz. All of the resistors shown in the figure are for dc biasing purposes. The input and output capacitors (1.0 μF) are the coupling capacitors that provide a low-impedance path for the signal to get to and from the transistor. The remaining 1.0-μF capacitor is a bypass capacitor for the emitter resistor for stabilizing

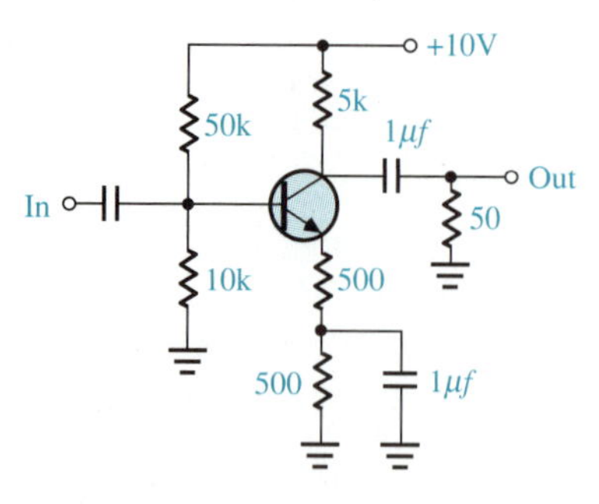

Figure 7.20 Low-frequency amplifier.

the circuit. There is nothing special about this circuit; you have probably seen such a circuit many times in electronics courses at lower levels. We include it here to remind you of its structure and to provide a basis for comparison with an rf amplifier.

To further compare a low-frequency amplifier with an rf amplifier, we present a typical rf amplifier circuit—shown in Figure 7.21. Once again, the resistors set up the bias conditions for the amplifier as in the case of the low-frequency amplifier. The major difference is with the output circuit: there is a transformer at the output rather than just a 50-Ω resistor, as in Figure 7.20. The transformer is substituted for the resistor because it is a much better impedance-matching device for high frequencies than a simple resistor would be. With the transformer, a class-A amplifier can achieve 50% collector efficiency at its full output-voltage swing. Figure 7.21 also shows a variable capacitor in the collector circuit. This is a tuning capacitor that can tune the amplifier for optimum performance at the desired frequency.

Thus you can see that the rf amplifier has a prominent place in communications equipment. Its primary application is for receiver front ends, but it can be used any place there is a need for rf amplification simply by changing bias levels or adjusting capacitances to enable the circuit to function as desired.

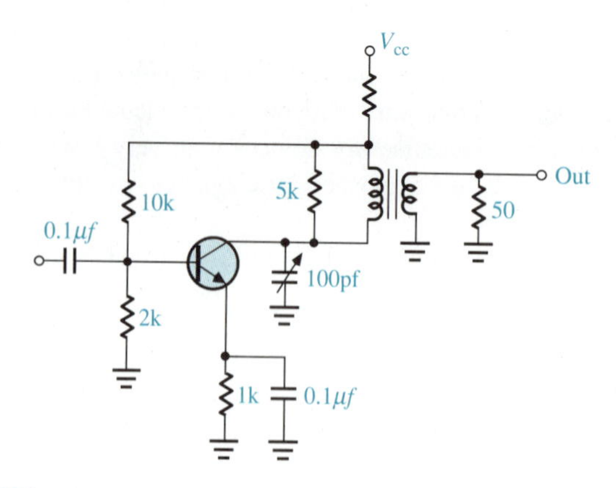

Figure 7.21 Rf amplifier.

7.4 Mixers

The mixing process consists of two signals entering a nonlinear device producing an output that is a combination of the two input signals. The most important term in this description is the word **nonlinear**. It is vital that the device exhibit nonlinear characteristics for mixing to occur. To understand the importance of nonlinearity in a device, refer to Figure 7.22. Part (a) is a typical P_{in}–P_{out} curve for a conventional amplifier, showing that a certain power input produces a corresponding output depending on the gain of the amplifier. If we were to put two signals into this circuit, we would have two output signals that were linear representations of the input, assuming that the frequencies of the two signals fell into the bandwidth of the amplifier. The two signals would not be mixed, only separately amplified.

If, on the other hand, the circuit were biased so that it was being overdriven, or placed into a nonlinear portion of the response curve (Figure 7.22b), an entirely different action would be performed on the two signals. Since the circuit is now a nonlinear device, it does not act on all of the signals in the same manner. If the original signal is designated as F_1, and the second signal as F_2, the end result at the output of this nonlinear stage would be F_1, F_2, $F_1 + F_2$, $F_1 - F_2$, $2F_1 + F_2$, and so on, just as we discussed in Chapter 6. For ordinary operation of an amplifier circuit, this is a very unacceptable condition. However, if we want a circuit to mix two signals together, it is ideal. We have used the term nonlinear many times thus far and shown response curves that illustrate the phenomenon, but to understand nonlinear we need to define linear.

Linear

A term where the output response of a system is directly proportional to an input

This definition exactly parallels our discussion of the performance and response of a conventional, properly biased, class-A amplifier. The response curve shown in Figure 7.22(a) substantiates the definition above. A specific input power results in a propor-

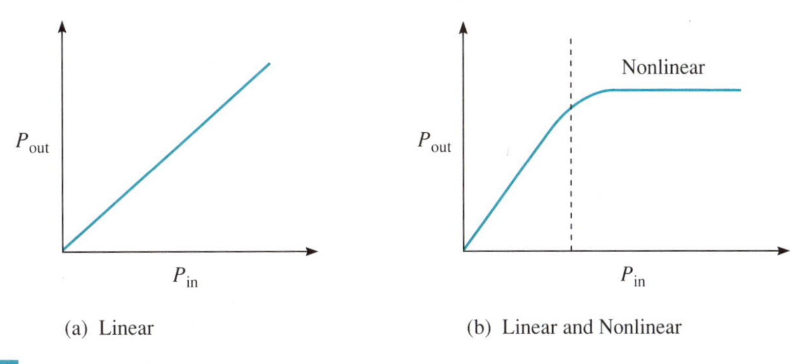

Figure 7.22 Response curves.

tional output power. The resulting plot is a straight line that exhibits linear characteristics.

Now for the contrast between linear and **nonlinear**. Since a linear system has an output directly proportional to the input, then nonlinear must be:

Nonlinear

The output response is something other than directly proportional to the input. This relationship may be proportional to the square, cube, or any other appropriate function.

A nonlinear response that should be familiar to most of you is that shown in Figure 7.23: the common *I–V* curve for a junction diode. For the first few volts of input on the curve there is a corresponding increase in current (*I*). This is a linear response. Then, at a certain point, a very small increase in voltage results in a very large value of current. The junction is now in a nonlinear region of operation because the output (*I*) is no longer directly proportional to the input (*V*). In actuality, the output is now the square of the input since this is now the portion of a diode junction termed the *square-law region*.

To summarize our discussions, we can say that a nonlinear circuit is needed in order to mix two signals since a linear system will not produce the necessary level of signals to cause the sum and difference of the frequencies to be produced. It should also be apparent that many systems are linear until a certain input level is obtained;

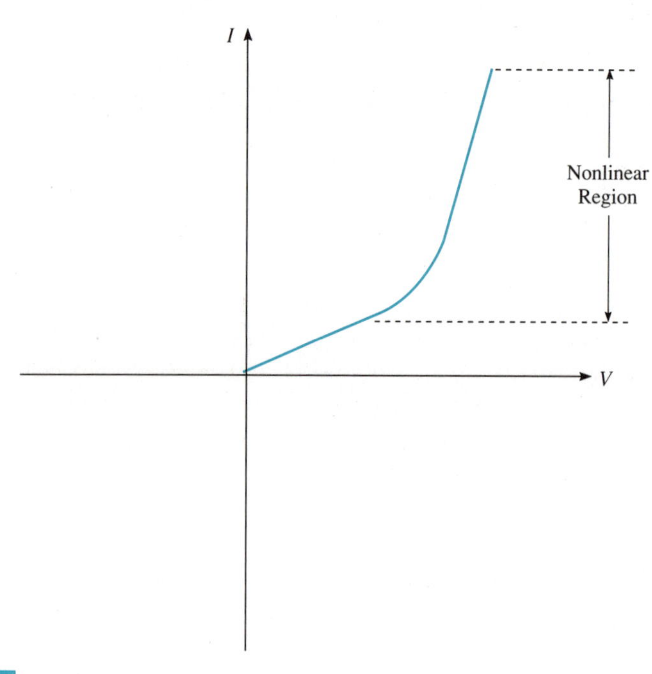

Figure 7.23 Nonlinear curve.

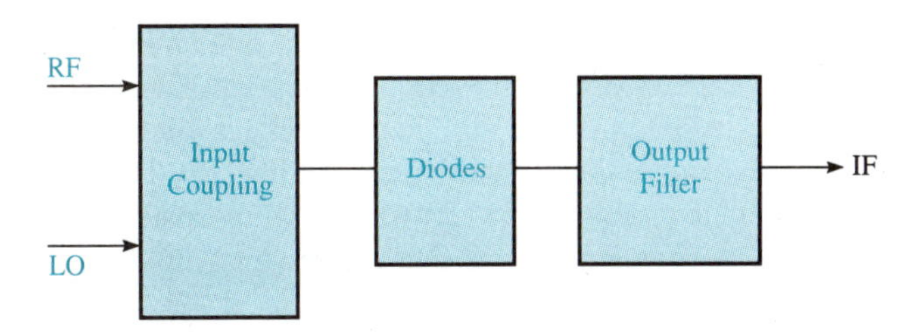

Figure 7.24 Basic mixer block diagram.

then the system becomes nonlinear. This was apparent when we talked about the junction diode in Figure 7.23. This should tell you that the condition that produces nonlinearity is an appropriately high input level. It is important to remember this about mixers: unless the proper input levels are attained, no mixing will take place.

Having discussed linear and nonlinear systems, let us now examine the circuits that can be used to perform the mixing operation. For a circuit to be a mixer, it must contain three basic blocks, as shown in Figure 7.24: an input-coupling network, a diode circuit, and an output filter. Let us now look at these three blocks and the mixers that can be used in them, and see how they apply to communications systems.

The first block, the input-coupling network, must meet two basic requirements: it must operate in the proper frequency range and must provide good isolation between the rf signal, which is a low-level signal, and the LO (local oscillator), which is a much higher level signal. Without this isolation, the signals cannot be separated properly and the mixing action will not occur. This network may be a transformer or the two gates of a dual gate FET (field-effect transistor), or it may be one signal applied to the base of a transistor while the other is applied to the emitter. Whatever the configuration, it is absolutely vital that the two signals be isolated from one another at the input to the mixer. The only place that the signals should get together is where they are intended to, at the diode circuitry.

The second block is the diode circuit. As mentioned above, this is where the two signals, the rf and local oscillator (LO), are intended to get together. This circuit may be a single *pn* diode, a pair of diodes, a diode bridge (four diodes), or the junctions of a FET or bipolar transistor. Regardless of the type of diode circuitry used, it must be able to exhibit the nonlinear properties needed for mixing. One point that you may not consider if you become too concerned with the rf properties of the circuit is that every diode needs a dc return to ground to operate properly. Without this dc path, the diode cannot conduct, and there is no chance of mixing occurring.

The final block in Figure 7.24 is the output filter. As we have said previously, there will be many frequencies at the output of a mixer circuit, and it is, therefore, vitally important there be some filtering to eliminate all unwanted signals in order to produce the desired mixed output frequency. Since most mixers are designed to produce an output that is the difference between the rf and LO signals, the logical choice for an output filter is a low-pass filter. This will ensure that only the difference frequency will be at the output. For example, suppose we had an rf frequency of 1.5

MHz and an LO frequency of 2.0 MHz. Some of the output frequencies before the filter would be 1.5 MHz, 2.0 MHz, 0.5 MHz, 3.5 MHz, to name only a few. A filter that passes everything below 0.5 MHz will do an excellent job of producing only the difference frequency at the output of the mixer. Another method of filtering at the output of a mixer is a tank circuit (L–C tank) which will resonate at the desired output frequency. This will, once again, ensure that the proper output is available for the circuits following the mixer.

Having explained the basic circuit blocks for a mixer, now let us examine the process that makes this newly constructed circuit mix two signals to produce a desired output. We have emphasized in this chapter that the circuit must be driven into a nonlinear mode for mixing to take place. We now need to discuss how to accomplish this task in a communications mixer circuit. It should be apparent from our early discussions that we can make a circuit nonlinear if we overdrive the junctions within the active devices. We saw this when we compared putting two signals into an amplifier with putting the same signals into an overdriven amplifier. In the first case we produced two amplified signals, assuming the bandwidth was sufficient; in the second case we ended up with many signals resulting from a nonlinear mixing process. We should, therefore, be able to produce a mixing process by overdriving the device with one signal. The signal used to overdrive the junctions of devices and cause mixing is the local oscillator (LO) signal. You will see from the specifications for a mixer that the power level of the LO is significantly higher than that of the rf input port. An example of the differences for communications mixers is that the rf input to a receiver mixer may be in the -60 dBm (1 μW) range, whereas the LO power may be at 0 dBm (1 mW) or higher. This wide difference in power is used to "turn on" the diode circuitry within the mixer and cause the nonlinear characteristics to work properly.

The following terms for mixers are important.

Conversion loss

This term is equivalent to an insertion loss in any other component. It is the loss from the rf input port to the i-f output port. These signals are at different frequencies and therefore cannot be measured with a power meter. This measurement must be made with a spectrum analyzer so that both the rf and i-f signals can be seen and their power levels measured.

Isolation

Three isolation figures are important to mixers: rf-to-LO isolation, rf-to-i-f isolation, and LO-to-i-f isolation. The critical isolation figures are those that involve the local oscillator. The power level of the LO signal is so much stronger than the other signals that it is sometimes a problem to attenuate the signal enough so that you do not get LO feedthrough. Thus, isolation that involves the LO must be high.

Noise figure

The noise characteristics of a mixer circuit follow very closely the characteristics of the conversion loss. If the conversion loss is 4 dB, generally the noise figure is 4 dB also. This parallels closely the theory in Chapter 5 that says that the noise and loss of an active system are related.

Example 7.7

Problem: The conversion loss of a mixer is specified as 6.5 dB. When the mixer is measured under actual operating conditions, the conversion loss is measured as 11 dB. What causes the difference?

Solution

Conversion loss depends on the operation of the diodes within the mixer. The loss encountered is due to the conversion technique that converts an rf signal to an i-f signal. Because this process is accomplished entirely within the diodes, the solution to this problem is to check the diodes within the mixer. At least one of them is probably bad.

Figure 7.25 shows some mixer circuits that can be used in communications systems. Figure 7.25(a) shows a single diode mixer, the most basic type of mixer built. Notice how it just barely meets all of the requirements of our block diagram for being called a mixer.

Figure 7.25(b) shows a balanced mixer that uses a series of transformers and a bridge configuration of diodes to accomplish the mixing process. Figure 7.25(c) shows a transistor mixer, consisting of a basic amplifier circuit with one input at the base and the other at the emitter. The power levels are in the neighborhood of 1 mV for the rf and 50 mV for the LO. The output takes one additional precaution that some other circuits do not take. You will recall that the LO signal, being very large, has a tendency to feed through into the i-f output. The circuit in Figure 7.25(c) has an additional trap filter designed to eliminate only the LO signal from the output. The remainder of the circuit is a low-pass filter that allows only the difference frequency to appear at the final output.

So it can be seen that the mixer circuit can produce a lower frequency signal from the mixing together of two high-frequency signals. This is an operation that is very helpful when dealing with communications receivers in which the signals must be "down converted" to a low frequency, or dc, signal that is much easier to process and analyze.

7.5 Filters

A **filter**, whether used in communications systems or in an air-conditioning system, has one purpose: it is designed to pass one thing and reject everything else. In the air-conditioning system the filter is designed to pass the cool air and to reject the dust and lint that can clog up the system. In a communications system a filter is designed to pass certain frequencies and to reject all others. The types of filters we will look at in this section are the bandpass, low-pass, and high-pass filters. Each of these will be presented with its important parameters and some basic applications.

Before we go any further we need to make one thing clear: there is no such thing as a *standard* filter. That is, we cannot get a filter for our communications

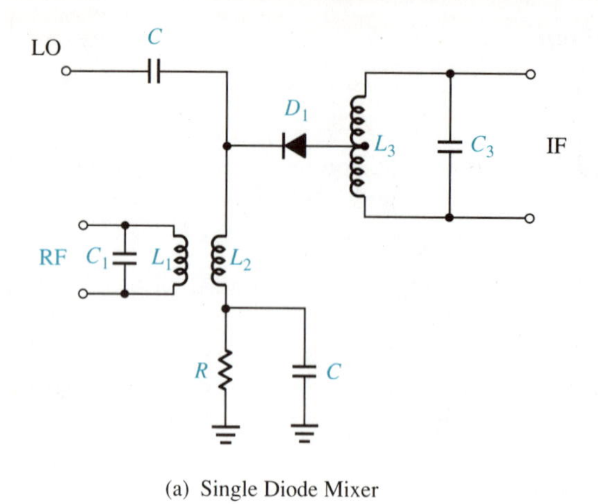

(a) Single Diode Mixer

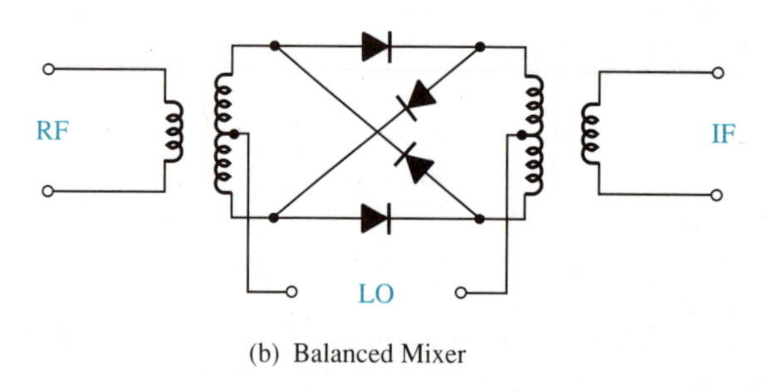

(b) Balanced Mixer

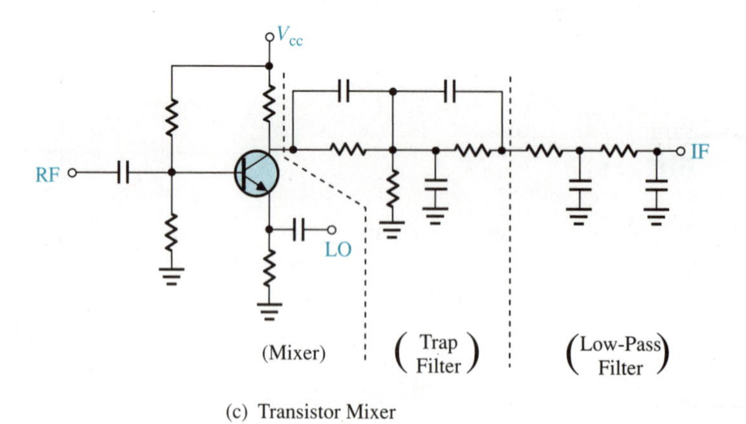

(c) Transistor Mixer

Figure 7.25 Mixer circuits.

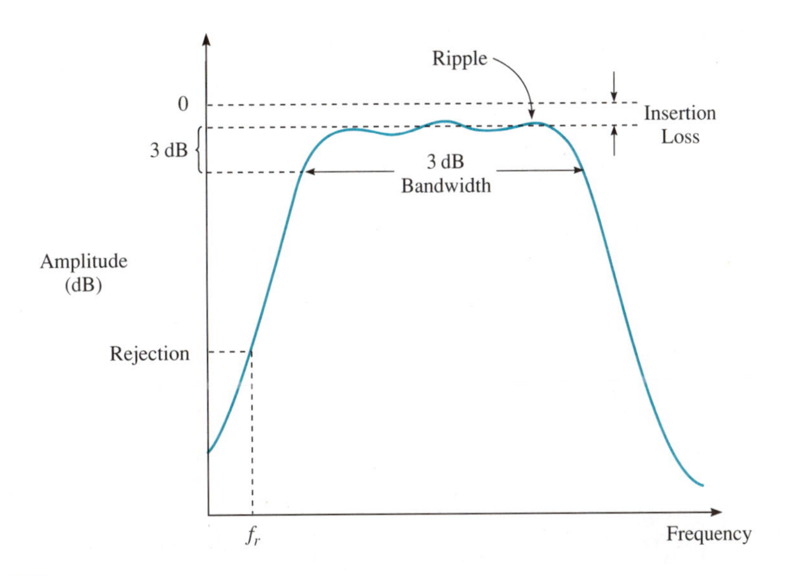

Figure 7.26 Bandpass filter.

system simply by calling a supplier and having them take a filter off the shelf. Each filter is a special component. As such, it is important to know the parameters of each filter and the meaning of each term used to describe filters. Knowledge of the filter terminology will enable you to create that special filter that will do the job for your system.

The first type of filter we will discuss is the **bandpass filter**. This filter, as the name suggests, will pass a certain band of frequencies and reject the frequencies on both sides of this band. A response curve for a bandpass filter is shown in Figure 7.26. The first term we will look at is **insertion loss**. This is the minimum loss of the filter due simply to rf energy propagating through a metallic conductor. There always is some small loss because of the number of sections, or poles, that make up the filter. The **poles** are sections of a filter that may consist of a single L–C network, R–C network, or parallel L–C tank circuit. These single poles are shown in Figure 7.27.

The ripple term shown in Figure 7.26 is the result of a series of poles being coupled together to make the entire filter. Each pole has a frequency-response curve of its own and when added together, as in Figure 7.28, they form the resulting response curve for the entire filter as shown by the dashed line in the figure.

One of the most difficult parameters to specify on a bandpass filter is the bandwidth because everyone has a different idea about which bandwidth is the most important. This parameter is also determined by system applications. Some applications call out the ripple bandwidth, the point where the ripple of the filter stops and the lower and upper skirt of the filter starts to fall off. Other applications call out a 1-dB or 3-dB bandwidth. In these cases the frequency is read where the response falls off by either 1 dB or 3 dB on both sides of the band, and the frequency difference is the

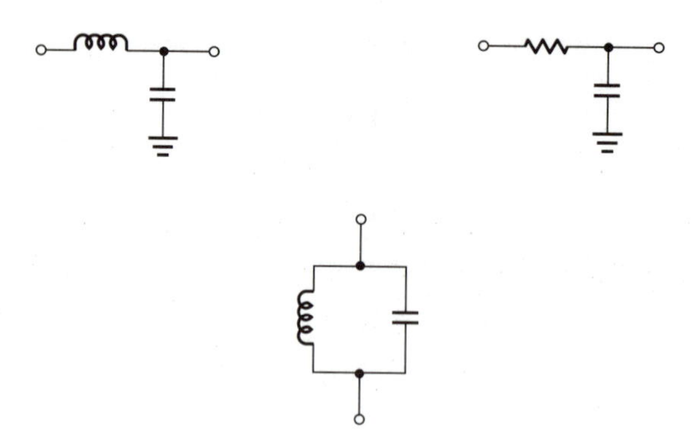

Figure 7.27 Examples of poles.

bandwidth. Regardless of which bandwidth is specified, it is the difference in frequency between the two points that the application calls out.

The final parameter in the figure is either a single or multiple rejection points. To specify these points you need to choose a particular frequency and specify the number of decibels that this frequency should be attenuated from the insertion loss. As an example, we may want a bandpass filter with a center frequency of 2 MHz, and we want a 3-MHz signal that is attenuated by 40 dB and a 0.5-MHz signal that is attenuated by 45 dB. The rejection frequencies would be 0.5 MHz and 3 MHz, whereas the rejection attenuations will be 45 dB and 40 dB, respectively.

The next type of filter to be discussed is the **low-pass filter**, which is designed to pass frequencies up to a certain cutoff frequency and attenuate everything from that point. The response curve for a low-pass filter is shown in Figure 7.29. Once again there is an insertion loss associated with the filter that is some number of decibels

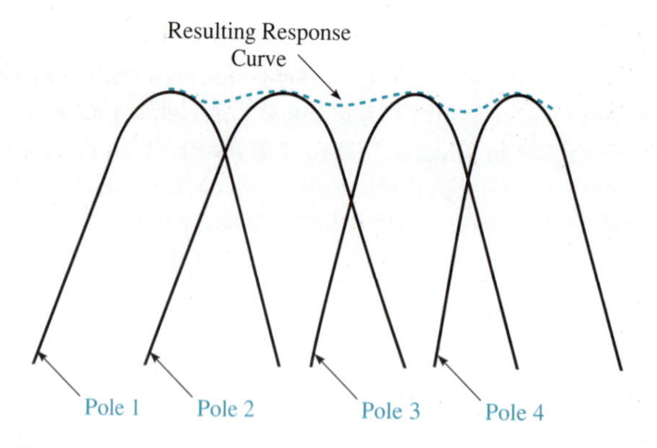

Figure 7.28 Result of poles being put together.

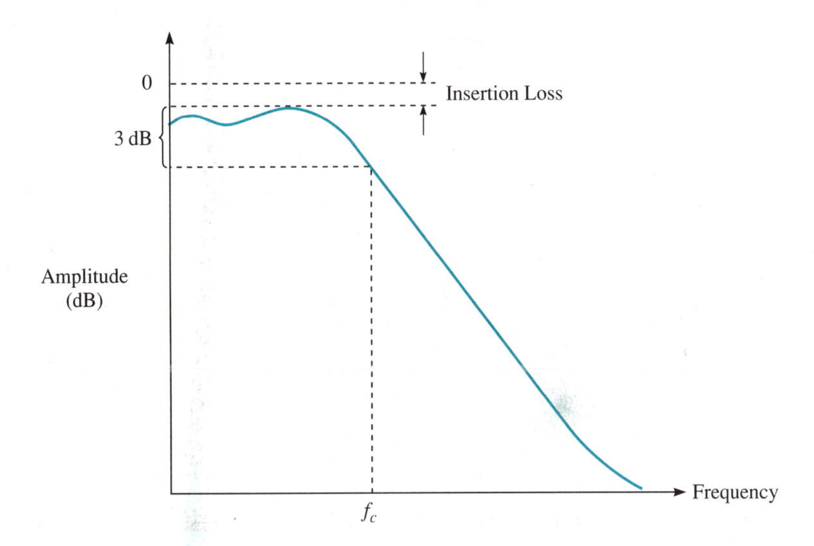

Figure 7.29 Low-pass filter.

down from 0. There is also a ripple term because of the number of poles used to fabricate the filter. The new term here is the cutoff frequency, F_c. This is the frequency where the response of the filter drops by 3 dB from the insertion-loss figure. This is probably the most important parameter of a low-pass filter because it defines the filter's area of operation. There are also rejection frequencies set up for low-pass filters similar to those for a bandpass filter. There needs to be a frequency and an attenuation figure associated with each point.

There is one term used with bandpass filters and low-pass filters that has not as yet been mentioned—ultimate attenuation. **Ultimate attenuation** ensures that when the response of a filter has gone to a high value of attenuation, it will remain there. This specification is necessary because when a filter is designed and a particular response is achieved, another response will begin, centered at $3F_0$ (or $3F_c$), that will allow any signals in that area to be passed. The designer can eliminate this condition if they know it is a requirement. Thus, an ultimate attenuation specification, such as −60 dB up to 5 GHz, tells the designer that responses up to 5 GHz should be taken care of. The specification shown above says that once we have gotten through the bandpass of the filter, or have reached the cutoff frequency of the low pass, all signals up to 5 GHz will be attenuated by at least 60 dB. This is a good specification to have in your requirements because it eliminates many problems that can occur from spurious signals or harmonics.

It should be pointed out that a low-pass filter will pass a dc signal. The construction of these filters allows the dc component of a signal or simply a dc signal to pass through the filter. Thus, if any problem can arise from the passage of dc in your circuit, precautions should be taken.

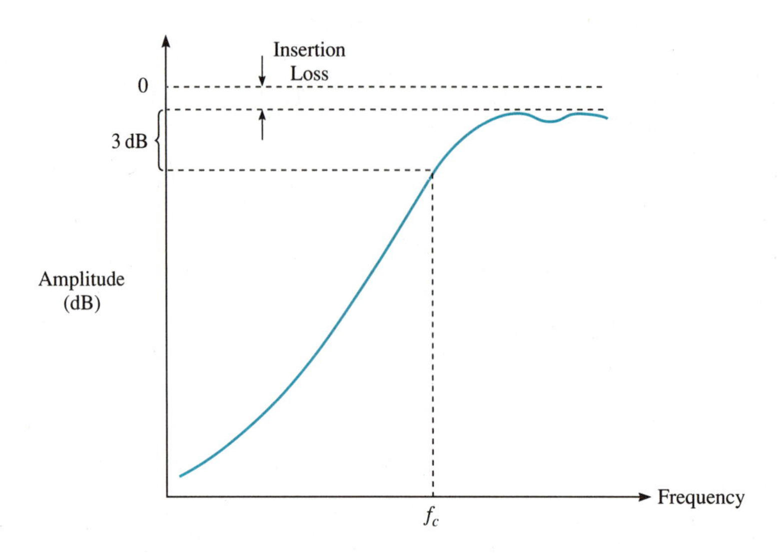

Figure 7.30 High-pass filter.

The last filter we will discuss is the reverse of the low-pass filter—the **high-pass filter**. Figure 7.30, the response curve for the high-pass filter, shows an insertion loss similar to those from other types of filters and a ripple due to the poles used. The cutoff frequency F_c is the frequency where the response is once again down 3 dB from the insertion loss of the filter.

Whereas a low-pass filter will pass dc, the opposite is not true of the high-pass filter. The high-pass filter does limit how high a frequency will exhibit the low insertion-loss characteristics. That is, it will not operate from F_c to blue-light frequencies with the same insertion loss. The point where it exhibits a higher loss does not concern us, however, because it is well beyond the useful range for which filters are normally used.

Before leaving the topic of filters, we should present a convenient way of changing a filter if the parameters need to be changed or if the original specifications do not actually allow the filter to perform its required functions. This method of changing, called **scaling**, enables you to change the frequency of a filter and still be able to retain the same characteristics as before. To accomplish this task you should take the following steps:

1. Find the multiplying factor K by dividing the new frequency by the old frequency.
2. Multiply all values of capacitances of the old filter by $1/K$.
3. Multiply all values of inductances of the old filter by $1/K$.
4. Do not do anything to resistor values. The resistance of the filters should remain the same.

Example 7.8 illustrates these steps.

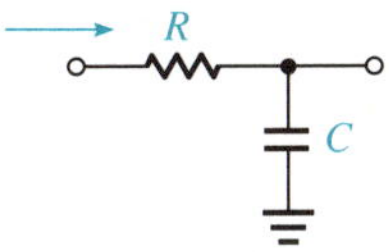

Figure 7.31 Low-pass filter example.

Example 7.8

Given: A low-pass filter consisting of a series resistor of 750 Ω and a shunt capacitor of 0.68 μF, as shown in Figure 7.31. The cutoff frequency for this filter is 312.1 Hz.

Find: The values of R and C to provide a cutoff frequency of 15 kHz.

Solution

The first step is to determine the multiplying factor K.

$$K = \frac{\text{New frequency}}{\text{Old frequency}}$$

$$= \frac{15{,}000}{312.1}$$

$$= 48.06$$

The factor $1/K$ is now $1/48.06 = 0.02$

Capacitor: New value $= 0.68\ \mu\text{F} \times 0.02$
$= 13\ \text{nF}$

Resistor: New value $= 750 \times 1$
$= 750\ \Omega$

(Recall that the resistors are not affected)

7.6 Summary

Many different types of components can be used in communications systems. Whether the components are oscillators, rf amplifiers, mixers, or specialized filters, they are all needed to make the overall communications system operate properly.

In this chapter we covered oscillators, including R–C, L–C, and crystal. The rf amplifier was also presented and compared with lower frequency devices. The mixer circuit, a very valuable part of many communications systems, was also discussed. Finally, the different types of filters (bandpass, low-pass, and high-pass) were discussed along with a frequency- and impedance-scaling method to facilitate the design of specific filters.

Questions

7.2 Oscillators

1. Define **oscillator.**
2. Name the four criteria needed for sustained oscillations.
3. What are the **Barkhausen criteria?**
4. Why is positive feedback needed in an oscillator?
5. Describe the operation of a lead–lag network.
6. Define **tank circuit.**
7. What are the feedback components in a Colpitts oscillator?
8. Why is the extra capacitor in a Clapp oscillator important?
9. Compare the operations of the Hartley and Colpitts oscillators.
10. Describe the **piezoelectric effect.**
11. Name the three axes of a crystal.
12. Name the two frequencies of a crystal.

7.3 Rf Amplifiers

13. Why are rf amplifiers used in communications systems?
14. Define **stray capacitance.**
15. How is an rf amplifier different from a low frequency amplifier?
16. What is an LNA?
17. Define **interelement capacitance.**

7.4 Mixers

18. Why must the circuit be nonlinear to have a mixing process?
19. Name the components necessary to have a mixer.
20. Define **conversion loss.**

7.5 Filters

21. Define **linear.**
22. Define **nonlinear.**
23. Why is isolation important in a mixer?
24. What is the relationship between conversion loss and noise figure in a mixer?
25. What are the **poles** of a filter?
26. Why do all filters have ripple?
27. Define **ultimate rejection.**
28. If the frequencies from 1 MHz to 1.5 MHz need to be passed and all other frequencies rejected, what type of filter should be used?
29. What type of filter should be used at the output of a signal generator to eliminate all harmonics?
30. Define **ripple.**
31. How is it possible for a low-pass filter to pass DC?
32. Define **scaling.**
33. Why is scaling valuable for filter design?

Problems

7.2 Oscillators

1. If we have an oscillator circuit with an overall system gain with feedback of 100, a forward system gain of 25, what is the feedback ratio?
2. An R–C phase-shift oscillator has resistors that are 75 Ω and capacitors that are 0.47 μF. What is the frequency of operation?
3. Suppose we want the frequency of the R–C oscillator in Problem 2 to be 3.2 kHz and we can only vary the resistors. What value must the resistors now have?
4. For the operational amplifier circuit shown in Figure 7.3, we have $R_i = 5\ \Omega$, $C = 0.005\ \mu\text{F}$, and $R = 10\ \text{k}\Omega$. Find the value of R_f required for oscillations and the frequency of oscillation.
5. Find the frequency of oscillation for a Colpitts oscillator, where $L = 0.1$ mH, $C_3 = 2\ \mu\text{F}$, and $C_4 = 5\ \mu\text{F}$. (Assume that the $Q > 10$.)
6. Repeat Problem 5 for a circuit with $Q = 4$.
7. A 6.8-MHz crystal has a positive temperature coefficient of 2 Hz/MHz/°C. What is the crystal frequency if the temperature decreases by 31°?

7.3 Rf Amplifiers

8. An rf amplifier in a communications system is experiencing very low gain. The base and collector voltages are checked and found to be near normal. The emitter voltage on the first stage is at 0 V. What is the problem?
9. The input stage of an rf amplifier is found to have 0 V on the base of the transistor. What is the probable cause of this lack of bias voltage?

7.4 Mixers

10. In the transistor mixer shown in Figure 7.25(c), the local oscillator signal is found to be very prominent at the output. Why?
11. The transistor mixer shown in Figure 7.25(c) has no i-f output. When the signal is traced through it is found that the sum, difference, rf, and LO frequencies are present at the collector of the transistor. It is also found that the trap circuit is eliminating the LO signal. What could be the cause of this lack of output?

7.5 Filters

12. We have a low-pass L–C filter with $L = 1.0$ mH, and $C = 1.5\ \mu\text{F}$ operating with $F_c = 4.1$ kHz. If we were to change this filter to operate at 2.5 kHz, what would be the new values of L and C?

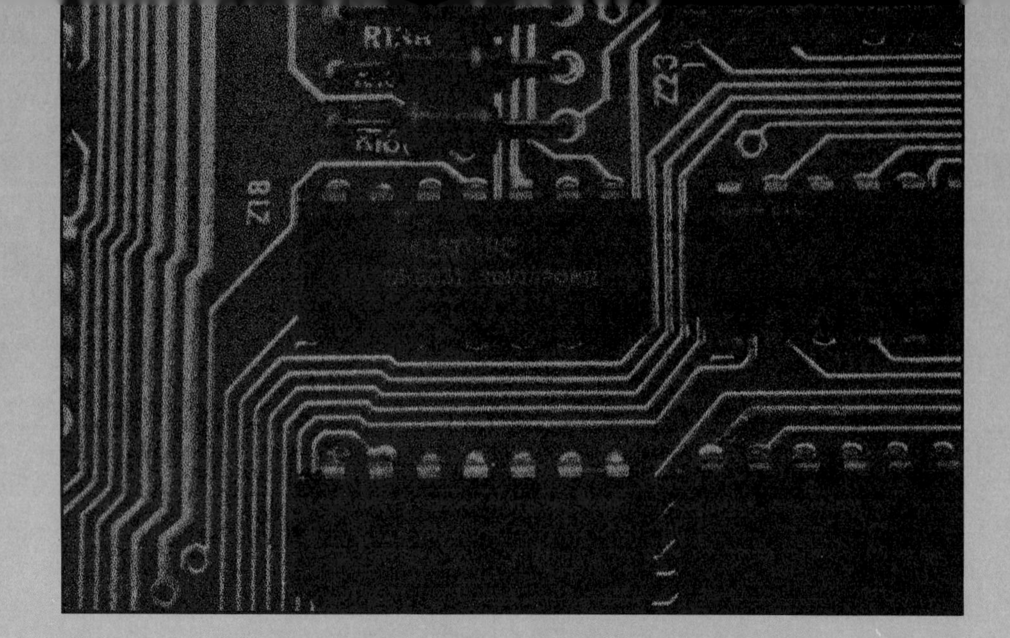

8

Outline

Objectives

- To introduce the concepts and applications of phase-locked loops and synthesizers in communications systems
- To discuss the individual parts of the phase-locked loop: the VCO, phase comparator, and the filter–amplifier
- To describe loop operation in order to enable troubleshooting any problems
- To discuss the concepts of direct and indirect synthesis, including the advantages and disadvantages of each type as well as their theory of operation

Key Terms

phase-locked loop
automatic gain control (AGC)
phase comparator
loop
pull-in range
direct synthesis
frequency synthesizer
voltage-controlled oscillator (VCO)
acquisition
acquisition time
capture range
hold-in range
indirect synthesis

Phase-Locked Loops and Synthesizers

8.1 Introduction

The phase-locked loop is generally thought of as a modern-day circuit that is only used in the most sophisticated applications. Although part of this statement is true, the phase-locked loop was used as far back as 1932 in France for synchronous reception of radio signals. It was the phase-locked loop that was the limiting factor in determining whether frequency modulation or phase modulation should be used for radio communications.

As we know, frequency modulation (FM) was chosen because demodulation of a phase-modulated signal required a phase-locked loop, and this could not be made small enough for some of the PM applications. You therefore carry around an FM rather than a PM radio. The necessity for a phase-locked loop in phase modulation is not a problem today because large-scale integration (LSI) techniques have miniaturized the phase-locked loop, producing a circuit that has all the needed accuracy but is considerably smaller than earlier versions.

The frequency synthesizer is a circuit that has found increased applications over the past few years. Many generators, particularly sweep generators, use the principles of the synthesizer to produce very accurate frequencies, which are used either for a single-frequency local oscillator or for a circuit that requires a series of precise frequencies for its operations. As with the phase-locked loop, the frequency synthesizer has become increasingly smaller because of improvements in technology. As such, the synthesizer is finding more and more uses throughout the electronics industry wherever a stable, accurate source is needed.

Both of the circuits presented above have been advanced through technology. We will now discuss each of them individually to show their design principles and operation, providing examples and applications of each.

8.2 Phase-Locked Loops

We can understand the term **phase-locked loop** (PLL) by dissecting the name. *Phase-lock* means that the phase of one signal is locked to the phase of another; that is, the frequencies of the two signals are the same, or are locked together. *Loop* refers to the method whereby the circuit is going to lock the two frequencies together—through the use of a feedback loop. Thus, we basically have a feedback loop designed to take a reference frequency and lock an incoming signal to it. We have a phase-locked loop.

This explanation is very basic, to say the least, but it serves to introduce you to the principles involved in phase-locked loops. There must be some sort of feedback system that will compare the frequency at the input to the circuit to a fixed reference frequency and have a means of locking on to that input frequency and holding it in the locked condition.

To aid in understanding phase-locked loops, we will review the concept of a feedback system. The basic block diagram for a simple feedback system is shown in Figure 8.1. Recall that we discussed feedback in Chapter 7 when discussing oscillators. In that instance we were after a positive, or regenerative, feedback to enable the circuit to oscillate. Here we need a negative, or degenerative, feedback to accomplish our purpose.

In Figure 8.1 we see that some input signal at the input of a circuit has a specific gain or loss characteristic. A designer needs to have the output of this circuit remain at a constant level at all times. This is very similar to an AGC (**automatic gain-control**) circuit used in some communications receivers. To maintain a specific level, the output must be constantly sampled to obtain its status relative to a certain reference level set prior to the circuit being turned on. The sampler sends the appropriate information back to a control circuit that either tells the circuit it is producing a signal that is too high or is not producing enough output. If the signal is too high, the control circuit decreases the gain of the circuit to comply with the requirements. When the sampler detects that the required level is once again attained, the control circuit is instructed to cease its correction. If the signal is too low, the opposite action occurs. This operation relies on a closed loop that feeds back a signal to a control circuit to control the actions of a specific circuit. This is a typical, basic feedback system.

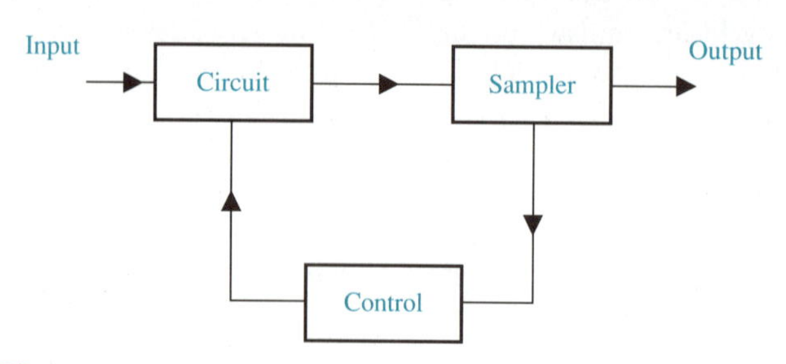

Figure 8.1 Basic feedback system.

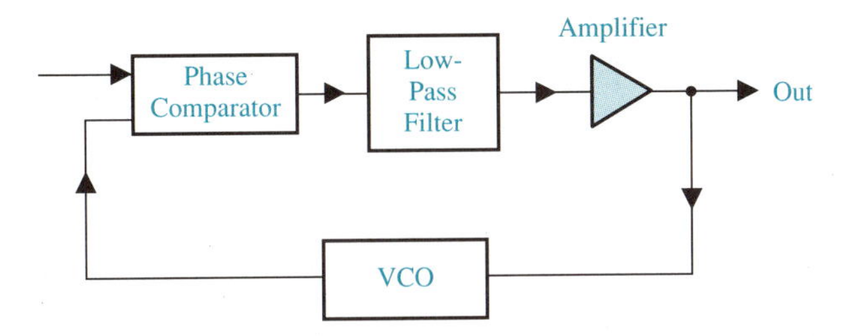

Figure 8.2 Phase-locked loop block diagram.

Having quickly reviewed feedback principles, we can now relate this theory to the phase-locked loop. A definition of a PLL would be:

Phase-locked loop

A feedback circuit designed to compare the frequency of a reference oscillator with that of an incoming signal. The combination of a phase comparator, voltage-controlled oscillator (VCO), and low-pass filter provide an efficient feedback loop to lock the phase of the VCO to the phase of the incoming signal.

Figure 8.2 illustrates the definition above. The figure is a block diagram of a PLL with all the terms referred to in the definition shown in it. The only block appearing in the figure that is not in the definition is the amplifier, the operation of which will be explained shortly. The PLL in Figure 8.2 consists of a phase comparator, a VCO (voltage-controlled oscillator), a low-pass filter, and an amplifier. Each of these blocks has a specific function as detailed in the following sections.

8.2.1 Voltage-Controlled Oscillator (VCO)

You will recall from Chapter 7, that an oscillator needs an input power source, a certain amount of gain, frequency-determining components, and a positive feedback system to produce self-sustaining oscillations. Also, if we want to change the frequency of the oscillations, we need to vary one of the frequency-determining components. The voltage-controlled oscillator must meet the above criteria for self-sustaining oscillations, but the frequency of operation is changed by changing a dc bias voltage to the circuit. This dc bias voltage is in addition to the input dc power required to originate oscillations.

To understand how an oscillator changes frequency proportional to a dc bias, consider the curve in Figure 8.3. This figure shows the output frequency as a function of the input dc bias voltage. This characteristic is called the **transfer function** of the oscillator, and we will designate it as H_0. Expressed mathematically it is

$$H_0 = \frac{\Delta F}{\Delta V} \tag{8.1}$$

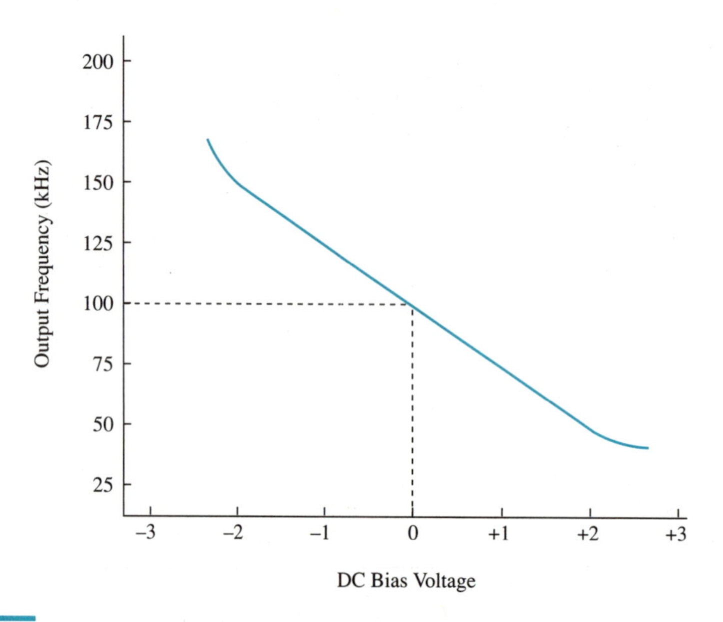

Figure 8.3 Input/output curve for VCO.

Where: H_0 = Oscillator transfer function (in hertz per volt)
ΔF = Frequency deviation (in hertz)
ΔV = Input bias voltage (in volts)

The dashed lines in the figure correspond to a frequency produced when the dc bias voltage is 0. This frequency is called the **natural frequency** of the oscillator and is designated as F_n. Any variation from this frequency by the application of a dc bias results in a different frequency at the output; this change in frequency is termed **frequency deviation** ΔF (see Equation 8.1). It's obvious that for a VCO to be predictable and accurate, the circuit should be operated within the linear range, that is, with a dc bias ranging from $+2$ V to -2 V. This is an important condition for the VCO to maintain so that it will operate properly within the entire PLL system.

8.2.2 Phase Comparator

The phase comparator is the very heart of the PLL system. Without a properly designed and operating phase comparator, there would be no such thing as a phase-locked loop. The **phase comparator** is actually a nonlinear mixer circuit in which one signal is an external signal (usually called the **incoming signal** (F_i), and the output of the VCO circuit is the other signal (designated F_o). As you recall from the discussions of mixer circuits in Chapter 6, if two signals are placed into a nonlinear circuit, the output will be the sum and difference of the two signals ($F_i + F_o$ and $F_i - F_o$), as well as a series of combinations of the two signals. The purpose of the phase comparator is to use only the difference in frequency of these combinations, and to ensure that the

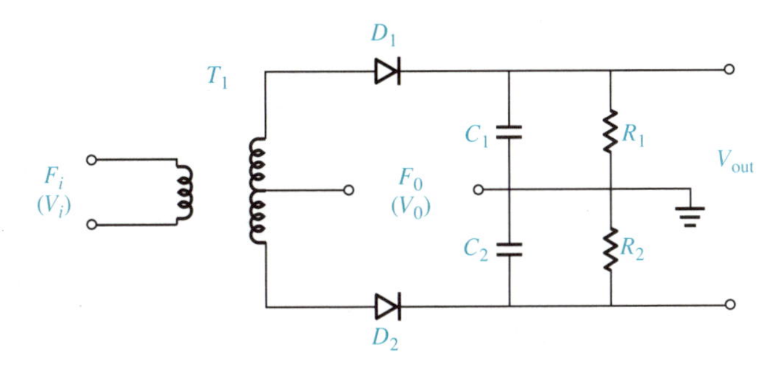

Figure 8.4 Phase comparator.

low-pass filter is placed at the output. This filter will pass only the difference of the two frequencies and reject the sum and all other combinations. Ultimately, for the loop to operate properly, F_i will have to equal F_o. This operation will be discussed later in this section.

A basic phase comparator is shown in Figure 8.4. Operation of the circuit depends on the forward or reverse biasing of diodes D_1 and D_2, and the subsequent charge and discharge of the associated capacitors C_1 and C_2. Basic operation of the circuit is as follows:

1. If we apply only the VCO voltage (V_o), which will probably be a sinusoidal voltage, to the comparator, we see that it is simultaneously applied to both halves of the transformer. With $C_1 = C_2$ and $R_1 = R_2$, it follows that the combination of D_1, C_1, and R_1 make up a half-wave rectifier circuit, as do D_2, C_2, and R_2.
2. On the positive half cycle of V_o, the diodes are forward-biased and the capacitors begin charging to equal values and opposite polarities. The net result is a 0-V output from the comparator.
3. On the reverse cycle, the diodes are reverse-biased and the capacitors can now discharge. The resulting output voltage is once again 0 V.

Thus, the output from the VCO voltage is constant and is equal to 0 V. This is important because it represents the reference we will be comparing any external input signal to. If the reference signal did not produce a constant output, there would be nothing to compare the signal to.

4. We now apply an input signal V_i at the same time as the VCO signal; we will be able to compare that input signal directly. The output of the comparator will depend on the phase relationship of the two signals. Figure 8.5 illustrates this concept showing four discrete phase relationships and the corresponding voltage at the output. Figure 8.5(a) shows the case where both V_o and V_i are in phase. During the ON time, V_i is positive. This results in an output voltage that is also positive and is equal to 0.636 V. This value is 0.636 V instead of the typical value of $1/\pi$ (0.318) because the two output states add together since the entire half cycle is the same.

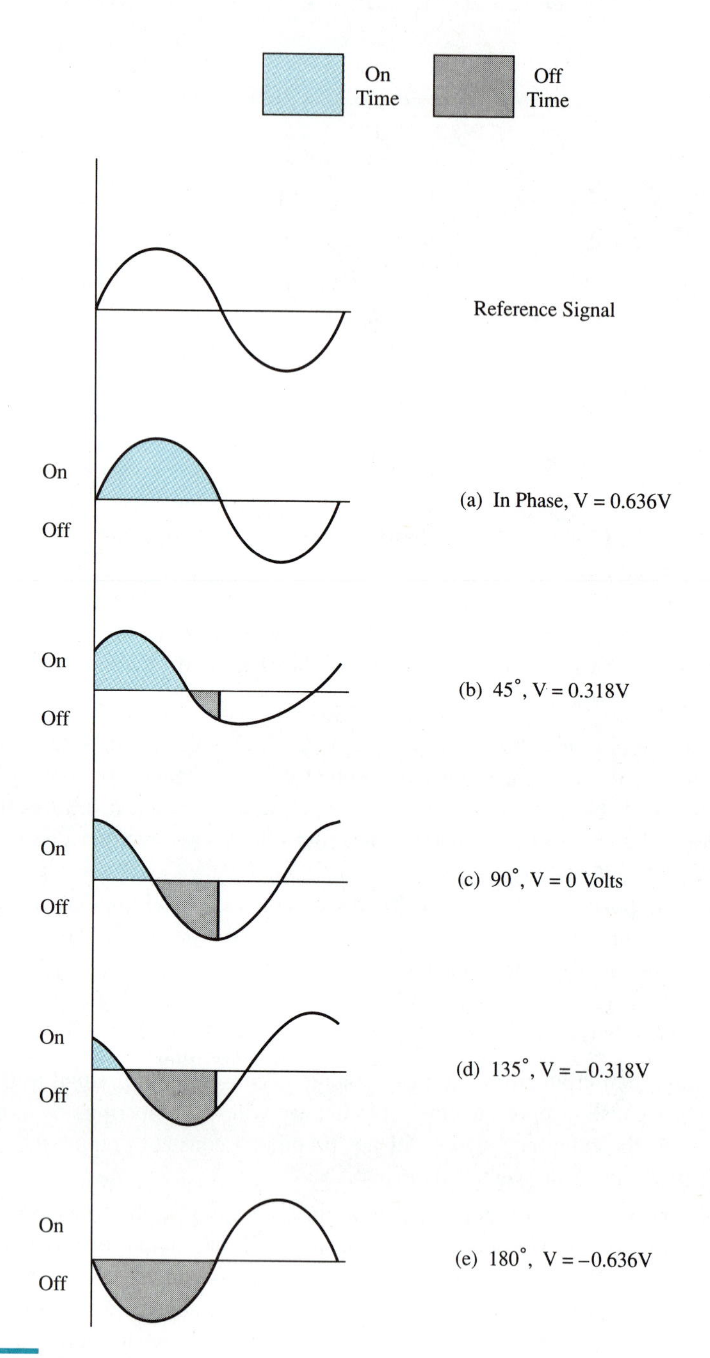

Figure 8.5 Phase comparator output.

5. Figure 8.5(b) shows V_o leading V_i by 45°. In this condition the circuit is ON for 75% of the time and OFF for 25% of the time. This results in a voltage output of 0.318 V (the true average value of a signal). For Figure 8.5(c), the signals are 90° out of phase and the circuit is ON for 50% of the time and OFF for 50% of the time. The result is 0 V at the output.
6. Figure 8.5(d) shows V_o leading V_i by 135°. In this condition, the circuit is ON for 25% of the time and OFF for 75% of the time. The output voltage is now negative because the circuit is OFF longer than it is ON. Thus, a −0.318 V is found at the output.
7. The final case is shown in Figure 8.5(e) for the condition where V_o leads V_i by 180°. In this case the circuit is OFF for 100% of the time. Thus, there is a negative voltage at the output: −0.636 V.

The conditions shown in Figure 8.5 are discrete relationships and have a corresponding output voltage associated with them. There are, of course, intermediate values of phase relationships between V_o and V_i that occur throughout a system. There are also corresponding voltages associated with these values. This correspondence results in a truly analog representation of the phase characteristics between the two signals. These characteristics enable the phase comparator to be an integral part of the PLL and to compare an input signal to a VCO signal, allowing the system to lock onto a specific signal.

8.2.3 Filter–Amplifier

From our discussions in previous sections, you can see that the low-pass filter shown in Figure 8.2 is necessary for cleaning up the signals in the loop. You will recall that the comparator is driven into a nonlinear region of operation in order to generate a variety of signal outputs, including the two signals, the sum, difference, and products of the signals involved. To ensure that there is a clean output, a low-pass filter is inserted in the loop as shown in Figure 8.2. This filter allows only the difference of the two signals (V_o and V_i) to proceed further around the loop, greatly increasing the accuracy of the loop because only a single frequency has to be processed rather than the multitude of signals present without the low-pass filter.

The amplifier in the loop is inserted to compensate for any losses that occur around the loop. This is especially needed to reinstate the level after the comparator action takes place. Anytime a system mixes two signals together in a nonlinear mixer, there will be losses in signal level. The amplifier, usually an operational amplifier, is employed to compensate for these losses and keep the loop gain constant during operations.

8.2.4 Loop Operation

Two conditions are important when speaking of phase-locked loops: acquisition and lock. **Acquisition** refers to acquiring, or capturing, a signal before we can use it. **Lock** means that once we acquire the signal we must be able to lock on to it and hold it so that we can maintain a stable, accurate frequency.

With these two terms in mind, let us look at a PLL and describe its operation with a specific input signal. Figure 8.6 will be used to trace the operation through the circuit.

Using the circuit shown in Figure 8.6, we apply an input signal to the phase comparator input (F_i) and mix it with the VCO output signal (F_o). This operation is similar to that of the phase comparator alone. In this initial condition F_i does *not* equal F_o and the system is considered to be **unlocked**. Since there is a difference in frequency, there will be a mixing action, and at the comparator output there will be frequencies F_i, F_o, $F_i + F_o$, and $F_i - F_o$. The low-pass filter now removes the original signals and the sum of the signals. What remains is a resultant signal F_r, which is the difference frequency, sometimes called the **beat frequency**.

The amplifier increases the level to compensate for losses and applies the frequency to the VCO. The output frequency F_o of the VCO is changed by an amount proportional to the polarity of the amplifier output frequency F_r. This is a frequency that changes the oscillator from its natural rest frequency F_n. Thus, $F_o = F_n + F$.

The process above continues, with F_o causing the amplitude and frequency of the resulting frequency F_r to change proportionally until $F_o = F_i$. This condition, termed **lock**, basically occurs when the frequency of the loop is the same as that of the input frequency. At this point, $F_o - F_i = 0$ Hz (dc voltage). This dc voltage keeps the two frequencies equal. At this point the comparator, which was acting as a frequency comparator, now becomes a phase comparator, and any difference in phase between F_o and F_i, from here on is converted to a dc bias voltage and fed back to the VCO to hold the circuit in its locked condition.

A certain amount of time is necessary for a PLL to acquire lock; this time is called the acquisition, or **pull-in, time**. With no loop filter in the circuit (uncompensated loop) this time is approximately equal to the reciprocal of the open-loop gain of the PLL. That is,

$$\text{Acquisition time} = \frac{1}{A_v} \tag{8.2}$$

Where: A_v = Open-loop gain of the PLL

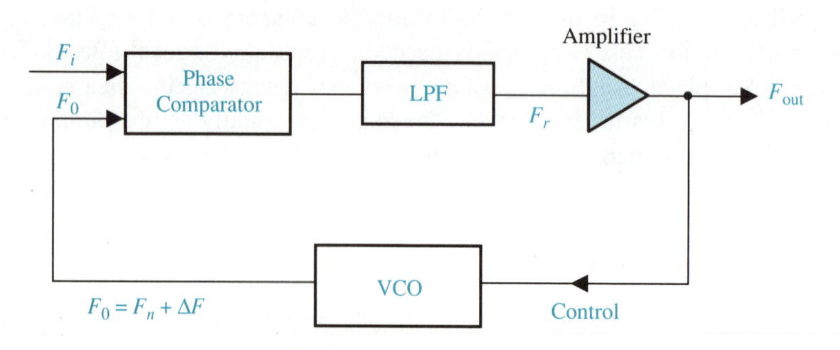

Figure 8.6 PLL operation block diagram.

For every PLL circuit there is a definite range within which the loop will lock onto the incoming signal. This range, called the **capture range**, is very definite because the loop does not have an infinite bandwidth. This range is usually between 1.1 and 1.7 times the natural frequency F_n of the VCO. Also associated with the capture range is the **pull-in range** of the circuit which is equal to $\frac{1}{2}$ the capture range of the PLL, meaning that the loop can acquire a signal, within the pull-in range, but will only lock if the signal is within the capture range.

Example 8.1

A PLL is set up with its VCO free-running at 5 MHz. The VCO will not change frequency until the input is within 6.5 kHz of the 5-MHz free-running frequency. When this occurs, the VCO follows the input to ±150 kHz. Determine the capture and lock range of this PLL.

Solution

The capture range occurs at 6.5 kHz from the free-running VCO. If we assume there is symmetrical operation, the capture range is

$$6.5 \text{ kHz} \times 2 = 13 \text{ kHz}$$

The lock range is determined by considering how far the circuit can follow the 5-MHz signal. This circuit can follow the input to 150 kHz. Thus

$$\text{lock range} = 150 \times 2 = 300 \text{ kHz}$$

We mentioned the loop gain of a PLL in Equation (8.2) and now need to pursue this term further. The loop gain is actually the product of each of the individual gains around the loop. That is

$$A_v = A_d L_f A_a H_0 \text{ (in hertz per rad)} \tag{8.3}$$

Where: A_v = Open-loop gain of the PLL
A_d = Comparator gain
L_f = Filter loss
A_a = Amplifier gain
H_0 = VCO transfer function

(The closed-loop gain of a locked PLL is always 1, or 0 dB, because there is always 100% feedback in a closed-loop PLL.)

The value of open-loop gain in decibles is given as

$$A_v \text{ (in decibels)} = 20 \log A_v \tag{8.4}$$

Once the PLL has achieved a locked condition, the next task is to keep the system locked. The range of frequencies that the input may vary and still maintain the locked condition is the **hold-in**, or **tracking, range**. This range is governed by the range of the VCO because the VCO must be able to track the input frequency to maintain lock. Since the maximum phase error allowable while maintaining lock in the PLL is $\pm\pi/2$, the maximum change in the VCO natural frequency is calculated as

$$\begin{aligned} F_{\max} &= \pm\frac{\pi}{2} A_d L_f A_a H_0 \\ &= \pm\frac{\pi}{2} A_v \end{aligned} \qquad (8.5)$$

To illustrate the material presented thus far on the PLL, consider the following example.

Example 8.2

Given: A PLL with F_n = 150 kHz, F_1 = 155 kHz, A_d = 0.5 V/rad, L_f = 1.3, A_a = 8, and H_0 = 10 kHz/v

Find:
1. ΔF to achieve lock
2. V_{out}
3. Open-loop gain A_v
4. Hold-in range

Solution

1. To find ΔF to achieve lock we use

$$\begin{aligned} \Delta F &= F_i - F_o \\ &= 155 \text{ kHz} - 150 \text{ kHz} \\ &= 5 \text{ kHz} \end{aligned}$$

2. To find V_{out}, use Equation (7.1)

$$\begin{aligned} V_{\text{out}} &= \frac{F}{H_0} \\ &= \frac{5 \text{ kHz}}{10 \text{ kHz/V}} \\ &= 0.5 \text{ V} \end{aligned}$$

3. Use Equation (8.3)

$$\begin{aligned} A_v &= A_d L_f A_a H_0 \\ &= (0.5)(1.3)(8)(10) \\ &= 52 \text{ kHz/rad} \end{aligned}$$

4. Use Equation (8.5)

$$F_{max} = \pm\frac{\pi}{2} A_d L_f A_a H_0$$

$$= \pm\frac{\pi}{2} A_v$$

$$= \pm\frac{\pi}{2}\, 52 \text{ kHz/rad}$$

$$= \pm 26\pi$$

$$F_{max} = \pm 81.68 \text{ kHz}$$

Data sheets for the LM565 phase-locked loop chip are shown in Figure 8.7. Figure 8.8 is a working circuit that uses the 565 chip in a phase-locked loop application.

8.3 Synthesizers

8.3.1 Introduction

Synthesis is, in a very general sense, the combination of parts to form a whole. In other words, if we need to synthesize something we need to gather together all of the parts and put them together. This sounds rather basic, but in order to obtain the proper results, the parts must be compatible and fit together in the proper way. Thus, our synthesizer is made up of specific parts that produce the required output.

At this point, we need to introduce the type of synthesizer we are going to be concerned with—the **frequency synthesizer**. This device, generally built around a PLL, provides a switch-selectable choice of a large number of output frequencies, which may be derived from one or more crystal-controlled oscillators. There are several points that should be made about this definition just presented. The first is that the frequency synthesizer is built around a phase-locked loop circuit, like the ones discussed in Section 8.2. The theory behind their operation should therefore be relatively easy to understand. Second, a large number of frequencies are available at the output of the synthesizer. This means that the frequency synthesizer is actually a very accurate form of sweep generator. It is not a true sweep generator in that it has a continuous sweep of frequencies, but is a sweep generator in the sense that it produces many frequencies very close together that simulate a full sweep of frequencies. Finally, as you recall from Chapter 7, the crystal-controlled oscillator is the most accurate oscillator that can be built. With this in mind, we understand that the frequency synthesizer, with one or more crystal-controlled oscillators, can generate not

(*Text continues on page 199*)

National Semiconductor

April 1987

LM565/LM565C Phase Locked Loop

LM565/LM565C Phase Locked Loop

General Description

The LM565 and LM565C are general purpose phase locked loops containing a stable, highly linear voltage controlled oscillator for low distortion FM demodulation, and a double balanced phase detector with good carrier suppression. The VCO frequency is set with an external resistor and capacitor, and a tuning range of 10:1 can be obtained with the same capacitor. The characteristics of the closed loop system—bandwidth, response speed, capture and pull in range—may be adjusted over a wide range with an external resistor and capacitor. The loop may be broken between the VCO and the phase detector for insertion of a digital frequency divider to obtain frequency multiplication.

The LM565H is specified for operation over the −55°C to +125°C military temperature range. The LM565CH and LM565CN are specified for operation over the 0°C to +70°C temperature range.

Features

- 200 ppm/°C frequency stability of the VCO
- Power supply range of ±5 to ±12 volts with 100 ppm/% typical
- 0.2% linearity of demodulated output
- Linear triangle wave with in phase zero crossings available
- TTL and DTL compatible phase detector input and square wave output
- Adjustable hold in range from ±1% to > ±60%

Applications

- Data and tape synchronization
- Modems
- FSK demodulation
- FM demodulation
- Frequency synthesizer
- Tone decoding
- Frequency multiplication and division
- SCA demodulators
- Telemetry receivers
- Signal regeneration
- Coherent demodulators

Connection Diagrams

Metal Can Package

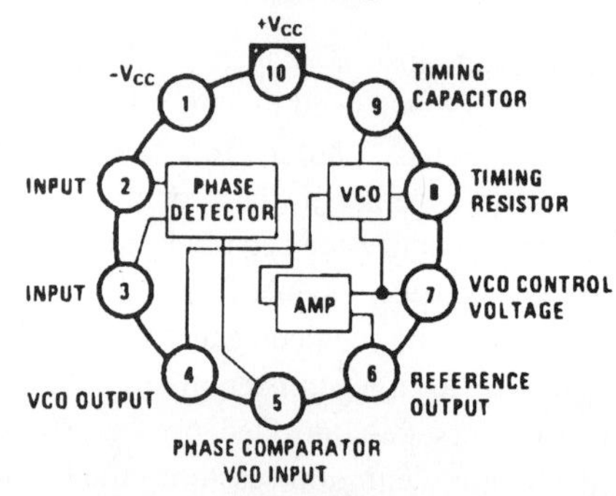

TL/H/7853-2

Order Number LM565H or LM565CH
See NS Package Number H10C

Dual-In-Line Package

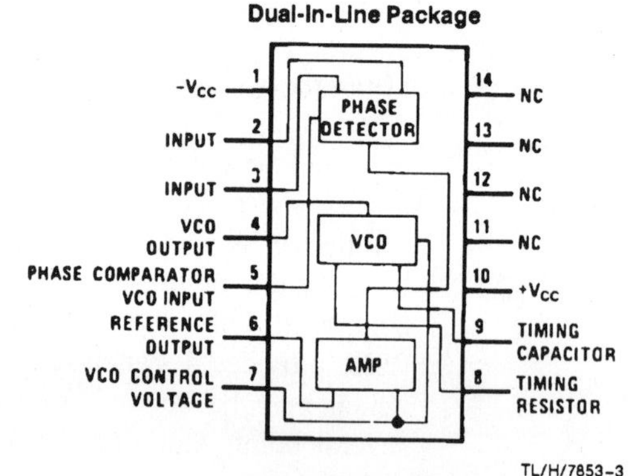

TL/H/7853-3

Order Number LM565CN
See NS Package Number N14A

Figure 8.7 PLL chip data sheet.

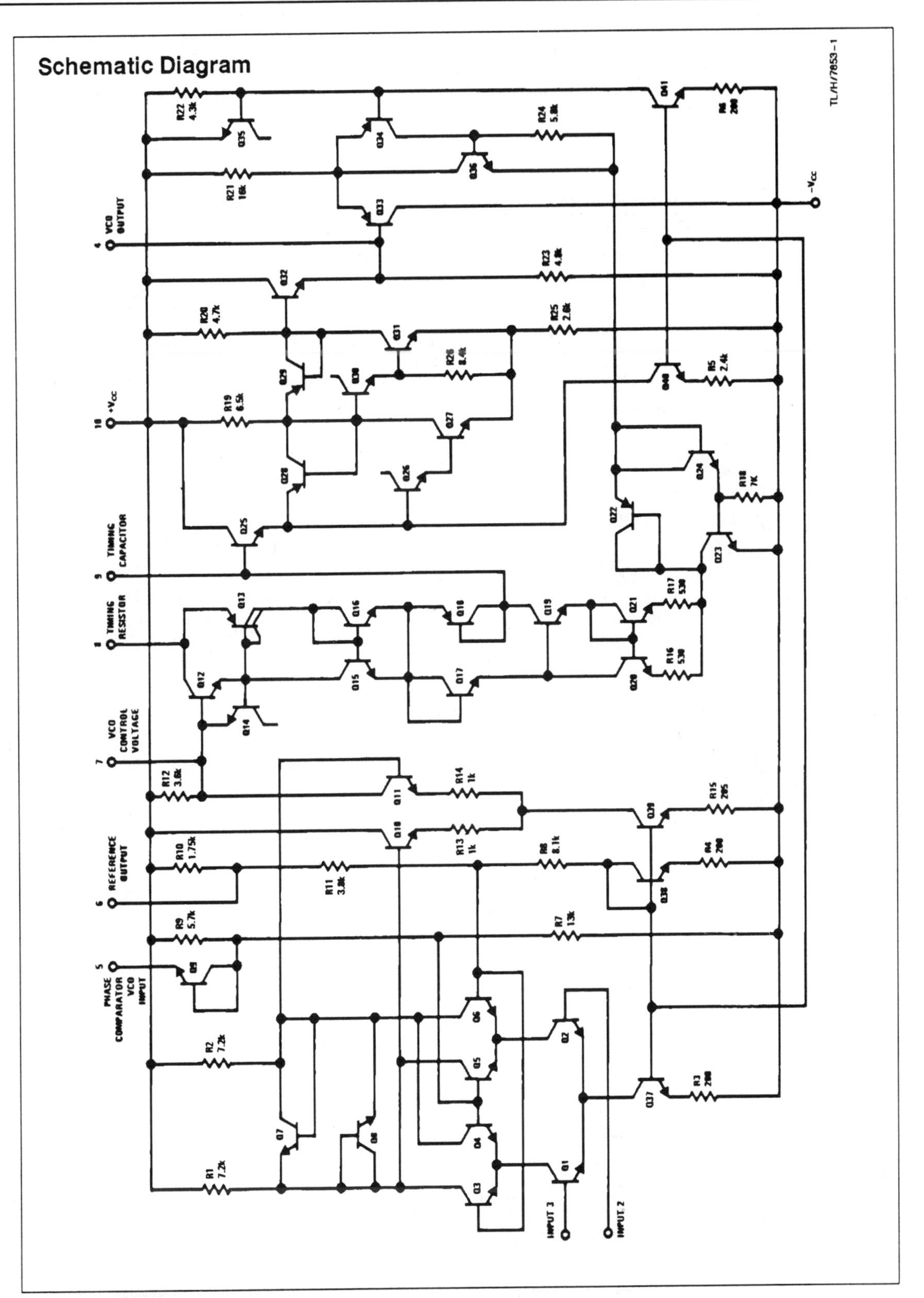

Figure 8.7 (*cont'd*)

Applications Information

In designing with phase locked loops such as the LM565, the important parameters of interest are:

FREE RUNNING FREQUENCY

$$f_o \cong \frac{0.3}{R_o C_o}$$

LOOP GAIN: relates the amount of phase change between the input signal and the VCO signal for a shift in input signal frequency (assuming the loop remains in lock). In servo theory, this is called the "velocity error coefficient."

$$\text{Loop gain} = K_o K_D \left(\frac{1}{\text{sec}}\right)$$

$$K_o = \text{oscillator sensitivity} \left(\frac{\text{radians/sec}}{\text{volt}}\right)$$

$$K_D = \text{phase detector sensitivity} \left(\frac{\text{volts}}{\text{radian}}\right)$$

The loop gain of the LM565 is dependent on supply voltage, and may be found from:

$$K_o K_D = \frac{33.6 f_o}{V_c}$$

f_o = VCO frequency in Hz

V_c = total supply voltage to circuit

Loop gain may be reduced by connecting a resistor between pins 6 and 7; this reduces the load impedance on the output amplifier and hence the loop gain.

HOLD IN RANGE: the range of frequencies that the loop will remain in lock after initially being locked.

$$f_H = \pm \frac{8 f_o}{V_c}$$

f_o = free running frequency of VCO

V_c = total supply voltage to the circuit

THE LOOP FILTER

In almost all applications, it will be desirable to filter the signal at the output of the phase detector (pin 7); this filter may take one of two forms:

Simple Lag Filter

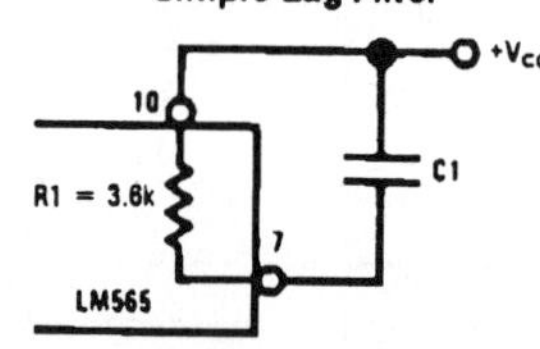

TL/H/7853-11

Lag-Lead Filter

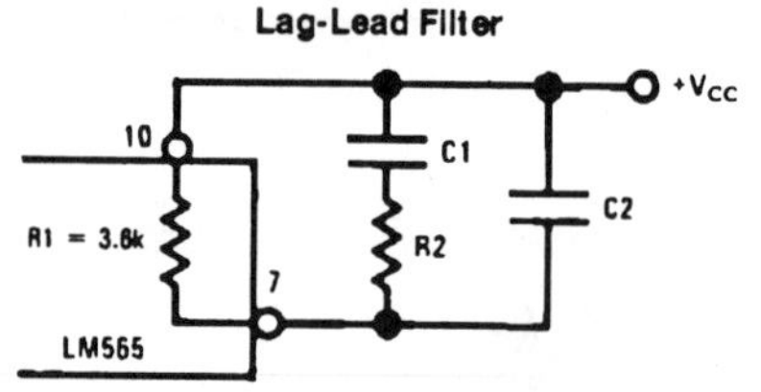

TL/H/7853-12

A simple lag filter may be used for wide closed loop bandwidth applications such as modulation following where the frequency deviation of the carrier is fairly high (greater than 10%), or where wideband modulating signals must be followed.

The natural bandwidth of the closed loop response may be found from:

$$f_n = \frac{1}{2\pi}\sqrt{\frac{K_o K_D}{R_1 C_1}}$$

Associated with this is a damping factor:

$$\delta = \frac{1}{2}\sqrt{\frac{1}{R_1 C_1 K_o K_D}}$$

For narrow band applications where a narrow noise bandwidth is desired, such as applications involving tracking a slowly varying carrier, a lead lag filter should be used. In general, if $1/R_1C_1 < K_o K_D$, the damping factor for the loop becomes quite small resulting in large overshoot and possible instability in the transient response of the loop. In this case, the natural frequency of the loop may be found from

$$f_n = \frac{1}{2\pi}\sqrt{\frac{K_o K_D}{\tau_1 + \tau_2}}$$

$$\tau_1 + \tau_2 = (R_1 + R_2) C_1$$

R_2 is selected to produce a desired damping factor δ, usually between 0.5 and 1.0. The damping factor is found from the approximation:

$$\delta \approx \pi \tau_2 f_n$$

These two equations are plotted for convenience.

Filter Time Constant vs Natural Frequency

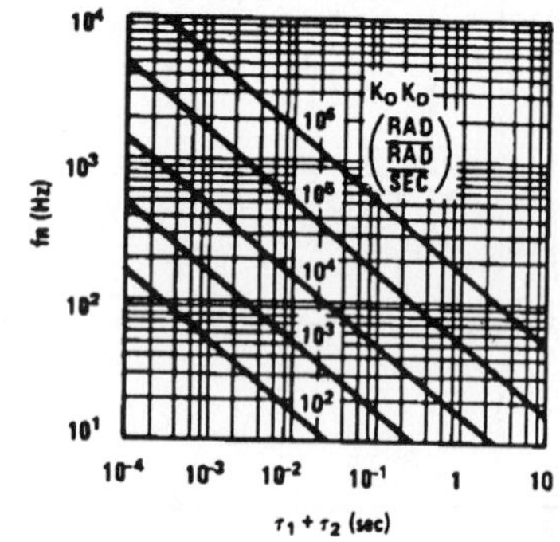

TL/H/7853-13

Damping Time Constant vs Natural Frequency

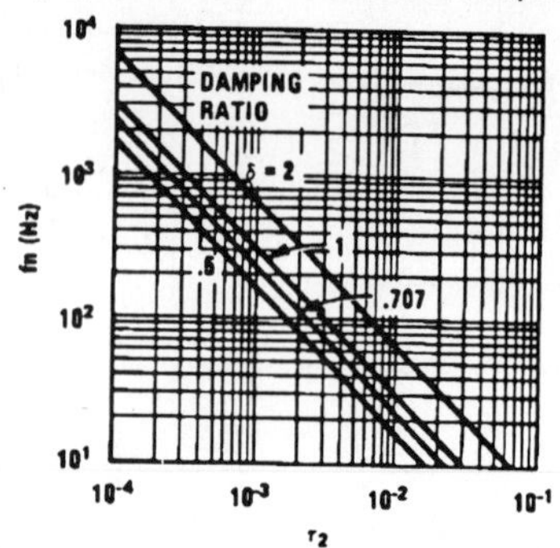

TL/H/7853-14

Capacitor C_2 should be much smaller than C_1 since its function is to provide filtering of carrier. In general $C_2 \leq 0.1\, C_1$.

Figure 8.7 *(cont'd)*

Absolute Maximum Ratings

If Military/Aerospace specified devices are required, please contact the National Semiconductor Sales Office/Distributors for availability and specifications.

Supply Voltage	±12V
Power Dissipation (Note 1)	1400 mW
Differential Input Voltage	±1V
Operating Temperature Range	
LM565H	−55°C to +125°C
LM565CH, LM565CN	0°C to +70°C
Storage Temperature Range	−65°C to +150°C
Lead Temperature (Soldering, 10 sec.)	260°C

Electrical Characteristics AC Test Circuit, $T_A = 25°C$, $V_{CC} = \pm 6V$

Parameter	Conditions	LM565			LM565C			Units
		Min	Typ	Max	Min	Typ	Max	
Power Supply Current			8.0	12.5		8.0	12.5	mA
Input Impedance (Pins 2, 3)	$-4V < V_2, V_3 < 0V$	7	10			5		kΩ
VCO Maximum Operating Frequency	$C_o = 2.7$ pF	300	500		250	500		kHz
VCO Free-Running Frequency	$C_o = 1.5$ nF $R_o = 20$ kΩ $f_o = 10$ kHz	−10	0	+10	−30	0	+30	%
Operating Frequency Temperature Coefficient			−100			−200		ppm/°C
Frequency Drift with Supply Voltage			0.1	1.0		0.2	1.5	%/V
Triangle Wave Output Voltage		2	2.4	3	2	2.4	3	$V_{p\text{-}p}$
Triangle Wave Output Linearity			0.2			0.5		%
Square Wave Output Level		4.7	5.4		4.7	5.4		$V_{p\text{-}p}$
Output Impedance (Pin 4)			5			5		kΩ
Square Wave Duty Cycle		45	50	55	40	50	60	%
Square Wave Rise Time			20			20		ns
Square Wave Fall Time			50			50		ns
Output Current Sink (Pin 4)		0.6	1		0.6	1		mA
VCO Sensitivity	$f_o = 10$ kHz		6600			6600		Hz/V
Demodulated Output Voltage (Pin 7)	±10% Frequency Deviation	250	300	400	200	300	450	$mV_{p\text{-}p}$
Total Harmonic Distortion	±10% Frequency Deviation		0.2	0.75		0.2	1.5	%
Output Impedance (Pin 7)			3.5			3.5		kΩ
DC Level (Pin 7)		4.25	4.5	4.75	4.0	4.5	5.0	V
Output Offset Voltage $\lvert V_7 - V_6 \rvert$			30	100		50	200	mV
Temperature Drift of $\lvert V_7 - V_6 \rvert$			500			500		μV/°C
AM Rejection		30	40			40		dB
Phase Detector Sensitivity K_D			.68			.68		V/radian

Note 1: The maximum junction temperature of the LM565 and LM565C is +150°C. For operation at elevated temperatures, devices in the TO-5 package must be derated based on a thermal resistance of +150°C/W junction to ambient or +45°C/W junction to case. Thermal resistance of the dual-in-line package is +85°C/W.

Figure 8.7 *(cont'd)*

Typical Performance Characteristics

Power Supply Current as a Function of Supply Voltage

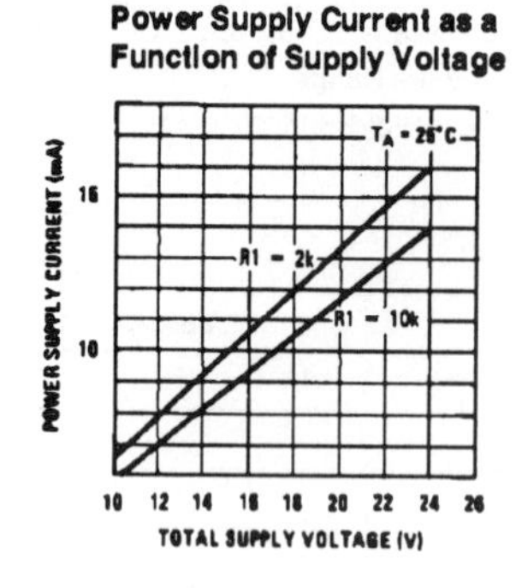

Lock Range as a Function of Input Voltage

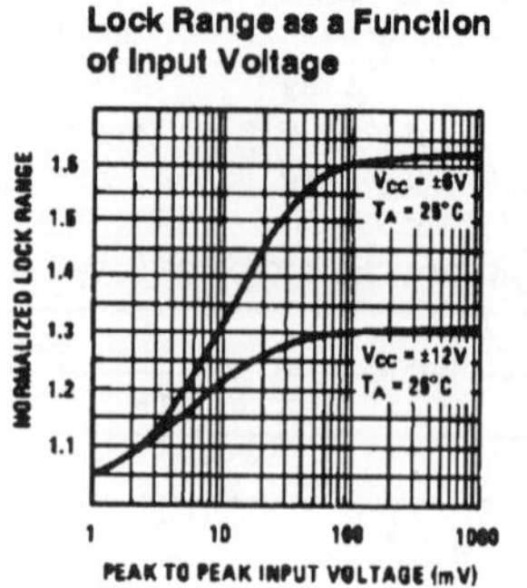

VCO Frequency

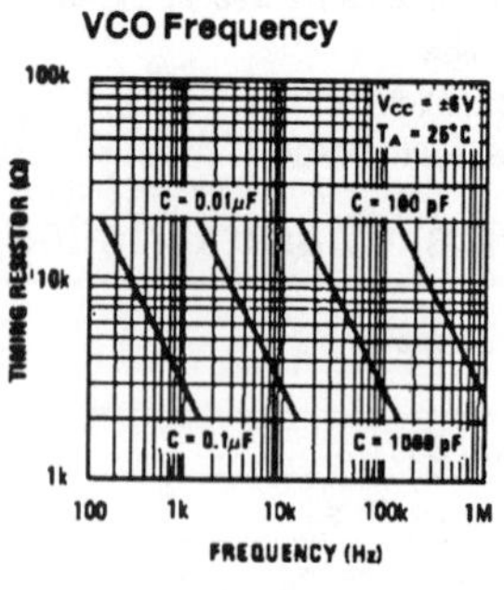

Oscillator Output Waveforms

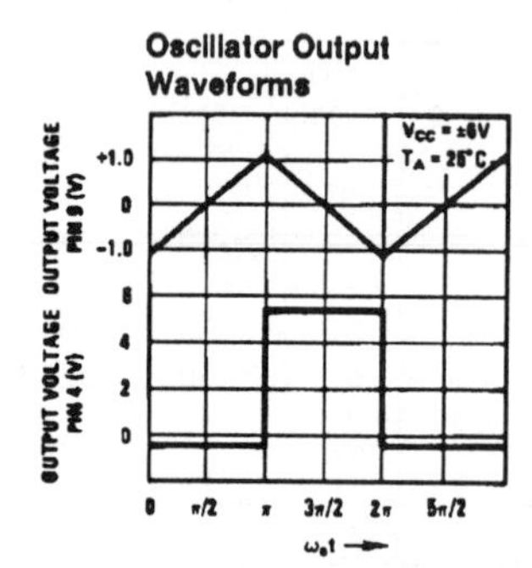

Phase Shift vs Frequency

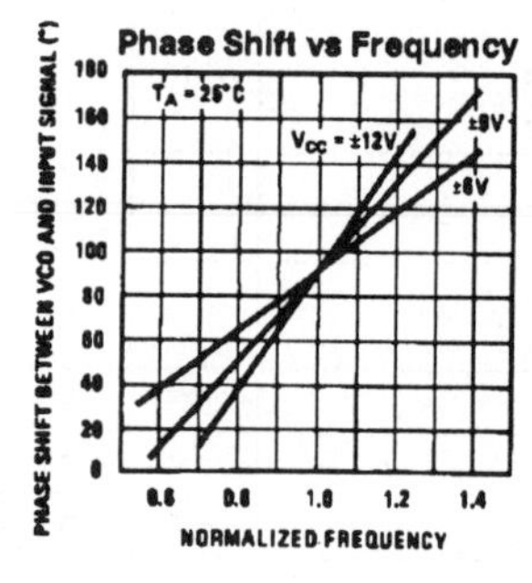

VCO Frequency as a Function of Temperature

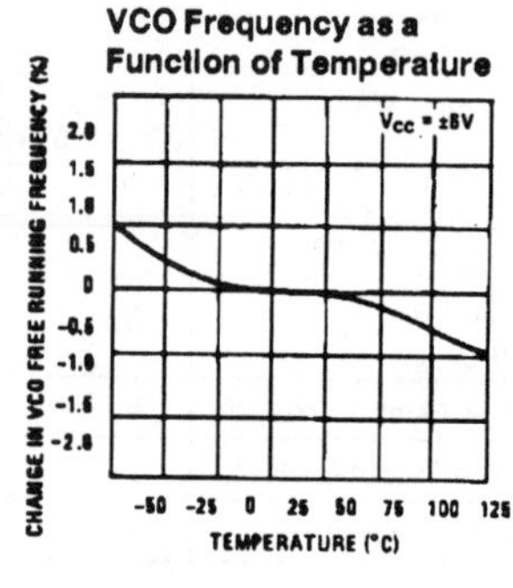

Loop Gain vs Load Resistance

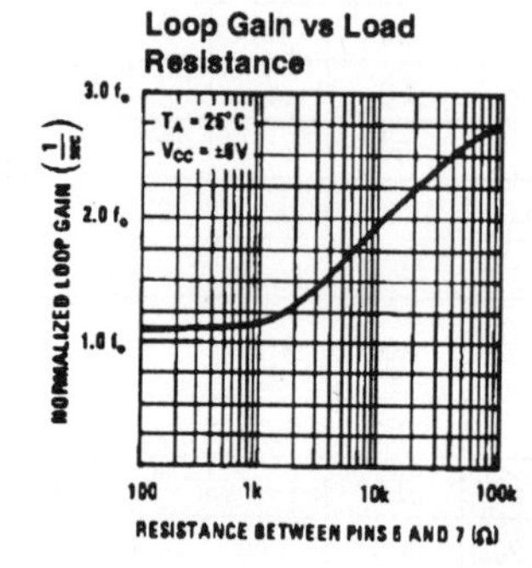

Hold In Range as a Function of R_{6-7}

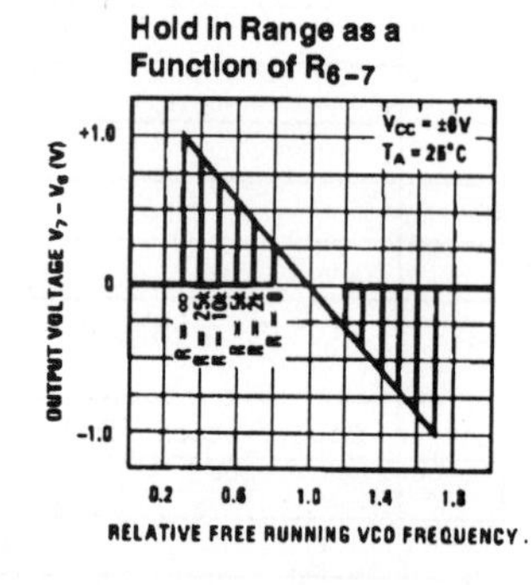

TL/H/7853-4

Figure 8.7 *(cont'd)*

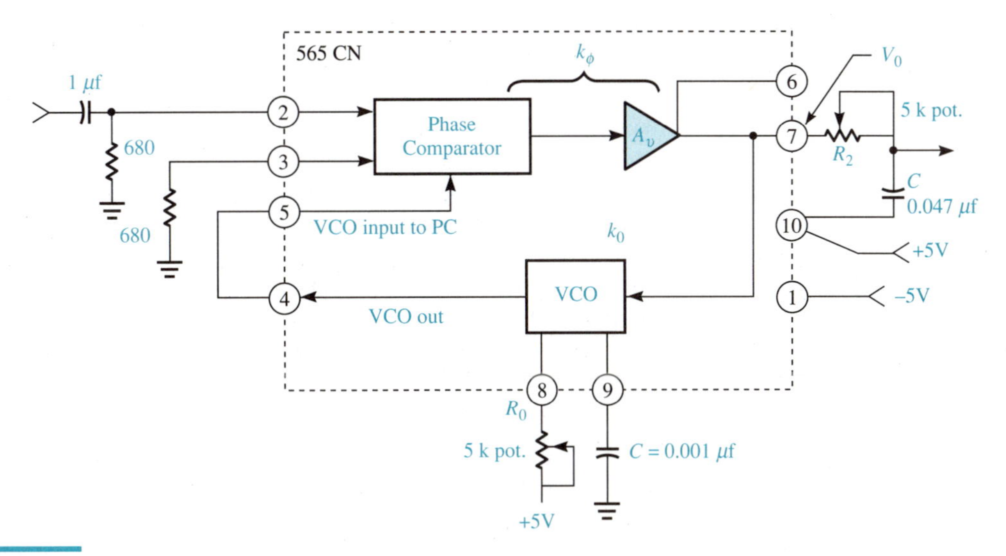

Figure 8.8 PLL circuit.

only a large number of frequencies, but very accurate and stable frequencies. Thus, the frequency synthesizer is a device with many applications in communications systems in both transmitter and receiver sections.

A basic block diagram of a frequency synthesizer, shown in Figure 8.9, is similar to the PLL block diagrams presented in Section 8.2. This is because the frequency synthesizer uses the same theory and operation as PLL's to produce the necessary and accurate frequencies that are required. The only block in the frequency synthesizer that is different from a PLL is the programmable divider. Under normal PLL operations, the output of the VCO would be sent to the phase comparator for comparison with the incoming signal. In the case for the synthesizer shown in Figure 8.9, the output of the VCO is the output but is also fed back to the divider circuit. By varying the division range in the circuit (N), the frequency compared at the input to the comparator is varied. Thus, the circuit now locks to a new frequency. This is the case for as many division ranges as the operator wants to place in the circuit. There is, of course, a limit to the frequency range but this limit is not one that is small. So, by varying the divider parameters (namely the N-division ratio), many very accurate, stable output frequencies can be generated.

Two methods are used in frequency synthesizers: direct and indirect synthesis.

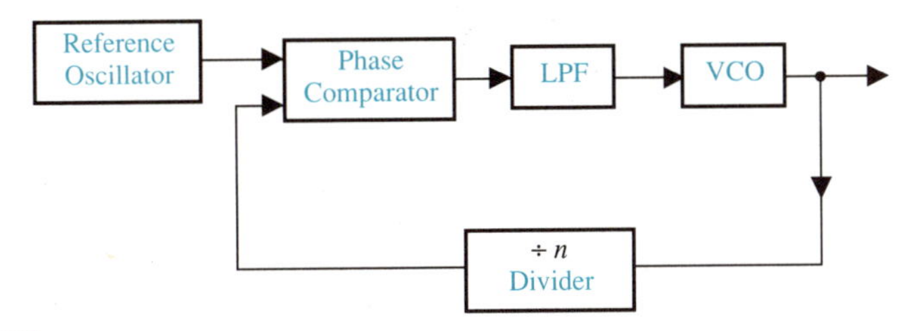

Figure 8.9 Basic frequency synthesizer.

8.3.2 Direct Synthesis

In **direct synthesis** the final output signal is derived from multiple-crystal-controlled oscillators or a single-crystal oscillator with multiple divider sections. Figure 8.10 shows a basic block diagram for a direct-frequency synthesizer. Figure 8.10(a) shows an arrangement using multiple crystals to produce individual frequencies, and Figure 8.10(b) shows a single-crystal oscillator with multiple dividers.

In the multiple-crystal system, some specified number of crystals usually supply a frequency to an oscillator module. The two banks of crystals shown in Figure 8.10(a) will produce two signals out of the two oscillator modules. These signals are then

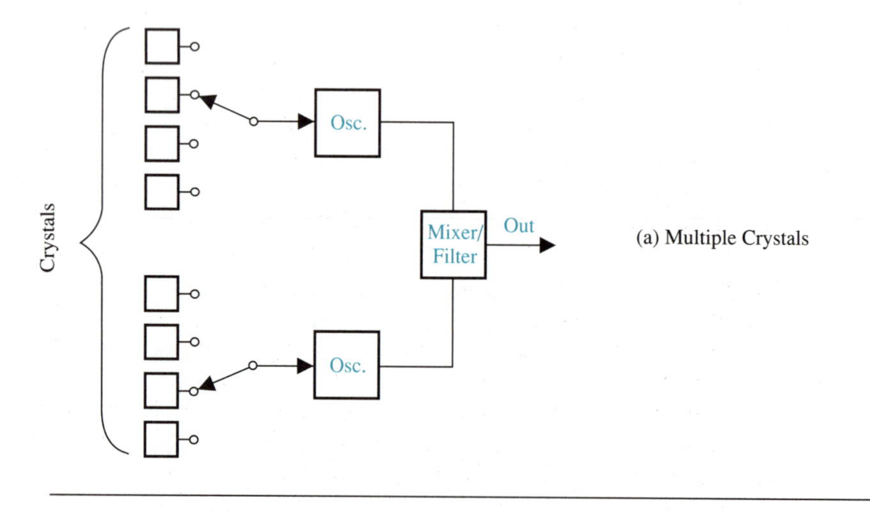

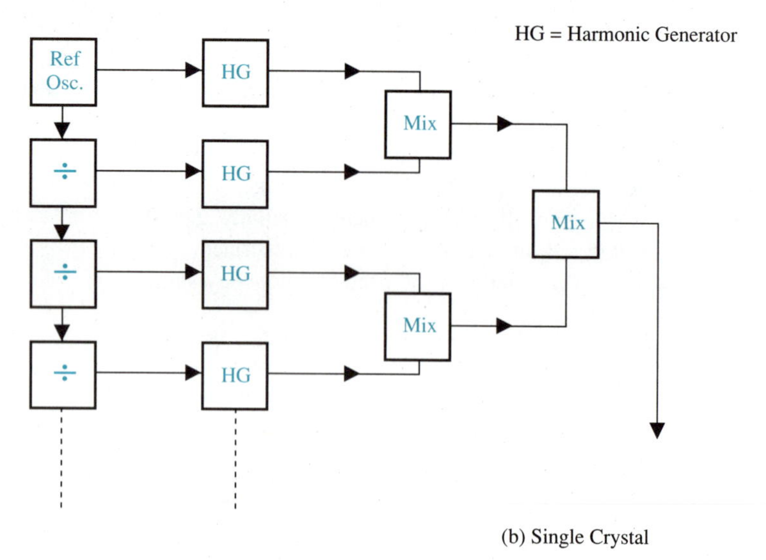

Figure 8.10 Direct synthesis.

mixed together and, with the proper filtering, a series of frequencies are available at the output. As an example, in a synthesizer with 20 crystals and two oscillator modules, a total of 128 frequencies are possible. Thus, a wide variety of frequencies can be obtained with this type of circuitry. There are, however, drawbacks to this type of system: there must be a switching system to insert the crystals into the circuit at the proper time to obtain the desired frequency, and 20 or more crystals can be rather expensive and time-consuming to integrate into a circuit for synthesizer applications. As with any other circuit, the trick is to compromise between good and bad features.

The single-crystal system shown in Figure 8.10(b) eliminates one of the disadvantages of the multiple-crystal system—it operates with only one crystal oscillator. This is the reference oscillator for all of the following circuitry. The secret to this circuit is the fact that the reference frequency can be divided by the ''divide-by-10'' dividers and can also be used as an input to a series of harmonic generators, allowing the single-crystal synthesizer to produce a large number of frequencies at a predetermined step size.

Figure 8.10(b) illustrates the reference oscillator beginning the sequence of events by producing a stable and accurate reference frequency. If we assume that this frequency is in the order of 100 kHz, the oscillator would send 100 kHz to a divide-by-10 circuit and to a harmonic generator. The **divide-by-10 circuit** does just what the name implies, it divides the 100-kHz signal by 10 to produce a 10-kHz signal. This process can be continued to produce signals of 1 kHz, 100 Hz, 10 Hz, 1 Hz, and so on, to whatever resolution is needed for a particular application.

The harmonic generators produce a series of harmonics of the input signal. For example, the 100-kHz input from the reference oscillator would produce 100-kHz, 200-kHz, 300-kHz, 400-kHz, 500-kHz, 600-kHz, 700-kHz, 800-kHz, and 900-kHz signals at its output. Similarly, the 10-kHz input to its harmonic generator would produce harmonics of the 10-kHz signal (in 10-kHz steps up to 90 kHz). This process is repeated for each division step needed for a particular synthesizer. The outputs of the harmonic generators are now sent to a series of mixer–summer circuits, where they are combined for an output. The operator determines the harmonic generator and mixer–summer section being used by a front panel control that selects the desired frequency. This particular circuit, if the dividers go down to a 1-Hz resolution, has a range of from 1 to 999,999 Hz in 1-Hz steps. This should give the operator considerable flexibility when choosing a frequency.

An obvious advantage of this type of circuit is that only one crystal oscillator is needed, greatly simplifying the stability requirements for the entire circuit. Also, the range and resolution of this circuit is a great advantage. The only disadvantage of this type of synthesizer is that more circuitry must be set up and calibrated to keep the circuit on frequency. This problem is often overlooked in order to obtain the single-crystal synthesizer's many advantages.

8.3.3 Indirect Synthesis

Probably the most widely used form of frequency synthesizer used in recent years incorporates a PLL circuit. This form of synthesizer, called an **indirect synthesizer**,

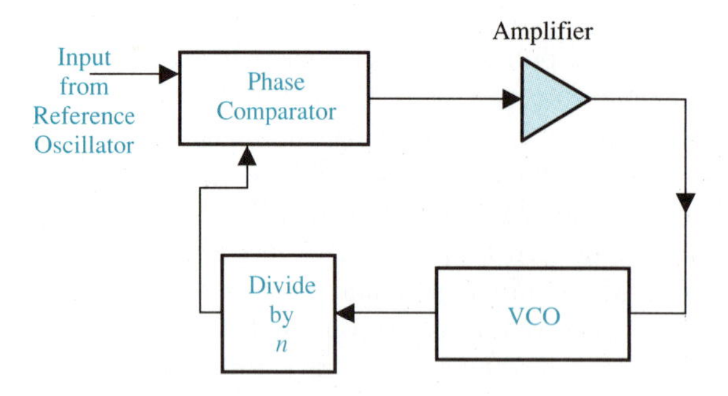

Figure 8.11 Indirect synthesis.

is shown in Figure 8.11. Note how it is similar to the basic PLL from Section 8.2 and exactly duplicates that in Figure 8.9 (the most typical synthesizer in use).

The circuit in Figure 8.11 is called a **single-loop synthesizer** because it contains only one PLL. More complex synthesizers are available, but we will look only at the single-loop device because if you understand the theory of the single-loop synthesizer, you should be able to relate it to other systems.

The single-loop synthesizer has a reference which is a single-crystal-controlled oscillator. This is the portion that was referred to as F_i in the phase-lock loop section of this text. The range of frequencies and the resolution of the system depends on the divider network and the gain of the loop (open-loop gain). The divider is a divide-by-n circuit, the frequency of which is usually controlled by an operator or by a control circuit that automatically sets a frequency or range. The synthesizer can basically be considered to be a times-n multiplier circuit, because the output of the synthesizer is actually n times F_{ref}.

As previously stated, the frequency range and resolution of the single-loop synthesizer depend on the n-divider and the open-loop gain of the system. Recall from Equation (8.3) that the open-loop gain A_v is equal to the product of the gain of the comparator A_d, the gain of the amplifier A_a, and the transfer function of the VCO (H_0). The other term in Equation (8.3), L_f, is the loss through the low-pass filter. Because there is no low-pass filter in the single-loop synthesizer, there is no L_f term. For the synthesizer in Figure 8.11, the open-loop gain is

$$A_v = \frac{A_d A_a H_0}{n} \tag{8.6}$$

Example 8.3 An indirect synthesizer exhibits the following parameters: $A_d = 0.7$ V/rad, $A_a = 8$, $H_0 = 12$ kz/V, and $n = 2$. The n-factor is inadvertently changed so that the open-loop gain of the synthesizer is 16.8. What is the new n-factor?

Solution To determine the new n-factor, the initial open-loop gain must be found. This is found by using Equation (8.6).

$$A_v = \frac{A_d A_a H_0}{n}$$
$$= \frac{(0.7)(8)(12)}{2}$$
$$= \frac{67.2}{2}$$
$$= 33.6$$

We can use the same equation to find the new n-factor.

The new gain is 16.8 and the numerator of Equation (8.6) is 67.2. Thus, we find the new n-factor as follows:

$$A_v = \frac{67.2}{n}$$
$$16.8 = \frac{67.2}{n}$$
$$n = \frac{67.2}{16.8}$$
$$n = 4$$

From Equation (8.6) we see that n is inversely proportional to the open-loop gain. To eliminate the reliance of open-loop gain on the n-factor of the synthesizer, a common practice is to program the amplifier gain to the divider ratio as well. This makes the gain of the amplifier equal to $n(A_a)$. The open-loop gain of the synthesizer is now

$$A_v = A_d A_a H_0 \tag{8.7}$$

Equation (8.7) ensures that changing the divider ratio in the synthesizer will not affect the open-loop gain of the system. Thus, only the divider ratio is used to control the frequency range and resolution.

Example 8.4 If we take the circuit in Figure 8.11 and apply a 10-kHz signal from the reference oscillator, and our divider ratio is equal to 10, the frequency range of the synthesizer will be

$$F_o = nF_{\text{ref}}$$
$$= (10)(1\text{ kHz})$$
$$= 10\text{-kHz range}$$
$$= 1\text{ kHz–}10\text{ kHz in 1-kHz steps}$$

As mentioned previously, there are other methods of designing indirect-frequency synthesizers. Some use a scheme called **prescaling** to achieve frequency division, which employs two methods of sampling the input frequency on alternate pulses. Others use a series of harmonic generators and filters to create a specific range and resolution. The method used depends on the particular application of the synthesizer, and the synthesizer should be tailored and designed accordingly.

8.4 Summary

This chapter served to introduce the very useful phase-locked loop, as well as its operation, capture, and locking capabilities, and its applications in communications systems.

We also presented the theory and operation of both direct and indirect synthesis, both of which find wide use in communications systems.

Questions

8.1 Introduction

1. Draw a block diagram of a phase-locked loop and describe the functions of each block.

8.2 Phase-Locked Loops

2. What is the difference between a **compensated** and an **uncompensated** loop?
3. When does the comparator change from a frequency comparator to a phase comparator?
4. Define **loop lock**.
5. Define **lock range**.
6. How is an AGC circuit related to a basic feedback system?
7. Define PLL.
8. What is a VCO?
9. Explain the basic operation of a VCO.
10. Define **transfer function**.
11. How can the comparator in a PLL be used more efficiently?
12. What is the purpose of the filter/amplifier in a PLL?
13. Define **acquisition time**.
14. What does the loop gain consist of in a PLL?
15. Define **hold-in range**.

8.3 Synthesizers

16. What is a frequency synthesizer?
17. Distinguish between **direct** and **indirect** synthesis.
18. How is the phase-locked loop an integral part of a frequency synthesizer?
19. What is the **resolution** of a frequency synthesizer?

20. What is the basic operation of a prescaling synthesizer?
21. List advantages and disadvantages of direct synthesis.
22. List advantages and disadvantages of indirect synthesis.
23. Define **prescaling**.

Problems

8.2 Phase-Locked Loops

1. For the circuit shown in Figure 8.12, determine
   a. The frequency at an input voltage of +3 V
   b. The amount of frequency deviation for a ±2 V input
   c. The transfer function for the linear portion of the curve.
2. For the PLL shown in Figure 8.6, the VCO has a natural frequency F_n of 100 kHz, the input frequency F_i is 120 kHz, $A_d = 0.22$ V/rad, $L_f = 1.3$, $A_a = 5$, and $H_0 =$ 12 kHz/V. Find:
   a. ΔF
   b. V_{out}
   c. V_d
   d. A_v
3. Determine the change in frequency for a VCO transfer function of 1.55 and a dc input-voltage change of 0.77 V.
4. Determine the **hold-in** range for a PLL with a loop gain of 100 kHz/rad.

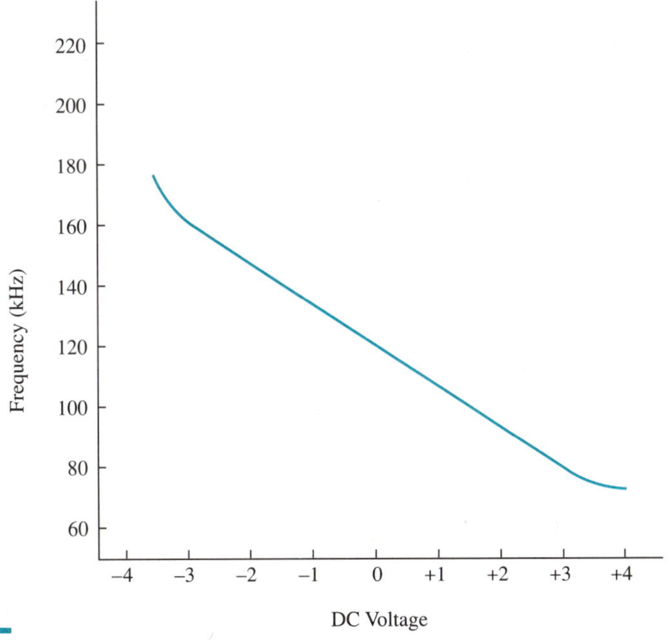

Figure 8.12 VCO characteristic.

8.3 Synthesizers

5. Determine the output frequency range of a synthesizer with a single-crystal oscillator at a frequency of 10 kHz and a dividing ratio of $n = 20$.
6. Determine the open-loop gain A_v of a synthesizer that has $A_d = 0.5$ V/rad, $A_a = 6$, and $H_0 = 25$ kHz/V.
7. We have an indirect-frequency synthesizer with a reference oscillator of 15 kHz and a divider ratio of 10. What is the frequency range of this synthesizer?
8. A frequency synthesizer has a comparator gain of 0.2 V/rad, an amplifier gain of 12, and a VCO transfer function of 25 kHz/V. What is the open-loop gain of this circuit?

9

Outline

9.1
Introduction

9.2
Theory

9.3
Transmission

9.4
Reception

9.5
Summary

Objectives

- To apply the preceding chapters on propagation, transmission lines, signal spectra, noise, and components to the amplitude-modulation scheme of electronic communications
- To introduce the study of modulation with a section on the theory behind amplitude modulation
- To describe the building of a transmitter to modulate a carrier and a receiver to demodulate the same carrier

Key Terms

modulate
modulating signal
frequency domain
overmodulation
double-sideband, full carrier (DSB-FC)
demodulation
trapezoidal pattern
preselector
rectifier distortion
carrier
time domain
modulation index
intelligence bandwidth
double-sideband, suppressed carrier (DSB-SC)
peak amplitude detector
carrier shift
intermediate frequency
diagonal clipping
sensitivity

Amplitude Modulation

9.1 Introduction

In Chapter 1 we used the term **modulate** to mean a method of applying a modulating signal (intelligence) to a carrier. How this modulating signal is applied to the carrier is what makes each modulation scheme unique. The primary goal of each scheme is to vary the carrier in some manner so that intelligence can be placed on it, transmitted to a receiver, and then removed to reproduce the original signal as closely as possible. There are three parameters of the carrier that can be varied to produce such a result: **amplitude**, **frequency**, and **phase**. This chapter will be concerned with the first of these parameters—amplitude.

Amplitude modulation is probably the most recognized type of modulation simply because of the commercial broadcast band (550–1600 kHz), which most of us grew up with and which provided the music that drove our parents crazy. This type of modulation is widely used because it is a relatively straightforward process to cause the amplitude of the carrier to vary at the modulating signal rate. It is also much easier to achieve a high degree of accuracy when reproducing the original signal since many demodulation methods consist of a diode detector and a low-pass filter.

Because of its relative simplicity, amplitude modulation serves as an excellent introduction to modulation techniques. All other forms of modulation can then be referenced from amplitude modulation.

9.2 Theory

In amplitude modulation an rf carrier and a modulating signal are joined together to produce a signal whose amplitude varies as a function of the modulating signal. Figure 9.1 shows a time-domain representation of an AM signal, with the rf carrier as the

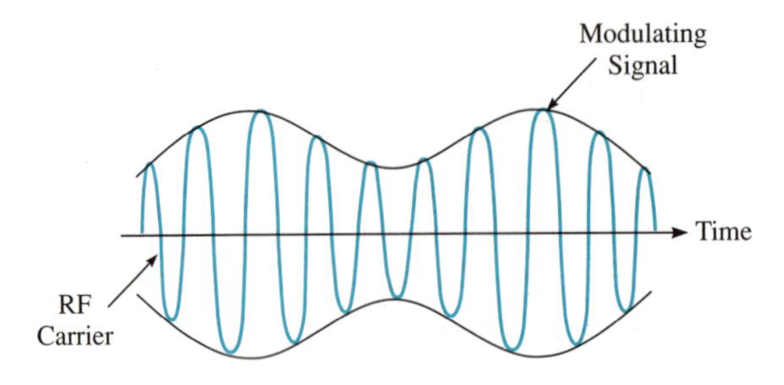

Figure 9.1 AM signal.

basic signal and a low-frequency modulating signal superimposed on the carrier to produce the amplitude variations.

To see how the signal in Figure 9.1 comes about, refer to Figure 9.2, which is a basic block diagram of an amplitude-modulation system. The associated signals, shown at each point where the modulation process takes place, are in the time domain as they would appear on an oscilloscope in the lab. The first signal, the carrier F_c, is the high-frequency component of the overall system. This signal may be in the megahertz or gigahertz range to enable the communications system to transmit its valuable intelligence over long distances. One point should be made about high-frequency signals: as the frequencies of the communications systems get higher and higher, certain precautions should be taken. These precautions basically involve keeping all leads very short and components close to the circuit board to decrease interelement and stray capacitances that were previously covered and to ensure that the circuit has a good chance of operating properly.

The second signal shown in Figure 9.2 is the modulating signal F_m. It is shown as a single-frequency signal but is usually an audio or video signal that is far from being a single frequency. For our discussions, however, we will use a single frequency in order to introduce the student to amplitude-modulation techniques. Always remember, however, that the signal is rarely a single frequency.

The resulting output signal is the same signal shown in Figure 9.1 as a picture of the signal in the time domain. A frequency-domain representation is shown in

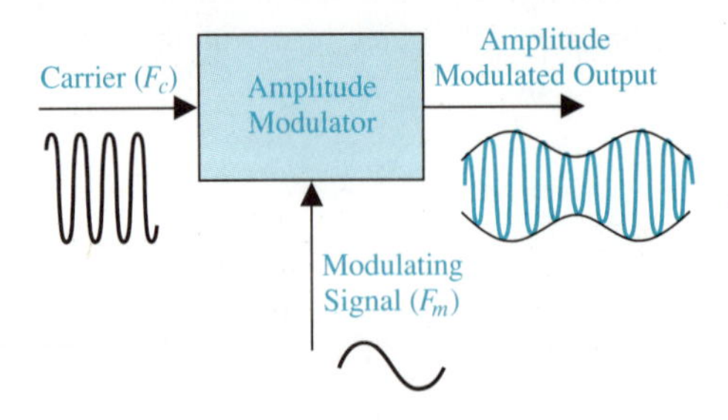

Figure 9.2 AM system block diagram.

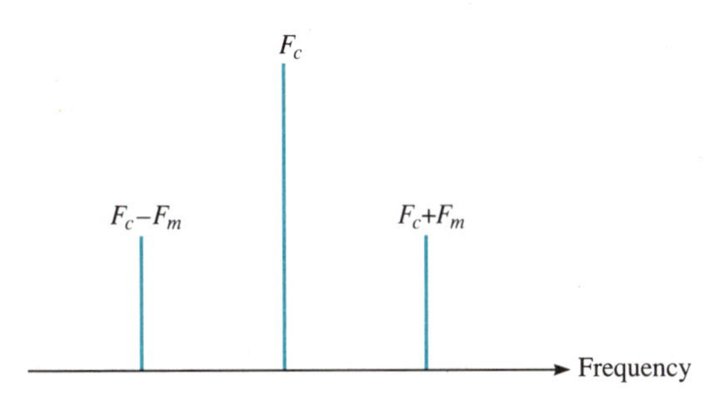

Figure 9.3 AM spectrum.

Figure 9.3, with the carrier signal F_c in the center of the frequency display, the upper sideband $F_c + F_m$ on one side, and the lower sideband $F_c - F_m$ on the other. This is a common picture for amplitude modulation.

To understand the modulation principles and how they apply to amplitude modulation, a mathematical representation of the process is needed. If we begin with the unmodulated signal, or the carrier, we can express this as

$$e = E_c \cos (2\pi F_c)t \quad \textbf{(9.1)}$$

or

$$e = E_c \cos \omega_c t$$

Where: E_c = The peak CW amplitude of the signal (in volts)
F_c = The frequency of the carrier (in hertz)
$\omega_c = 2\pi F_c$

Equation (9.1) is saying that in an unmodulated state the rf carrier is a sinusoidal signal with a frequency of F_c and a peak amplitude of E_c. Both the frequency and amplitude of the carrier are important to the modulating scheme because we need to know the frequency in order to distinguish mathematically between the carrier and the modulation signal, whereas the amplitude is an invaluable parameter when looking at amplitude modulation.

Similarly, the modulating signal can be expressed as

$$e = E_m \cos (2\pi F_m)t \quad \textbf{(9.2)}$$

or

$$e = E_m \cos \omega_m t$$

Where: E_m = The peak CW amplitude of the modulating signal (in volts)
F_m = The frequency of the modulating signal (in hertz)
$\omega_m = 2\pi F_m$

The objective of an amplitude-modulating scheme is to combine the signals shown in Equation (9.1) with that of Equation (9.2). When this combination is made, the carrier amplitude varies at a sinusoidal rate that is the same as the modulating signal. This expression now becomes

$$e = [E_c + E_m \cos(2\pi F_m)t] \cos(2\pi F_c)t \tag{9.3}$$

In this equation the following pieces may be broken out and identified:

1. E_c = The peak amplitude of the carrier sinusoid
2. $E_m \cos(2\pi F_m)t$ = The sinusoidal variation of the carrier amplitude due to the modulating signal
3. $\cos(2\pi F_c)t$ = The original carrier signal

Equation (9.3) is a good representation for a modulated signal with all of its parts identified. The equation assumes that the carrier is completely (100%) modulated; that is, the modulated signal has some maximum value at the peaks and a zero value in the minimum position. This complete modulation is shown in Figure 9.4. Notice that in the figure, the value for the peaks and the minimum points for both the upper and lower halves of the signal is the same—zero. This condition constitutes 100% modulation.

In most of the cases you will encounter, 100% of anything is a positive condition. In the case of amplitude modulation, however, it is not the best condition and not what the experienced designer aims for. Usually something in the range of 80% to 87% is an ideal modulation to strive for. We will clarify this statement as you become more familiar with the concepts of amplitude modulation and modulation index. To determine what this percentage of modulation really is in a system, the student must be familiar with the term **modulation index**. This term is a figure of merit for the modulated system and tells the designer what percentage of modulation is actually being used in the system (What percentage of the carrier is being affected by the modulating signal). The mathematical expression for the modulation index m is,

$$m = \frac{E_m}{E_c} \text{ (unitless)} \tag{9.4}$$

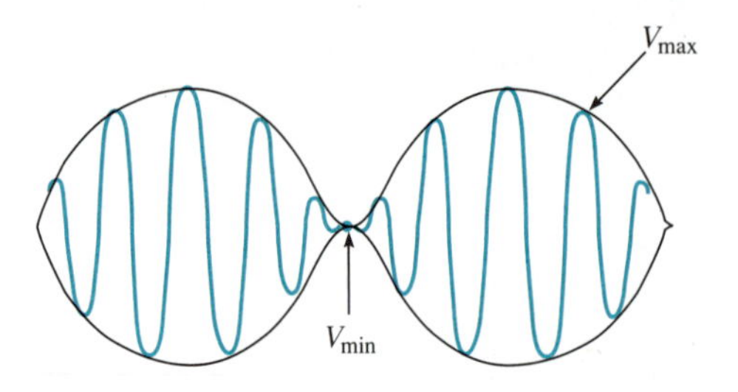

Figure 9.4 100% modulation.

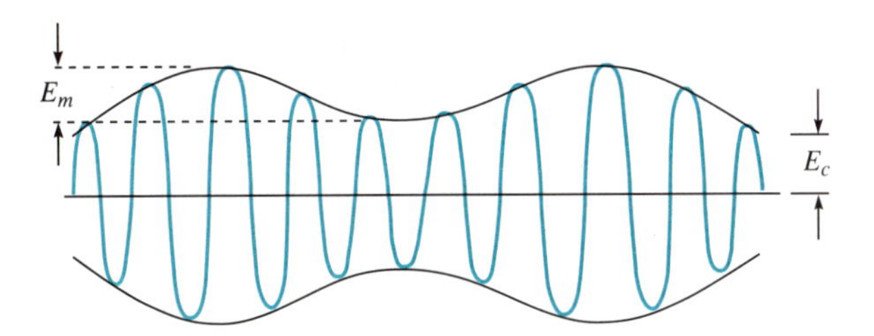

Figure 9.5 Modulation index reference.

Where: m = Modulation index ($m \leq 1$)
E_m = Peak change in the amplitude of the modulated signal (in volts)
E_c = Peak amplitude of the carrier (unmodulated) (in volts)

These values can be seen in Figure 9.5.

Relating Equation (9.4) to a percentage, we obtain

$$m = \frac{E_m}{E_c} \times 100 \tag{9.5}$$

To further clarify the equations above, let us see what E_m and E_c really are. An inspection of the figures presented thus far (Figures 9.4 and 9.5) will lead us to conclude that the following relationships are true:

$$E_m = \frac{1}{2}(V_{max} - V_{min}) \tag{9.6}$$

and

$$E_c = \frac{1}{2}(V_{max} + V_{min}) \tag{9.7}$$

If we now go back to Equation (9.4), we see a definite relationship between the modulation index and the presentation on an oscilloscope. That relationship is

$$\begin{aligned} m &= \frac{E_m}{E_c} \\ &= \frac{1/2(V_{max} - V_{min})}{1/2(V_{max} + V_{min})} \\ &= \frac{V_{max} - V_{min}}{V_{max} + V_{min}} \end{aligned} \tag{9.8}$$

Referring now to Figure 9.4, in which we said that we had 100% modulation, we can discover whether we actually have 100% modulation or not. Recall that we said that the peaks were some maximum value V_{max}, and the minimums were equal to zero. If we plug these values into Equation (9.8), we should be able to find out what the percentage of modulation really is.

$$m = \frac{V_{max} - V_{min}}{V_{max} + V_{min}}$$
$$= \frac{V_{max} - 0}{V_{max} + 0}$$
$$= \frac{V_{max}}{V_{max}}$$
$$= 1.0, \text{ or } 100\%$$

So we see that by observing the amplitude-modulated signal on an oscilloscope and measuring the maximum and minimum voltages, we obtain the percentage of modulation for the circuit being used.

Example 9.1

An amplitude-modulated signal is displayed on an oscilloscope and the following measurements are taken:

$$V_{max} = 4.2 \text{ V}$$
$$V_{min} = 1.1 \text{ V}$$

Find: The modulation index for this system.

Solution

Using Equation (9.8), we find

$$m = \frac{V_{max} - V_{min}}{V_{max} + V_{min}}$$
$$= \frac{4.2 - 1.1}{4.2 + 1.1}$$
$$= \frac{3.1}{5.3}$$
$$= 0.5849, \text{ or } 58.49\%$$

We will now explain why 100% modulation is not the number to be striving for. Recall from Figure 9.4 that 100% modulation was a signal with a maximum voltage and a zero voltage for a minimum. This zero term is what can cause problems in a modulation system. Looking back at the ultimate objective of any communications system—to modulate a carrier with some form of intelligence, transmit it to a receiver, and demodulate it to reproduce the original intelligence as closely as possible—we realize that the modulating signal must remain intact. If we now look at Figure 9.6(a), we see that this restriction is true for a 100% modulated signal. If the system relies on the fact that the minimum signal must always go to zero, there is the possibility that, because of variations in the system (in temperature, for instance), the signal will go to this zero level sooner than expected, as shown in Figure 9.6(b). It should be obvious that this condition will not reproduce the original intelligence, but will cause "gaps" in the signal where there is a flat portion in the response. This condition is called **overmodulation**. If we now look at the condition in Figure 9.6(c), we see that no matter what drifts around in the system, the intelligence will be preserved and the modulating signal will be reproduced. Thus, the figure of 80–87% is used to ensure as much modulation as possible, but not so much as to cause overmodulation.

Having explained the equations and concepts for modulation index, we can now rewrite Equation (9.3) in terms of the signals and the modulation index—the usual way these expressions are presented. The new equation is

$$e = E_c[1 + m \cos(2\pi F_m)t] \cos(2\pi F_c)t \tag{9.9}$$

It can be seen that this is the same expression that we had in Equation (9.3) except that E_m has been replaced with its equivalent expression, mE_c. (This comes about from Equation (9.4) which said that $m = E_m/E_c$. Therefore, E_m will equal mE_c.)

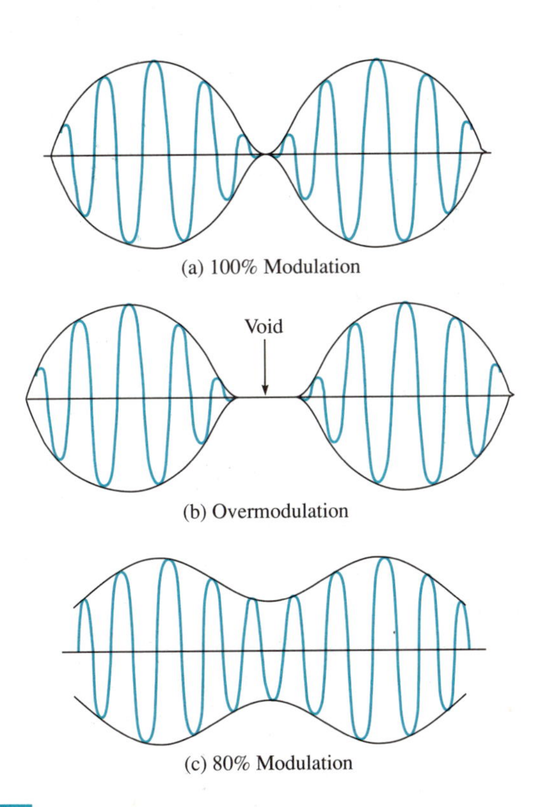

Figure 9.6 Modulation.

The equations above assume that there is only one signal modulating the carrier frequency. If more than one tone is modulating the carrier, as is usually the case, each signal will have a modulation index. Thus, the resulting index is used to calculate the overall modulation index. This is expressed as,

$$m = \sqrt{m_1^2 + m_2^2 + m_3^2 + \cdots} \quad \textbf{(9.10)}$$

Where: $m, m_1, m_2, m_3,$ = Modulating indices of each signal in the system

Example 9.2

Given: Modulation indexes for four signals within a spectrum that is modulating an rf carrier are:

$$m_1 = 45\%$$
$$m_2 = 37\%$$
$$m_3 = 21\%$$
$$m_4 = 49\%$$

Find: The total modulation index for the system

Solution

Using Equation (9.10) and plugging in the proper numbers, we obtain

$$\begin{aligned} m &= \sqrt{0.2025 + 0.1369 + 0.0441 + 0.2401} \\ &= \sqrt{0.6236} \\ &= 0.789, \text{ or } 78.9\% \end{aligned}$$

As previously stated, amplitude-modulated signals may be presented either on an oscilloscope (time-domain) or a spectrum analyzer (frequency-domain). Each of these displays give a certain amount of information about the signal but not all that may be needed. There are certain times when a time-domain representation will do very nicely for a particular application. Other times a frequency-domain picture is an absolute necessity. It is up to you to determine which presentation, or combination of presentations, is needed for the application being considered. To aid in the decision making, we will now present both time- and frequency-domain displays and indicate the information that can be obtained from each. From these discussions you should be able to determine which display will give the desired information.

Figure 9.7 shows the common time-domain presentation for an amplitude-modulated signal. This representation has been used more than once throughout this chapter, but is repeated here since the previous presentations have concentrated on one particular aspect of the display. The figure describes the entire display and the four parameters that can be determined directly from it. These are

1. E_m: The peak change in amplitude of the output wave
2. E_c: The peak amplitude of the unmodulated carrier
3. V_{max}: Which is equal to $E_c + E_m$
4. V_{min}: which is equal to $E_c - E_m$

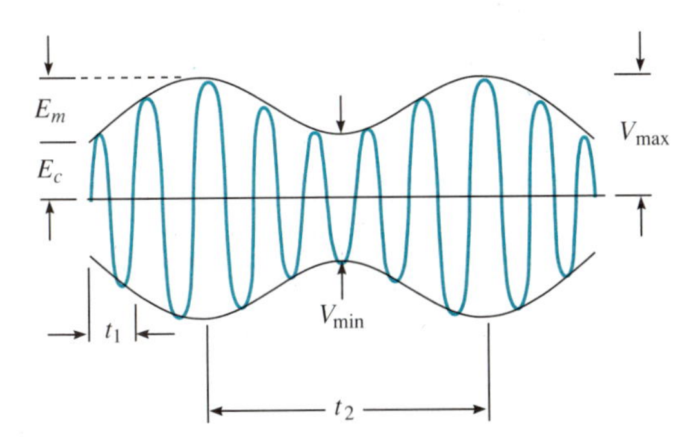

Figure 9.7 Time domain representation.

These parameters are included in Figure 9.7 to show you where each of the values is and to enable you to obtain readings for each one. Once these parameters are read from the screen, they can be used to calculate the coefficient of modulation and the percentage of modulation using the previous equations.

Two parameters can be determined indirectly from the time-domain presentation: the frequency of the carrier F_c, and the frequency of the modulating signal F_m. These parameters are considered indirect because a certain amount of calculations are needed to obtain them. Figure 9.7 designates t_1 and t_2 as specific times on the display. The first, t_1, is the time for one cycle of the rf carrier. Recall from previous courses and discussions in this text that we can calculate the frequency of a signal if we know the time duration of one cycle by the relationship $F = 1/T$. Therefore, $F_c = 1/t_1$. Similarly, the value t_2 is the time for one cycle of the modulation signal. This means that $F_m = 1/t_2$. It should be pointed out that this example once again uses a single frequency for the modulating signal for analysis purposes. It is obviously not an easy task to determine F_m for a signal with a voice-modulation component using the time-domain presentation. There must be other means of determining this parameter.

Example 9.3 A time-domain display shows that the $V_{max} = 6.4$ V, $V_{min} = 1.5$ V, $t_1 =$ 1.1 μs, and $t_2 = 2.2$ ms. Find the percentage modulation mF_c, and F_m.

Solution From Equation (9.8) we obtain

$$m = \frac{V_{max} - V_{min}}{V_{max} + V_{min}}$$

$$= \frac{6.4 - 1.5}{6.4 + 1.5}$$

$$= \frac{4.9}{7.9}$$

$$= 0.62 \text{ or } 62\%$$

$$F_c = \frac{1}{t_1}$$

$$= \frac{1}{1.1} \times 10^{-6}$$

$$= 909 \text{ kHz}$$

$$F_m = \frac{1}{t_2}$$

$$= \frac{1}{2.2} \times 10^{-3}$$

$$= 454 \text{ Hz}$$

Now we investigate amplitude modulation in the frequency domain, pictured in Figure 9.8. As shown before, there is a carrier frequency F_c in the center of the display with an upper sideband ($F_c + F_m$) and a lower sideband ($F_c - F_m$). These points were made previously when this display was first introduced; now we need to know what measurements can be obtained from this display. The first parameter that can be obtained is the obvious one, that is, the carrier and modulation frequencies. These can be read directly from the spectrum-analyzer frequency dial and recorded as needed. Thus, an accurate reading of both the carrier and modulating signal frequencies can be obtained.

Having determined the carrier and sideband frequencies, we can calculate the bandwidth of the system. Recalling the definition of bandwidth, you will note the similarity to Figure 9.9. This figure shows a center frequency F_c and two frequencies at the edges of the designated bandwidth F_1 and F_2. The definition of a bandwidth for this arrangement was $F_2 - F_1$. This is the same arrangement used for the spectrum-analyzer display in Figure 9.8, the difference in the two sideband frequencies.

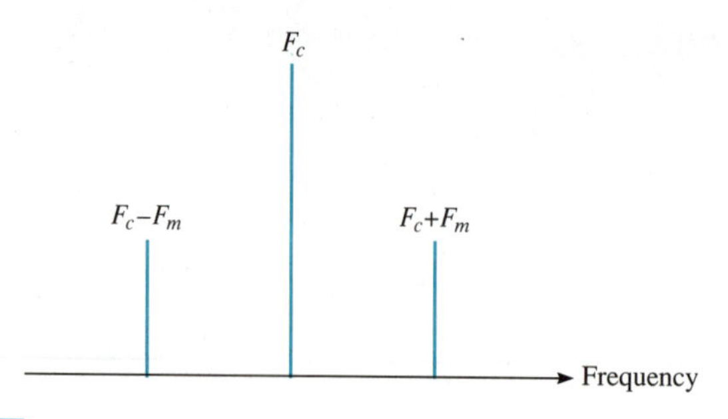

Figure 9.8 Frequency domain.

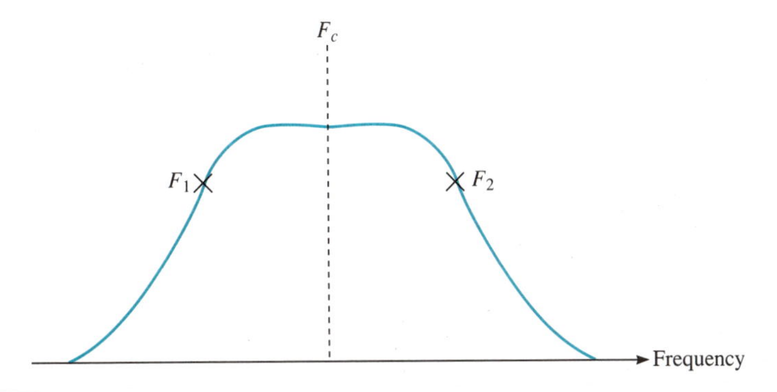

Figure 9.9 Bandwidth example.

To determine the actual bandwidth of the system we subtract the higher sideband from the lower sideband. That is

$$\begin{aligned} BW &= F_2 - F_1 \\ &= (F_c + F_m) - (F_c - F_m) \\ &= F_c + F_m - F_c + F_m \\ &= F_m + F_m \\ &= 2F_m \text{ (Hz)} \end{aligned} \tag{9.11}$$

Thus, for an amplitude-modulated signal, the bandwidth of the system is equal to twice the modulating frequency. This bandwidth is called the **intelligence bandwidth** because it is the bandwidth in which all of the transmitted intelligence is contained.

Example 9.4 A spectrum-analyzer display shows a carrier frequency of 2 MHz, an upper sideband (USB) of 2.01 MHz, and a lower sideband (LSB) of 1.99 MHz. What is the intelligence bandwidth?

Solution To find the intelligence bandwidth, we need the value of F_m, which can be found by taking the difference between the carrier and either sideband. This is

$$\begin{aligned} F_m &= F_{\text{USB}} - F_c \\ &= 2.01 \times 10^{-6} - 2.00 \times 10^{-6} \\ &= 10 \text{ kHz} \end{aligned}$$

Thus,

$$\begin{aligned} BW &= 2F_m \\ &= 20 \text{ kHz} \end{aligned}$$

Another parameter that can be taken from the spectrum-analyzer display is the amplitudes of the frequency components displayed. As we see in Figure 9.8, the carrier frequency is at a higher level than the two sideband frequencies. By checking the calibration and reference points on the spectrum-analyzer front panel, we can attach an absolute-power number to each component in the display to determine how much power is in each part of the total spectrum. The carrier may be at a level of 0 dBm, for example, with the sidebands at −12 dBm.

The difference in the power level of the individual frequency components can be used to find another parameter—the percentage of modulation. As was the case for finding frequencies of the carrier and modulating signal with a time display, we can similarly find the percent of modulation by doing some minor calculations. To illustrate this, refer to Figure 9.10, where the carrier signal F_c is at 0 dBm and each of the sidebands is at a −12 dBm level, the same as our sample values above. To use this display, we must use the following relationship:

$$m = \frac{2E_s}{E_c} \tag{9.12}$$

Where: E_s = Voltage of the sideband
E_c = Voltage of the carrier

This calculation appears to be straightforward until we realize that Equation (9.12) calls out voltages and the parameter that is read off the spectrum analyzer is power. We must, therefore, convert the units. To accomplish this we first make note of the *difference* in the power levels of the signals: the carrier is at 0 dBm and the sidebands are at −12 dBm (a difference of 12 dB). This number can now be translated into a voltage ratio in the following manner:

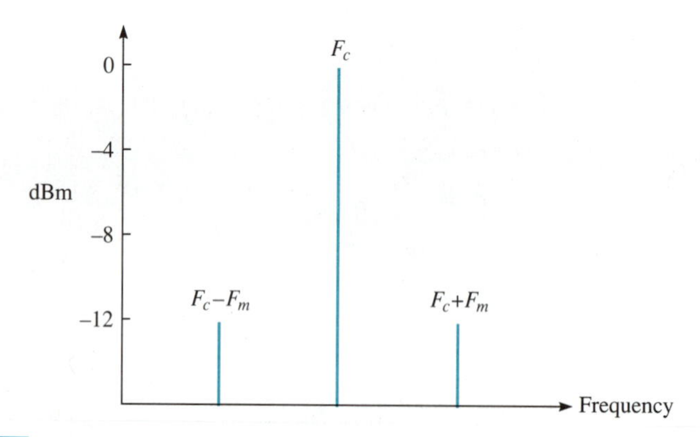

Figure 9.10 Percent modulation example.

$$\text{dB} = 20 \log \text{voltage ratio}$$
$$12 = 20 \log \text{voltage ratio}$$
$$0.6 = \log \text{voltage ratio}$$
$$\text{Ratio} = \log 0.6$$
$$\text{Ratio} = 4$$

The ratio of 4 is established. If we assign a value of 1 to the sideband signals, then the carrier will have a value of 4. Using these values, the percentage of modulation can be found as follows:

$$\text{Percentage modulation} = \frac{2E_s}{E_c}$$
$$= \frac{2(1)}{4}$$
$$= \frac{2}{4}$$
$$= 0.5, \text{ or } 50\%$$

Table 9.1 shows a series of amplitude differences and their associated percentages of modulation

Table 9.1

$E_c - E_s$ (dB)	Percentage Modulation
10	60
12	50
16	30
26	10
46	1
60	0.2

To determine the amount of power in the amplitude-modulated signal, we must expand Equation (9.3) to obtain both the upper and lower sidebands. The result is

$$e = E_c \cos (2\pi F_c)t + \frac{E_m}{2 \cos (2\pi)(F_c - F_m)} + \frac{E_m}{2 \cos (2\pi)(F_c + F_m)} \tag{9.13}$$

If this signal is placed on an antenna with an effective real resistance R, we can determine the power in each sideband and in the carrier. Thus, we are able to determine the total power within the system. The values obtained are

$$P_c = \frac{V_{\text{pk}}^2}{2R} \quad \text{(in watts)} \tag{9.14}$$

$$P(\text{lower}) = P(\text{upper}) = \frac{m^2 P_c}{4} \tag{9.15}$$

The total power in the entire AM signal will now be

$$\begin{aligned} P &= P_c + P\ (\text{lower}) + P\ (\text{upper}) \\ &= P_c + \frac{mP_c}{4} + \frac{mP_c}{4} \qquad (9.16) \\ &= P_c\left(1 + \frac{m^2}{2}\right) \quad \text{(in watts)} \end{aligned}$$

Example 9.5 With a peak carrier voltage of 35 V and a percentage modulation of 58%, what is the power for each AM spectral component and the total power if an effective resistance of 72 Ω is available?

Solution

1. Using Equation (9.14), we find P_c

$$\begin{aligned} P_c &= \frac{E_c^2}{2R} \\ &= \frac{(35)^2}{2(72)} \\ &= \frac{1225}{144} \\ &= 8.506 \text{ W} \end{aligned}$$

2. Equation (9.15) will give the sideband powers.

$$\begin{aligned} P\ (\text{lower}) &= \frac{m^2 P_c}{4} \\ &= \frac{(.58)^2(8.506)}{4} \\ &= \frac{(.3364)(8.506)}{4} \\ &= 0.715 \text{ W} \end{aligned}$$

3. The total power in the signal is

$$P = P_c + P(\text{lower}) + P(\text{upper})$$
$$= 8.506 + 0.715 + 0.715$$
$$= 9.936 \text{ W}$$

An alternative method is to use Equation (9.16).

$$P = P_c\left(1 + \frac{m^2}{2}\right)$$
$$= 8.506\left(1 + \frac{0.58^2}{2}\right)$$
$$= 8.506(1.1682)$$
$$= 9.936 \text{ W}$$

Thus, the theory of amplitude modulation has shown how a high-frequency carrier F_c varies its amplitude by a modulating signal F_m to produce a modulated waveform. The important parameters, such as carrier level, modulation index (percentage), maximum and minimum envelope voltages, and sideband levels, all contribute to making the system operate as intended.

Having explained why 100% modulation is not desirable, we now present theory and methods for transmitting and receiving AM signals.

9.3 Transmission

We use ''AM Transmission'' to mean ''What does it take to generate an amplitude-modulated signal?'' That is, how do we modulate a carrier to cause a constant-level rf carrier to become an amplitude-modulated signal. When we speak of an amplitude-modulated signal we are referring to a double-sideband, full carrier (DSB-FC) signal. These are the signals displayed in both the time domain and frequency domain throughout this text. Later in this section we will present a different type of AM—double-sideband, suppressed carrier AM (DSB-SC)—and look at it for possible use in specific systems. For our discussions here, however, we will be looking at the AM signal with a full carrier.

To understand how a circuit must look to produce amplitude modulation, you must understand what function must be performed to cause the two frequencies F_c and F_m to combine. Recall from our discussions of mixers in Chapter 7 that if we apply two signals to an amplifier operating in a linear region, one at 1 kHz and the other at 100 kHz, for example, we will end up with the two signals at the output.

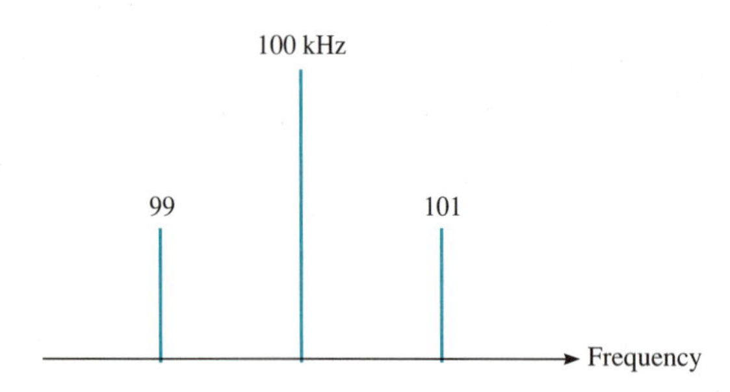

Figure 9.11 Modulation example.

One may be higher in amplitude than the other because of the difference in the frequency characteristics of the amplifier, but they will both be at the output, one at 1 kHz and one at 100 kHz. These will be the only two there. If we apply these same signals at the input to a nonlinear device, we have the original signals (1 kHz and 100 kHz), the sum of the two (101 kHz) and the difference of the two (99 kHz). This is the basic mixing process that we have come to know and understand.

The process for amplitude modulation is much the same procedure in that a nonlinear device is needed to produce a combined signal at the output. If we were to take the same two signals (1 kHz and 100 kHz) and modulate the two into an AM signal, we would produce the spectrum in Figure 9.11, which shows the carrier (100 kHz), the upper sideband (101 kHz), and the lower sideband (99 kHz). This is a typical picture, but it's missing one more signal to connect the modulator circuit to the mixing circuit.

Recall that when we combined two signals in a mixer circuit, we ended up with the two original signals and the sum and difference of the original signals. Examining Figure 9.11 we see the sum of the two signals (101 kHz), the difference of the two signals (99 kHz), and one of the original signals (100 kHz). What is missing, as mentioned above, is one of the original signals—the 1-kHz signal that was present at the input (the modulating signal). If we were to expand the presentation on the spectrum analyzer, we would see that another signal is actually present (see Fig. 9.12). Thus, the modulation process is actually the same as the mixing process.

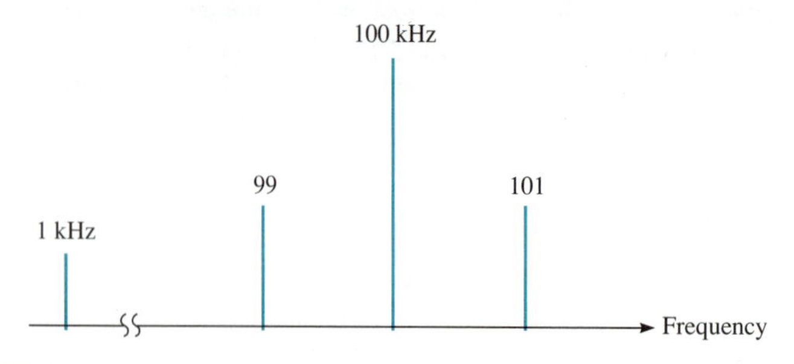

Figure 9.12 Entire spectrum.

One area of the modulation process that is not the same as the mixing process is the level of signal used. Recall that in the mixing process the rf signal was low level and the local oscillator was a higher level to enable it to drive the circuit into a nonlinear region. In some cases the rf signal may be 0.1 V with the LO ranging between 3 and 5 V.

In the modulation process the circuit is already in a nonlinear condition, accomplished by using the proper biasing techniques (biased class C). With the circuit in a nonlinear mode of operation, both the carrier F_c and the modulating signal F_m can be low-level signals and still accomplish the proper results. When both signals are at a low level, there is a savings in both design effort and energy.

There are two methods for modulating an rf carrier with a modulating signal to produce an AM signal: a discrete transistor circuit and an integrated circuit. Figure 9.13 is an example of a discrete transistor circuit used to modulate the amplitude of an rf carrier. The circuit consists of a single-stage amplifier with both the carrier and modulating signals applied to the amplifier. The carrier is applied to the base, and the modulating signal is applied to the collector. The one area of the figure that makes this a modulator rather than just a simple amplifier of two signals is in the base circuit where the circuit is biased class C.

Recall from previous circuit course that a class-C biased amplifier operates differently from one that is the normal class A. In the class-A operation the input signal is applied to the input of the transistor and the transistor conducts over the entire 360° of the cycle. In this case the individual signals are amplified and no modulation occurs.

In class-C biasing the signal is applied to the input of the transistor, and the device conducts for less than half the period of the cycle, usually 120 to 150° of the cycle. This condition causes the circuit to be in a nonlinear state and results in the two signals being mixed (or modulated) together to obtain the proper results. You can

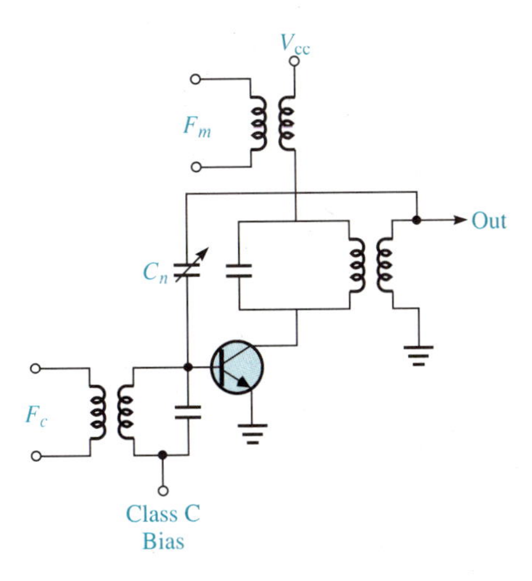

Figure 9.13 Discrete transistor modulator.

see how the transistor has a different biasing scheme from those shown in Chapter 7 (Figs. 7.20 and 7.21, for example). In these circuits there is a resistive voltage divider that takes a portion of the supply voltage and biases the base of the transistor for class-A operation. In Figure 9.13 no such biasing scheme exists. A condition exists wherein a bias is applied far below cut-off at the base of the transistor. This means that there must be a certain level of signal present at the base to get the device to conduct. When this level is reached, the transistor will conduct, and the modulation process occurs for only that portion of the conduction cycle (120–150°). This is a much different situation from the class-A operation of the low-frequency or rf amplifiers discussed in Chapter 7.

Notice also how the two signals F_c and F_m are transformer-coupled to the modulator circuit, making this an efficient use of a modulator circuit. The tank circuit in the collector and any impedance-matching components that may be used are tuned to the carrier frequency because it is the primary signal that must be passed. Once this signal is successfully reproduced the modulation signal, which is riding on the carrier, will follow. Thus, by applying the carrier to the base and the modulating signal to the collector of the device, the carrier and modulating signal are both present at the collector during the conduction cycle and result in the modulating signal modulating the amplitude of the carrier. This is similar to improperly bypassing the dc power-supply lines on a conventional rf amplifier or opening a bypass capacitor, thereby causing a 60-Hz component to be present on the bias lines. The result is an rf output signal whose amplitude varies at a rate of 60 Hz. That is, the rf is amplitude-modulated with 60 Hz, which is not the desired output. In the case where we do want amplitude modulation, this process is excellent.

The second type of modulator, shown in Figure 9.14, uses an integrated circuit. This particular device is the 1496 14-pin balanced modulator–demodulator chip. Figure 9.15 is the full data sheet for the device. The carrier is applied to the V_c input

(*Text continues on page 235*)

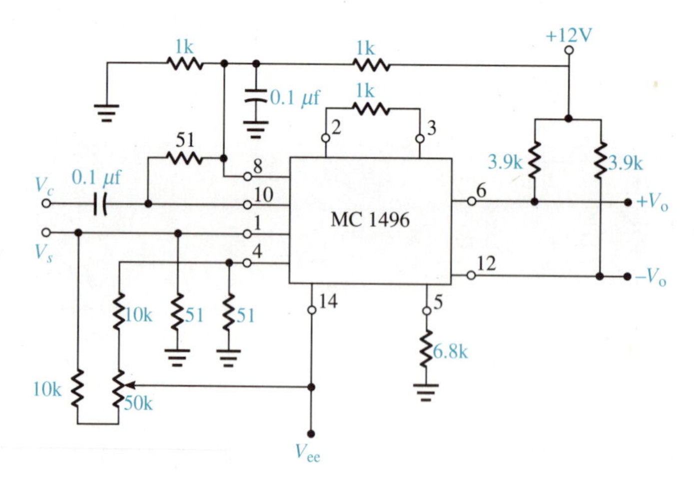

Figure 9.14 Integrated circuit modulator.

MC1496
MC1596

Specifications and Applications Information

BALANCED MODULATOR/DEMODULATOR

SILICON MONOLITHIC INTEGRATED CIRCUIT

BALANCED MODULATOR/ DEMODULATOR

. . . designed for use where the output voltage is a product of an input voltage (signal) and a switching function (carrier). Typical applications include suppressed carrier and amplitude modulation, synchronous detection, FM detection, phase detection, and chopper applications. See Motorola Application Note AN-531 for additional design information.

- Excellent Carrier Suppression – 65 dB typ @ 0.5 MHz
  – 50 dB typ @ 10 MHz
- Adjustable Gain and Signal Handling
- Balanced Inputs and Outputs
- High Common Mode Rejection – 85 dB typ

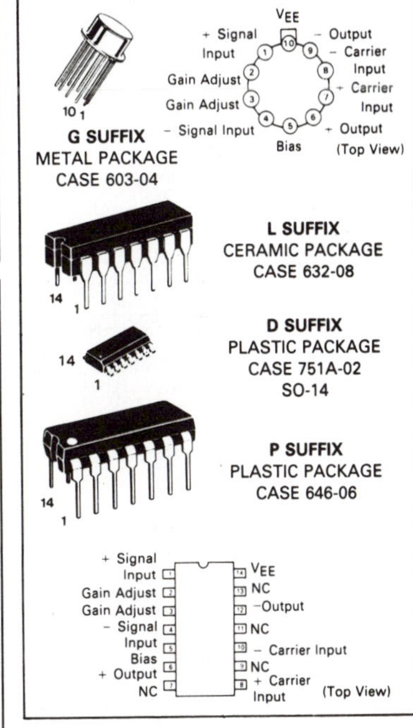

ORDERING INFORMATION

Device	Temperature Range	Package
MC1496D	0°C to +70°C	SO-14
MC1496G		Metal Can
MC1496L		Ceramic DIP
MC1496P		Plastic DIP
MC1596G	–55°C to +125°C	Metal Can
MC1596L		Ceramic DIP

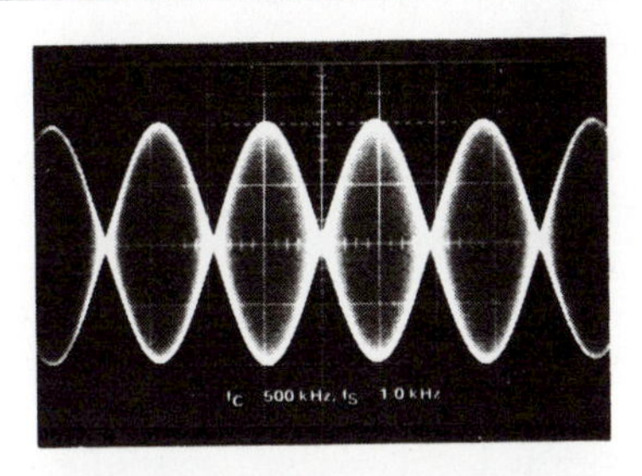

FIGURE 1 – SUPPRESSED-CARRIER OUTPUT WAVEFORM

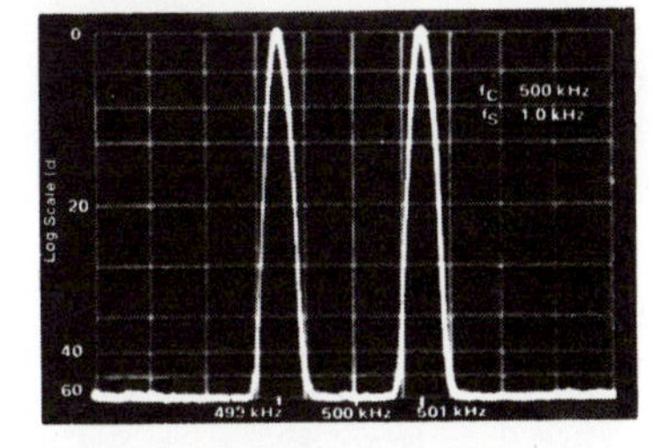

FIGURE 2 – SUPPRESSED-CARRIER SPECTRUM

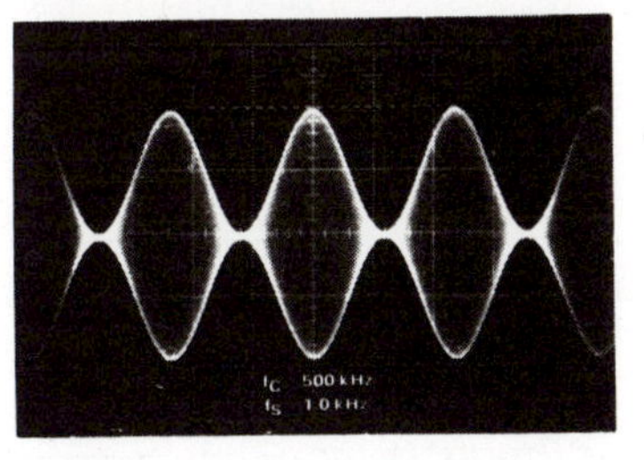

FIGURE 3 – AMPLITUDE-MODULATION OUTPUT WAVEFORM

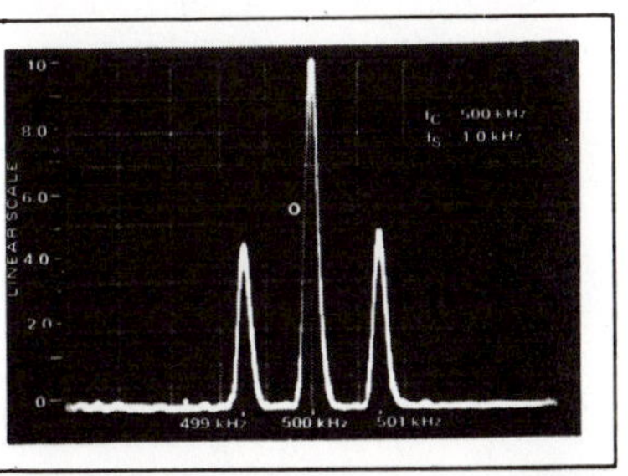

FIGURE 4 – AMPLITUDE-MODULATION SPECTRUM

 DS9132R3

Figure 9.15 MC1496 data sheet.

MAXIMUM RATINGS* (T_A = +25°C unless otherwise noted)

Rating	Symbol	Value	Unit
Applied Voltage ($V_6 - V_7$, $V_8 - V_1$, $V_9 - V_7$, $V_9 - V_8$, $V_7 - V_4$, $V_7 - V_1$, $V_8 - V_4$, $V_6 - V_8$, $V_2 - V_5$, $V_3 - V_5$)	ΔV	30	Vdc
Differential Input Signal	$V_7 - V_8$ $V_4 - V_1$	+5.0 $\pm(5 + I_5 R_e)$	Vdc
Maximum Bias Current	I_5	10	mA
Thermal Resistance, Junction to Air Ceramic Dual In-Line Package Plastic Dual In-Line Package Metal Package	$R_{\theta JA}$	 100 100 160	°C/W
Operating Temperature Range MC1496 MC1596	T_A	 0 to +70 −55 to +125	°C
Storage Temperature Range	T_{stg}	−65 to +150	°C

ELECTRICAL CHARACTERISTICS* (V_{CC} = +12 Vdc, V_{EE} = −8.0 Vdc, I_5 = 1.0 mAdc, R_L = 3.9 kΩ, R_e = 1.0 kΩ, T_A = +25°C unless otherwise noted) (All input and output characteristics are single-ended unless otherwise noted.)

Characteristic	Fig.	Note	Symbol	MC1596			MC1496			Unit
				Min	Typ	Max	Min	Typ	Max	
Carrier Feedthrough V_C = 60 mV(rms) sine wave and offset adjusted to zero, f_C = 1.0 kHz f_C = 10 MHz V_C = 300 mVp-p square wave: offset adjusted to zero, f_C = 1.0 kHz offset not adjusted, f_C = 1.0 kHz	5	1	V_{CFT}	 — — — —	 40 140 0.04 20	 — — 0.2 100	 — — — —	 40 140 0.04 20	 — — 0.4 200	μV(rms) mV(rms)
Carrier Suppression f_S = 10 kHz, 300 mV(rms) f_C = 500 kHz, 60 mV(rms) sine wave f_C = 10 MHz, 60 mV(rms) sine wave	5	2	V_{CS}	 50 —	 65 50	 — —	 40 —	 65 50	 — —	dB k
Transadmittance Bandwidth (Magnitude) (R_L = 50 ohms) Carrier Input Port, V_C = 60 mV(rms) sine wave f_S = 1.0 kHz, 300 mV(rms) sine wave Signal Input Port, V_S = 300 mV(rms) sine wave $\|V_C\|$ = 0.5 Vdc	8	8	BW_{3dB}	 — —	 300 80	 — —	 — —	 300 80	 — —	MHz
Signal Gain V_S = 100 mV(rms), f = 1.0 kHz; $\|V_C\|$ = 0.5 Vdc	10	3	A_{VS}	2.5	3.5	—	2.5	3.5	—	V/V
Single-Ended Input Impedance, Signal Port, f = 5.0 MHz Parallel Input Resistance Parallel Input Capacitance	6	—	 r_{ip} c_{ip}	 — —	 200 2.0	 — —	 — —	 200 2.0	 — —	 kΩ pF
Single-Ended Output Impedance, f = 10 MHz Parallel Output Resistance Parallel Output Capacitance	6	—	 r_{op} c_{oo}	 — —	 40 5.0	 — —	 — —	 40 5.0	 — —	 kΩ pF
Input Bias Current $I_{bS} = \frac{I_1 + I_4}{2}$; $I_{bC} = \frac{I_7 + I_8}{2}$	7	—	 I_{bS} I_{bC}	 — —	 12 12	 25 25	 — —	 12 12	 30 30	μA
Input Offset Current $I_{ioS} = I_1 - I_4$; $I_{ioC} = I_7 - I_8$	7	—	 $\|I_{ioS}\|$ $\|I_{ioC}\|$	 — —	 0.7 0.7	 5.0 5.0	 — —	 0.7 0.7	 7.0 7.0	μA
Average Temperature Coefficient of Input Offset Current (T_A = −55°C to +125°C)	7	—	$\|TC_{Iio}\|$	—	2.0	—	—	2.0	—	nA/°C
Output Offset Current ($I_6 - I_9$)	7	—	$\|I_{oo}\|$	—	14	50	—	14	80	μA
Average Temperature Coefficient of Output Offset Current (T_A = −55°C to +125°C)	7	—	$\|TC_{Ioo}\|$	—	90	—	—	90	—	nA/°C
Common-Mode Input Swing, Signal Port, f_S = 1.0 kHz	9	4	CMV	—	5.0	—	—	5.0	—	Vp-p
Common-Mode Gain, Signal Port, f_S = 1.0 kHz, $\|V_C\|$ = 0.5 Vdc	9	—	ACM	—	−85	—	—	−85	—	dB
Common-Mode Quiescent Output Voltage (Pin 6 or Pin 9)	10	—	V_{out}	—	8.0	—	—	8.0	—	Vp-p
Differential Output Voltage Swing Capability	10	—	V_{out}	—	8.0	—	—	8.0	—	Vp-p
Power Supply Current $I_6 + I_9$ I_{10}	7	6	 I_{CC} I_{EE}	 — —	 2.0 3.0	 3.0 4.0	 — —	 2.0 3.0	 4.0 5.0	mAdc
DC Power Dissipation	7	5	P_D	—	33	—	—	33	—	mW

* Pin number references pertain to this device when packaged in a metal can. To ascertain the corresponding pin numbers for plastic or ceramic packaged devices refer to the first page of this specification sheet.

MOTOROLA *Semiconductor Products Inc.*

Figure 9.15 *(cont'd)*

GENERAL OPERATING INFORMATION *

Note 1 – Carrier Feedthrough

Carrier feedthrough is defined as the output voltage at carrier frequency with only the carrier applied (signal voltage = 0).

Carrier null is achieved by balancing the currents in the differential amplifier by means of a bias trim potentiometer (R_1 of Figure 5).

Note 2 – Carrier Suppression

Carrier suppression is defined as the ratio of each sideband output to carrier output for the carrier and signal voltage levels specified.

Carrier suppression is very dependent on carrier input level, as shown in Figure 22. A low value of the carrier does not fully switch the upper switching devices, and results in lower signal gain, hence lower carrier suppression. A higher than optimum carrier level results in unnecessary device and circuit carrier feedthrough, which again degenerates the suppression figure. The MC1596 has been characterized with a 60 mV(rms) sinewave carrier input signal. This level provides optimum carrier suppression at carrier frequencies in the vicinity of 500 kHz, and is generally recommended for balanced modulator applications.

Carrier feedthrough is independent of signal level, V_S. Thus carrier suppression can be maximized by operating with large signal levels. However, a linear operating mode must be maintained in the signal-input transistor pair – or harmonics of the modulating signal will be generated and appear in the device output as spurious sidebands of the suppressed carrier. This requirement places an upper limit on input-signal amplitude (see Note 3 and Figure 20). Note also that an optimum carrier level is recommended in Figure 22 for good carrier suppression and minimum spurious sideband generation.

At higher frequencies circuit layout is very important in order to minimize carrier feedthrough. Shielding may be necessary in order to prevent capacitive coupling between the carrier input leads and the output leads.

Note 3 – Signal Gain and Maximum Input Level

Signal gain (single-ended) at low frequencies is defined as the voltage gain,

$$A_{VS} = \frac{V_o}{V_S} = \frac{R_L}{R_e + 2r_e} \text{ where } r_e = \frac{26 \text{ mV}}{I_5 \text{ (mA)}}$$

A constant dc potential is applied to the carrier input terminals to fully switch two of the upper transistors "on" and two transistors "off" (V_C = 0.5 Vdc). This in effect forms a cascode differential amplifier.

Linear operation requires that the signal input be below a critical value determined by R_E and the bias current I_5

$$V_S \le I_5 R_E \text{ (Volts peak)}$$

Note that in the test circuit of Figure 10, V_S corresponds to a maximum value of 1 volt peak.

Note 4 – Common-Mode Swing

The common-mode swing is the voltage which may be applied to both bases of the signal differential amplifier, without saturating the current sources or without saturating the differential amplifier itself by swinging it into the upper switching devices. This swing is variable depending on the particular circuit and biasing conditions chosen (see Note 6).

Note 5 – Power Dissipation

Power dissipation, P_D, within the integrated circuit package should be calculated as the summation of the voltage-current products at each port, i.e. assuming $V_9 = V_6$, $I_5 = I_6 = I_9$ and ignoring base current, $P_D = 2\, I_5\, (V_6 - V_{10}) + I_5\, (V_5 - V_{10})$ where subscripts refer to pin numbers.

Note 6 – Design Equations

The following is a partial list of design equations needed to operate the circuit with other supply voltages and input conditions. See Note 3 for R_e equation.

A. Operating Current

The internal bias currents are set by the conditions at pin 5. Assume:

$$I_5 = I_6 = I_9$$

$$I_B \ll I_C \text{ for all transistors}$$

then:

$$R_5 = \frac{V^- - \phi}{I_5} - 500\ \Omega$$

where: R_5 is the resistor between pin 5 and ground; ϕ = 0.75 V at T_A = +25°C

The MC1596 has been characterized for the condition I_5 = 1.0 mA and is the generally recommended value.

B. Common-Mode Quiescent Output Voltage

$$V_6 = V_9 = V^+ - I_5 R_L$$

Note 7 – Biasing

The MC1596 requires three dc bias voltage levels which must be set externally. Guidelines for setting up these three levels include maintaining at least 2 volts collector-base bias on all transistors while not exceeding the voltages given in the absolute maximum rating table;

$$30 \text{ Vdc} \ge [(V_6, V_9) - (V_7, V_8)] \ge 2 \text{ Vdc}$$

$$30 \text{ Vdc} \ge [(V_7, V_8) - (V_1, V_4)] \ge 2.7 \text{ Vdc}$$

$$30 \text{ Vdc} \ge [(V_1, V_4) - (V_5)] \ge 2.7 \text{ Vdc}$$

The foregoing conditions are based on the following approximations:

$$V_6 = V_9, \quad V_7 = V_8, \quad V_1 = V_4$$

Bias currents flowing into pins 1, 4, 7, and 8 are transistor base currents and can normally be neglected if external bias dividers are designed to carry 1.0 mA or more.

Note 8 – Transadmittance Bandwidth

Carrier transadmittance bandwidth is the 3-dB bandwidth of the device forward transadmittance as defined by:

$$y_{21C} = \left.\frac{i_o \text{ (each sideband)}}{v_s \text{ (signal)}}\right|_{V_o = 0}$$

Signal transadmittance bandwidth is the 3-dB bandwidth of the device forward transadmittance as defined by:

$$y_{21S} = \left.\frac{i_o \text{ (signal)}}{v_s \text{ (signal)}}\right|_{V_c = 0.5 \text{ Vdc},\ V_o = 0}$$

*Pin number references pertain to this device when packaged in a metal can. To ascertain the corresponding pin numbers for plastic or ceramic packaged devices refer to the first page of this specification sheet.

MOTOROLA *Semiconductor Products Inc.*

Figure 9.15 *(cont'd)*

Note 9 – Coupling and Bypass Capacitors C_1 and C_2

Capacitors C_1 and C_2 (Figure 5) should be selected for a reactance of less than 5.0 ohms at the carrier frequency.

Note 10 – Output Signal, V_o

The output signal is taken from pins 6 and 9, either balanced or single-ended. Figure 12 shows the output levels of each of the two output sidebands resulting from variations in both the carrier and modulating signal inputs with a single-ended output connection.

Note 11 – Negative Supply, V_{EE}

V_{EE} should be dc only. The insertion of an RF choke in series with V_{EE} can enhance the stability of the internal current sources.

Note 12 – Signal Port Stability

Under certain values of driving source impedance, oscillation may occur. In this event, an RC suppression network should be connected directly to each input using short leads. This will reduce the Q of the source-tuned circuits that cause the oscillation.

SIGNAL INPUT (PINS 1 & 4) — 510 — 10 pF

An alternate method for low-frequency applications is to insert a 1 k-ohm resistor in series with the inputs, pins 1 and 4. In this case input current drift may cause serious degradation of carrier suppression.

TEST CIRCUITS

FIGURE 5 – CARRIER REJECTION AND SUPPRESSION

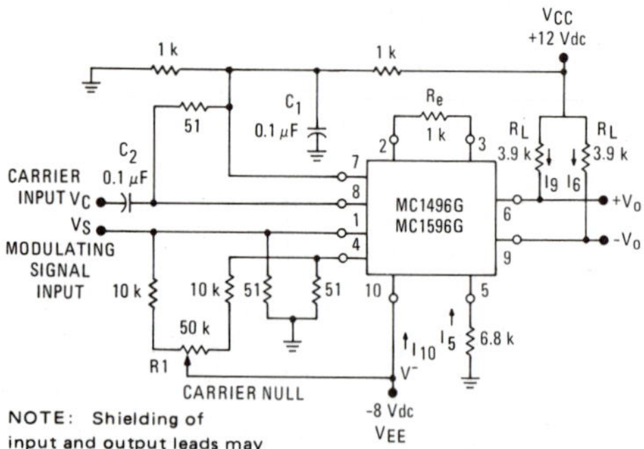

NOTE: Shielding of input and output leads may be needed to properly perform these tests.

FIGURE 6 – INPUT-OUTPUT IMPEDANCE

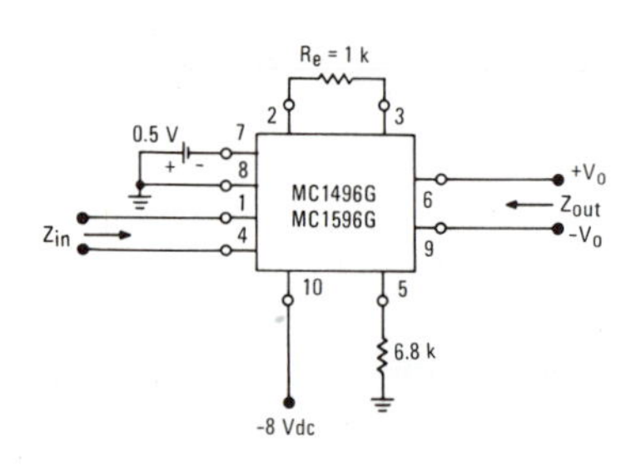

FIGURE 7 – BIAS AND OFFSET CURRENTS

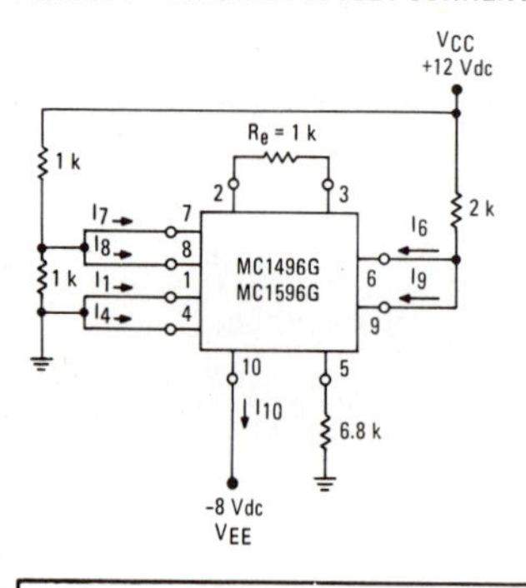

FIGURE 8 – TRANSCONDUCTANCE BANDWIDTH

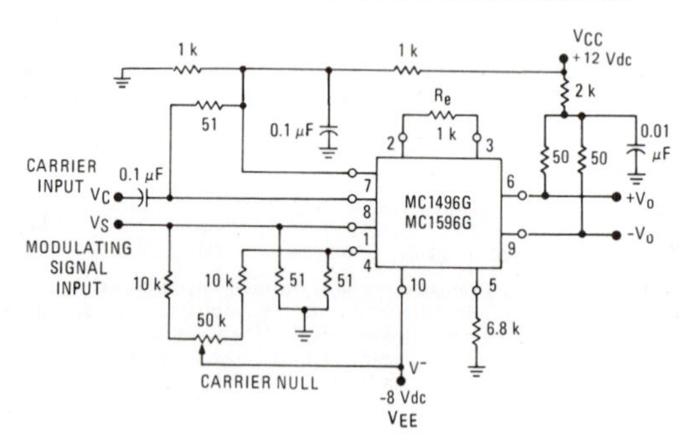

NOTE: Pin number references pertain to this device when packaged in a metal can. To ascertain the corresponding pin numbers for plastic or ceramic packaged devices refer to the first page of this specification sheet.

MOTOROLA *Semiconductor Products Inc.*

Figure 9.15 *(cont'd)*

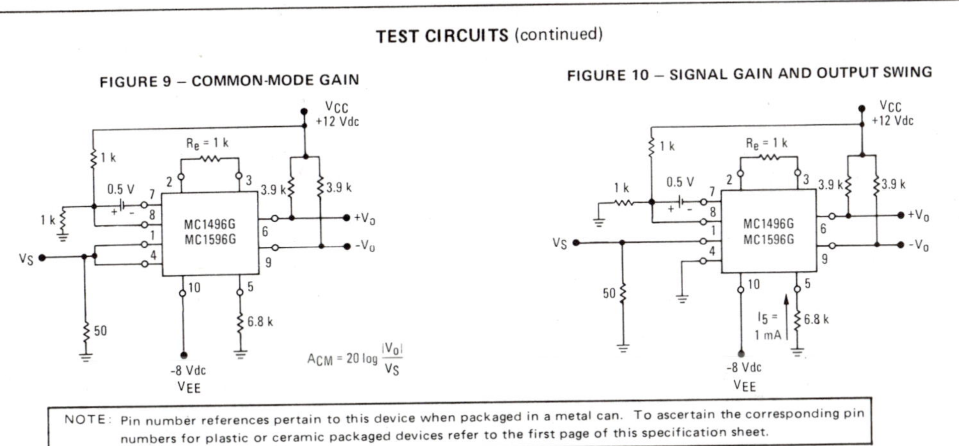

TYPICAL CHARACTERISTICS (continued)

Typical characteristics were obtained with circuit shown in Figure 5, f_C = 500 kHz (sine wave), V_C = 60 mV(rms), f_S = 1 kHz, V_S = 300 mV(rms), T_A = +25°C unless otherwise noted.

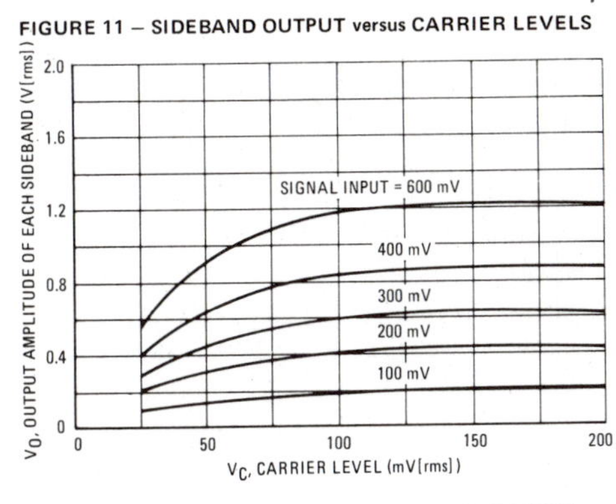

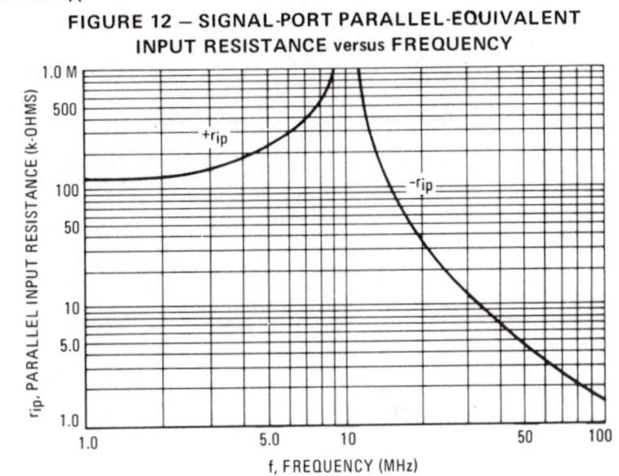

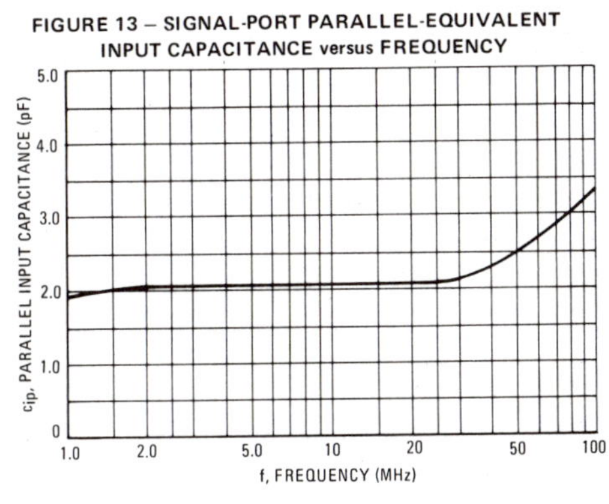

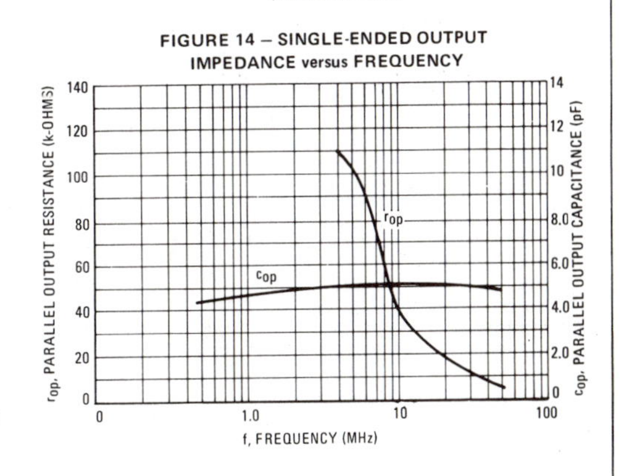

Figure 9.15 *(cont'd)*

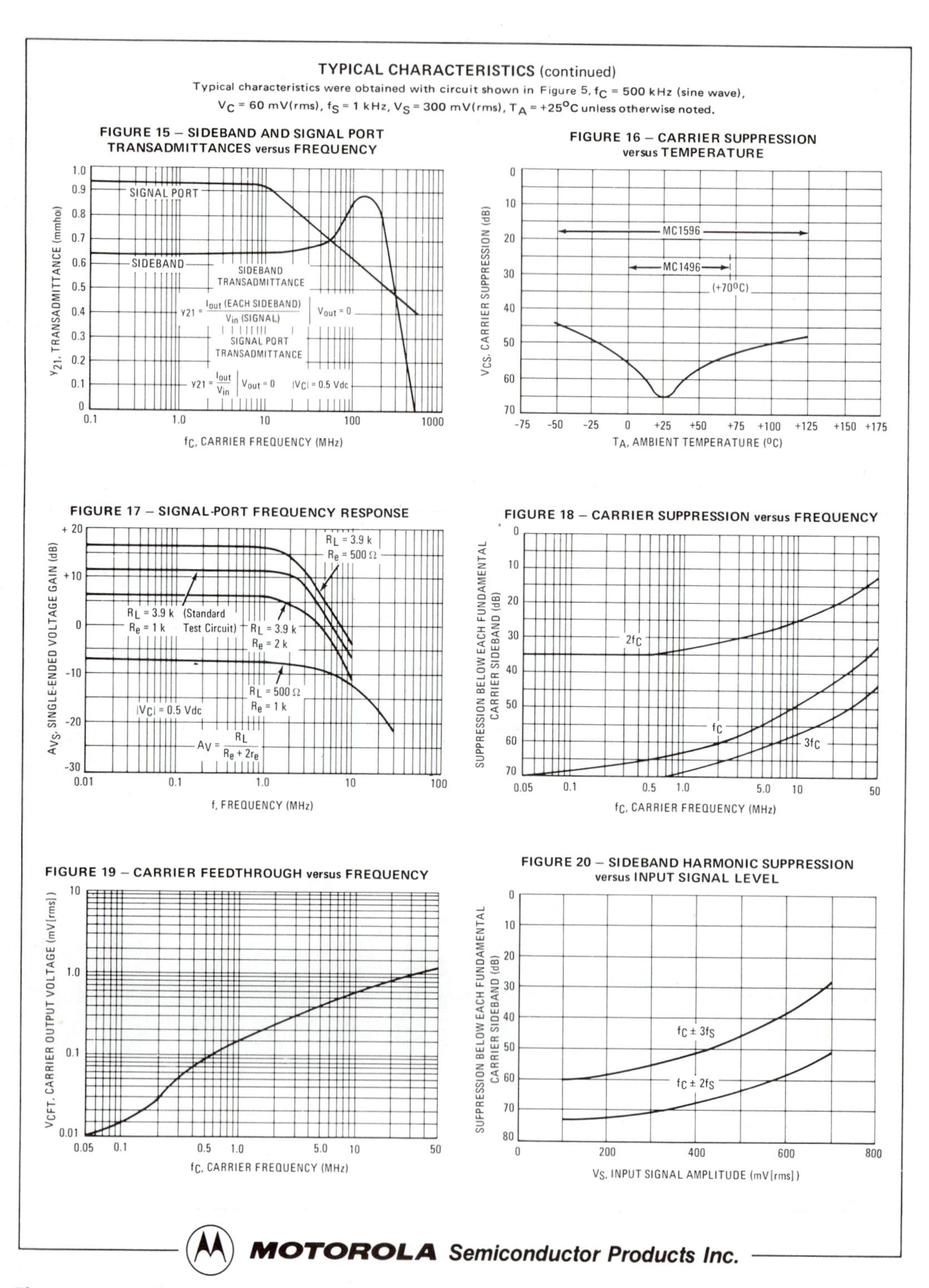

Figure 9.15 *(cont'd)*

TYPICAL CHARACTERISTICS (continued)

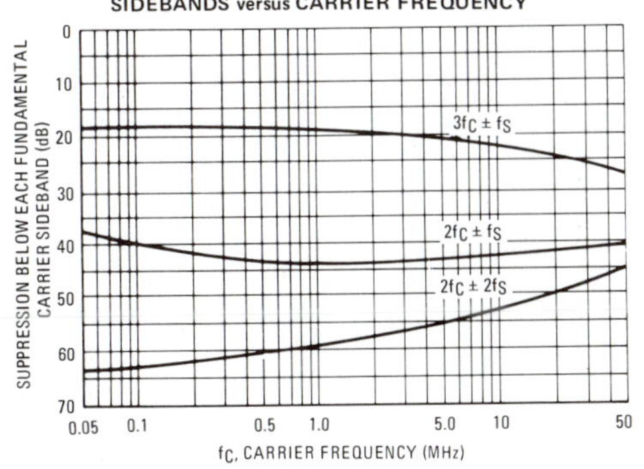

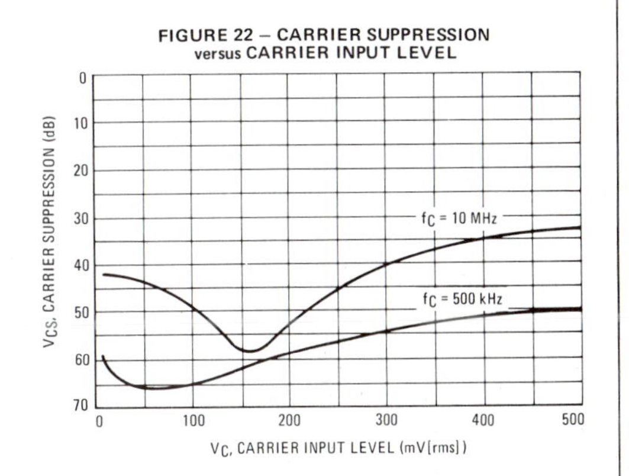

OPERATIONS INFORMATION

The MC1596/MC1496, a monolithic balanced modulator circuit, is shown in Figure 23.

This circuit consists of an upper quad differential amplifier driven by a standard differential amplifier with dual current sources. The output collectors are cross-coupled so that full-wave balanced multiplication of the two input voltages occurs. That is, the output signal is a constant times the product of the two input signals.

Mathematical analysis of linear ac signal multiplication indicates that the output spectrum will consist of only the sum and difference of the two input frequencies. Thus, the device may be used as a balanced modulator, doubly balanced mixer, product detector, frequency doubler, and other applications requiring these particular output signal characteristics.

The lower differential amplifier has its emitters connected to the package pins so that an external emitter resistance may be used. Also, external load resistors are employed at the device output.

Signal Levels

The upper quad differential amplifier may be operated either in a linear or a saturated mode. The lower differential amplifier is operated in a linear mode for most applications.

For low-level operation at both input ports, the output signal will contain sum and difference frequency components and have an amplitude which is a function of the product of the input signal amplitudes.

For high-level operation at the carrier input port and linear operation at the modulating signal port, the output signal will contain sum and difference frequency components of the modulating signal frequency and the fundamental and odd harmonics of the carrier frequency. The output amplitude will be a constant times the modulating signal amplitude. Any amplitude variations in the carrier signal will not appear in the output.

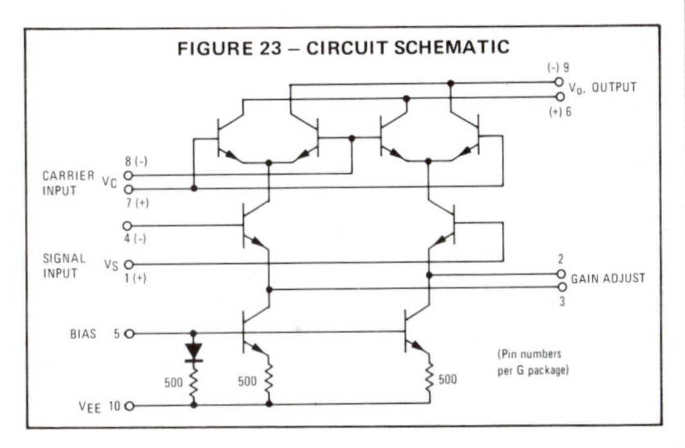

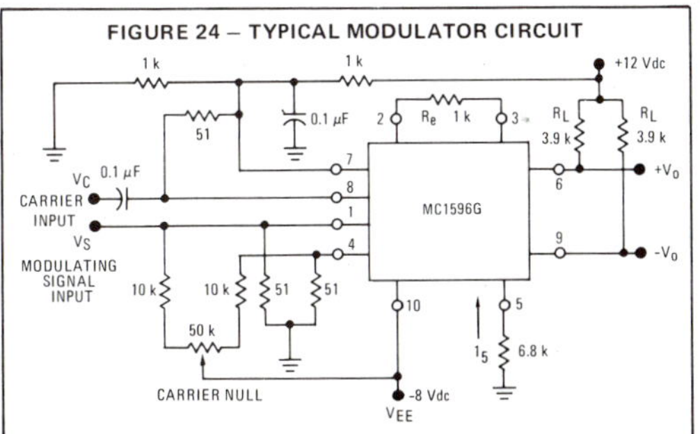

NOTE: Pin number references pertain to this device when packaged in a metal can. To ascertain the corresponding pin numbers for plastic or ceramic packaged devices refer to the first page of this specification sheet.

MOTOROLA Semiconductor Products Inc.

Figure 9.15 *(cont'd)*

OPERATIONS INFORMATION (continued)

The linear signal handling capabilities of a differential amplifier are well defined. With no emitter degeneration, the maximum input voltage for linear operation is approximately 25 mV peak. Since the upper differential amplifier has its emitters internally connected, this voltage applies to the carrier input port for all conditions.

Since the lower differential amplifier has provisions for an external emitter resistance, its linear signal handling range may be adjusted by the user. The maximum input voltage for linear operation may be approximated from the following expression:

$$V = (I_5)(R_E) \text{ volts peak.}$$

This expression may be used to compute the minimum value of R_E for a given input voltage amplitude.

FIGURE 25 – TABLE 1
VOLTAGE GAIN AND OUTPUT FREQUENCIES

Carrier Input Signal (V_C)	Approximate Voltage Gain	Output Signal Frequency(s)
Low-level dc	$\frac{R_L V_C}{2(R_E + 2r_e)\left(\frac{KT}{q}\right)}$	f_M
High-level dc	$\frac{R_L}{R_E + 2r_e}$	f_M
Low-level ac	$\frac{R_L V_C(rms)}{2\sqrt{2}\left(\frac{KT}{q}\right)(R_E + 2r_e)}$	$f_C \pm f_M$
High-level ac	$\frac{0.637 R_L}{R_E + 2r_e}$	$f_C \pm f_M$, $3f_C \pm f_M$, $5f_C \pm f_M$, . . .

The gain from the modulating signal input port to the output is the MC1596/MC1496 gain parameter which is most often of interest to the designer. This gain has significance only when the lower differential amplifier is operated in a linear mode, but this includes most applications of the device.

As previously mentioned, the upper quad differential amplifier may be operated either in a linear or a saturated mode. Approximate gain expressions have been developed for the MC1596/MC1496 for a low-level modulating signal input and the following carrier input conditions:

1) Low-level dc
2) High-level dc
3) Low-level ac
4) High-level ac

These gains are summarized in Table 1, along with the frequency components contained in the output signal.

NOTES:
1. Low-level Modulating Signal, V_M, assumed in all cases. V_C is Carrier Input Voltage.
2. When the output signal contains multiple frequencies, the gain expression given is for the output amplitude of each of the two desired outputs, $f_C + f_M$ and $f_C - f_M$.
3. All gain expressions are for a single-ended output. For a differential output connection, multiply each expression by two.
4. R_L = Load resistance.
5. R_E = Emitter resistance between pins 2 and 3.
6. r_e = Transistor dynamic emitter resistance, at +25°C;

$$r_e \approx \frac{26 \text{ mV}}{I_5 \text{ (mA)}}$$

7. K = Boltzmann's Constant, T = temperature in degrees Kelvin, q = the charge on an electron.

$$\frac{KT}{q} \approx 26 \text{ mV at room temperature}$$

APPLICATIONS INFORMATION

Double sideband suppressed carrier modulation is the basic application of the MC1596/MC1496. The suggested circuit for this application is shown on the front page of this data sheet.

In some applications, it may be necessary to operate the MC1596/MC1496 with a single dc supply voltage instead of dual supplies. Figure 26 shows a balanced modulator designed for operation with a single +12 Vdc supply. Performance of this circuit is similar to that of the dual supply modulator.

AM Modulator

The circuit shown in Figure 27 may be used as an amplitude modulator with a minor modification.

All that is required to shift from suppressed carrier to AM operation is to adjust the carrier null potentiometer for the proper amount of carrier insertion in the output signal.

However, the suppressed carrier null circuitry as shown in Figure 27 does not have sufficient adjustment range. Therefore, the modulator may be modified for AM operation by changing two resistor values in the null circuit as shown in Figure 28.

Product Detector

The MC1596/MC1496 makes an excellent SSB product detector (see Figure 29).

This product detector has a sensitivity of 3.0 microvolts and a dynamic range of 90 dB when operating at an intermediate frequency of 9 MHz.

The detector is broadband for the entire high frequency range. For operation at very low intermediate frequencies down to 50 kHz the 0.1 μF capacitors on pins 7 and 8 should be increased to 1.0 μF. Also, the output filter at pin 9 can be tailored to a specific intermediate frequency and audio amplifier input impedance.

As in all applications of the MC1596/MC1496, the emitter resistance between pins 2 and 3 may be increased or decreased to adjust circuit gain, sensitivity, and dynamic range.

This circuit may also be used as an AM detector by introducing carrier signal at the carrier input and an AM signal at the SSB input.

The carrier signal may be derived from the intermediate frequency signal or generated locally. The carrier signal may be introduced with or without modulation, provided its level is sufficiently high to saturate the upper quad differential amplifier. If the carrier signal is modulated, a 300 mV(rms) input level is recommended.

MOTOROLA *Semiconductor Products Inc.*

Figure 9.15 *(cont'd)*

port (pin 10), the modulating signal to the V_s port (pin 1), and the output is taken from the $+V_0$ or $-V_0$ port (pin 6 or 12). The pin numbers quoted are for the 14-pin ceramic package. Care must be taken to ensure that you have the proper package when building such a circuit. The 50-kΩ potentiometer is used to adjust the carrier level to achieve either a double sideband-full carrier output or a double sideband-suppressed carrier output. Being able to control the amplitude of the carrier signal with a single control is invaluable for the experimenter. Using an integrated circuit for modulation purposes allows the experimenter greater efficiency and provides a high degree of accuracy with only a few components. Integrated circuits are also an ideal choice because they allow the advantage of carrier-level control. Time-domain and frequency-domain (spectrum) pictures of DSB-SC are shown in figures 1 and 2 of the data sheet (Fig. 9.15), and DSB-FC is shown in figures 3 and 4.

All of the signals discussed thus far for amplitude modulation are for double-sideband full-carrier systems, meaning that two sidebands are present (the 101-kHz and 99-kHz components of the previous example) and a full carrier (100 kHz) that is not attenuated at all. Another method used in AM systems to reduce the amount of power used·in a system is the double-sideband suppressed carrier application.

Recall from our discussions on power within the carrier and sidebands, that the vast majority of required power is in the carrier. Thus, if we could reduce the level of the carrier, we would save a considerable amount of power. When suppressing the carrier, the only requirement is that enough power is left in the carrier to obtain the required range for the system. A DSB-SC presentation on a spectrum analyzer is shown in Figure 9.16. The term *suppressed* is used for these applications because the carrier is not totally eliminated, it is only reduced in amplitude. Some applications use this type of transmission to conserve power. The modulation process must, therefore, be one which combines the carrier and the modulating signal while reducing the amplitude of the carrier. This can be accomplished by a tuning or an attenuation process. Whichever method is used, the resulting signal must have the two sidebands and a highly reduced amplitude carrier as shown in Figure 9.16.

Having examined the basic method of modulating the amplitude of a carrier, you can now appreciate the basic requirements of a complete amplitude-modulated transmitter. Figure 9.17 is a simple block diagram of the circuits that could be used in such a transmitter. Two separate channels exist in this diagram which both come together in the modulator circuit. The first channel is the carrier. There is a carrier oscillator that may be any of the oscillator circuits discussed earlier in Chapter 7: a single-transistor Colpitts or Hartley. Or, if you need to tightly control frequency, a crystal-controlled oscillator is called for here. A typical circuit is shown in Figure

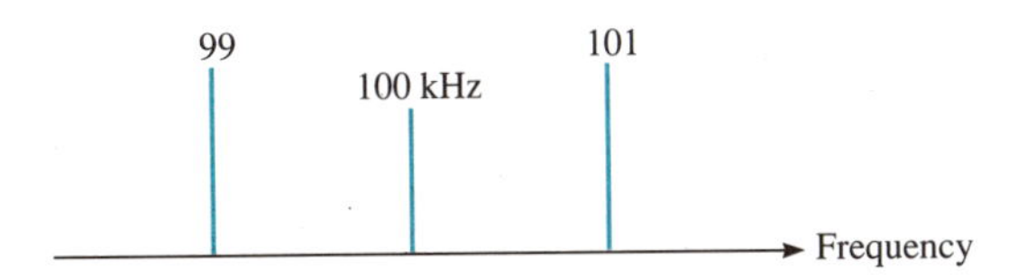

Figure 9.16 Double-sideband suppressed carrier (DSB-SC).

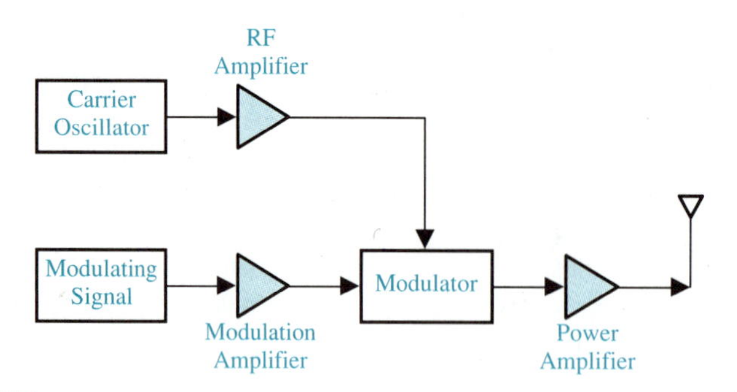

Figure 9.17 AM transmitter.

9.18. This crystal-controlled oscillator (discussed in Chapter 7) is operated in the parallel mode and gives our AM transmitter a stable source. We have chosen the crystal source for its excellent frequency stability. This circuit is of a basic crystal-controlled Colpitts construction.

The next block in the carrier chain is a rf amplifier. This circuit (also discussed in Chapter 7) would be placed in the chain if the oscillator did not have sufficient power to satisfy the carrier requirements or there was a danger of the modulator pulling the oscillator off frequency. There generally is an rf amplifier in most carrier circuits because the main function of the oscillator is to generate a stable frequency signal, not to produce a signal of any significant level. Figure 9.19 is a schematic of a typical rf amplifier that might be used for the application we are describing. It is a straightforward class-A amplifier, which requires great care during fabrication to avoid the effects of stray capacitance. This amplifier is actually serving two purposes: it provides a certain amount of gain to allow the entire transmitter to operate properly, and it acts as a buffer to isolate the oscillator output from the load (which is the modulator in this case). The amplifier thus ensures that the oscillator will not be "pulled off" frequency by any variations in impedance at the modulator, which is a nonlinear device. Frequency pulling is a phenomenon that must be considered and compensated

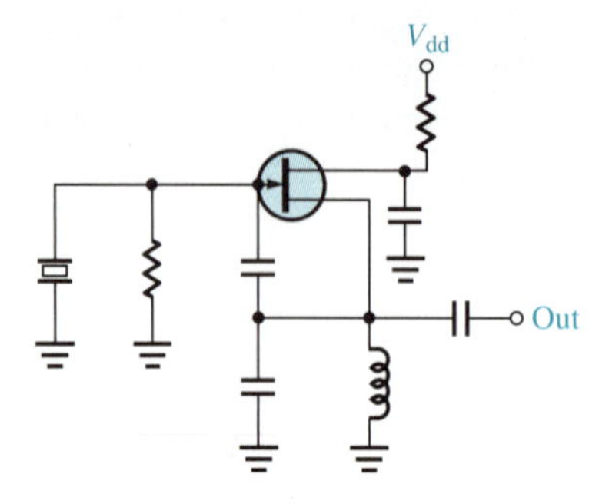

Figure 9.18 Crystal oscillator.

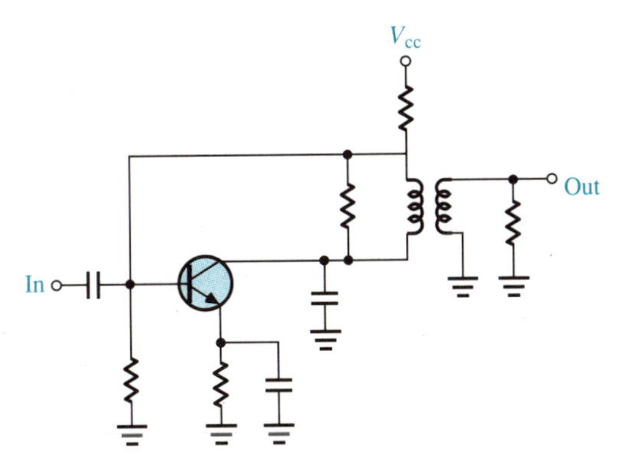

Figure 9.19 Rf amplifier.

for, which the rf amplifier does very nicely. Other circuits may be placed in this channel, such as monitoring or other testing circuits or a form of gain control for the amplifier, but in general the oscillator and rf amplifier are the main blocks.

The second channel is the modulating signal. The first block in this channel is called the **modulating signal**, meaning that generally some outside source supplies the signal to the transmitter. This outside source may be a microphone, a TV camera, or any number of sources containing intelligence to be broadcast through and out of the transmitter. Once again, there is an amplifier to increase the level. This time it is not an rf amplifier but an audio or video amplifier designed to amplify the necessary frequencies in the intelligence. Figure 9.20 is an example of the type circuit present in the modulating signal line. Notice that this amplifier is much less complex than the rf amplifier in the carrier-signal channel. Also, there need not be all of the construction precautions taken for this amplifier because it is a basic class-A, solid-state amplifier that was presented many times when you studied amplifiers and biasing methods in earlier electronic courses. The bias arrangement is the popular and efficient voltage-divider method with the input capacitively coupled.

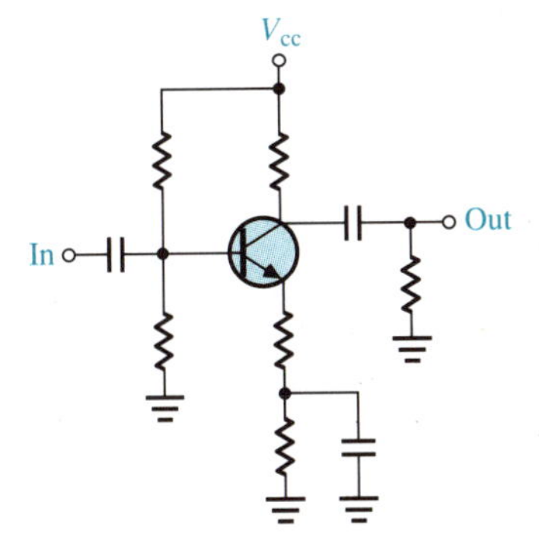

Figure 9.20 Modulation signal amplifier.

The two channels just described are brought together in the **modulator** block. As described earlier, this block may be a discrete transistor (bipolar or FET) or an integrated circuit. Its main function is to combine the carrier and modulating signal and produce the amplitude-modulated signal. As discussed previously, the modulator must be biased in such a way (class-C, preferably) as to always operate nonlinearly. With this arrangement, any two applied signals, such as a carrier and modulating signal, will result in a combined output that amplifies the original carrier and varies its amplitude at the modulating-frequency rate; that is, an amplitude-modulated signal. If we let the carrier be

$$e = E_c \cos (2\pi F_c)t \tag{9.1}$$

let the modulating signal be

$$e = E_m \cos (2\pi F_m)t \tag{9.2}$$

and let

$$m = \frac{E_m}{E_c} \tag{9.4}$$

the output of the modulator will be Equation (9.9), which is

$$e = E_c[1 + m \cos (2\pi F_m)t] \cos (2\pi F_c)t$$

We see from this equation that all the components are present for an amplitude-modulated signal. There is the carrier amplitude E_c, which is operated on by the modulating signal $\cos (2\pi F_m)t$, and there is a certain percentage modulation m for the signal. Were we to allow the percentage modulation to go to zero, the modulation equation would simply be $E_c \cos (2\pi F_c)t$, or the carrier signal alone. This makes sense because there would be no modulation with a zero percentage modulation, and only the carrier would be present. Any other value of percentage modulation between 0 and 1 will result in the appropriate amount of modulation for the system.

If the amplitude of the carrier E_c goes to zero, there will be no output. We covered this previously when discussing how the carrier ''turns on'' the modulator when it reaches a certain level. If the carrier level is zero, we understand that the modulator will not operate.

Thus, the modulator and the conditions it imposes on the circuits that drive it is the very heart of the AM transmitter as shown in Figure 9.17.

The final amplifier stage shown in Figure 9.17 can be varied according to the particular application. If we have a low-level transmitter (a few watts), the amplifier will be linear. If we have high-level transmission (tens of watts and up), the amplifier will be class-C in order to achieve the proper level of modulated output power. Whichever power level is needed, the transmitter in Figure 9.17 can be modified to provide it. Whatever the level, the power amplifier is a device that requires special considerations.

The mere name *power* amplifier indicates that this is a different device and should be treated as such. To this point we have considered circuits for which the power ratings of individual components is a secondary consideration. A resistor with a $\frac{1}{4}$-W or $\frac{1}{8}$-W rating will work very well for 99% of the applications we have discussed. Occasionally there will be an area where more current is to be drawn and a higher value resistor or stockier transistor is used. In general, however, most of the applications do not need to have power dissipation of components be a prime specification in their design. This is not the case for the power amplifier. Whether it is a linear amplifier, higher power amplifier, bipolar transistor, FET, integrated circuit, or vacuum tube, *the dissipation of power by each component that makes up the circuit must be considered.* This also includes the dissipation properties of the active devices (transistors and tubes), as well as the passive devices (resistors). Probably the most important devices to dissipate power are the active devices because the power put into a device does not come out as signal power, but rather as heat.

It is obvious by this point that nothing is 100% efficient. For any amount of power put into a circuit (dc and signal), some lesser value comes out as the signal. The difference between these two powers is heat, which must be efficiently transferred from the device so as not to destroy it.

As an example, consider a transistor amplifier operating with a V_{cc} of +24 V, has an I_c of 0.5 A, and an input signal power of $\frac{1}{2}$ W. The gain of the stage is 10, which means that there is a 5-W output. The total power input in this example is the dc power (24 V × 0.5 A = 12 W) and the signal power (0.5 W). Total input is 12.5 W. The output power is only 5 W. This says that 7.5 W, or 60% of the total input power, is now inside the transistor and must be dissipated or it will destroy the device. This is where heat sinking to metallic plates with fins going into the air are very useful. Also, sometimes air cooling (air blowing across a circuit) or water cooling (water or liquid circulating through pipes around a circuit) may be used to carry the excess heat away. Whatever the method, some cooling technique must be used to allow the transistor to operate at its rated temperature. This temperature is usually 200°C at the junctions of the devices. Consult the data sheet for each device for specific temperatures for the device being used.

Example 9.6 A transistor power amplifier is operating with the following parameters:

$$V_{cc} = +28 \text{ V}$$
$$I_c = 0.7 \text{ A}$$
$$\text{Signal in} = 0.6 \text{ W}$$
$$\text{Gain} = 15$$

Find:
1. What is the signal power output?
2. What is the efficiency of the stage?
3. How much heat is to be dissipated?

Solution

1. $P_{out} = P_{in} \text{ (gain)}$

$$= 0.6 \text{ W } (15)$$

$$= 9 \text{ W}$$

2. $\text{Efficiency} = \frac{P_{out}}{P_{in}}$

$$= \frac{9}{(28)(0.7) + 0.6}$$

$$= \frac{9}{17.4}$$

$$= 52\%$$

3. $P_{diss} = P_{in} - P_{out}$

$$= 17.4 - 9$$

$$= 8.4 \text{ W}$$

There are many applications where the power needed is only available through the use of a vacuum tube in the power-amplifier circuit. The theory and applications of vacuum tubes are uncommon in today's world of miniature solid-state circuits. The vacuum tube, however, has many applications in the realm of high-power communications transmitters. As an example, the high-power AM and FM commercial broadcast transmitters would not be possible without the vacuum tube. For this reason we will spend a short time here to describe vacuum-tube theory.

The vacuum tube relies on a phenomenon called **thermionic emission**. The term thermionic pertains to temperature or, more precisely, heat and emission, in the vacuum tube sense, is the release of electrons. Thus, thermionic emission is the release of electrons by a heating process. More precisely, it is the "boiling off" of electrons from a metallic structure. This process is also called the **Edison effect** because it was used to describe the electrons released from the filaments of Thomas Edison's incandescent light.

The first type of vacuum tube we will explore is the diode. Just as a solid-state diode consists of two elements, a cathode and an anode, so also does the vacuum-tube diode. The elements are also called the cathode and anode, although the anode is more commonly called the *plate*. The cathode in a vacuum tube is a metallic element heated by a filament, or it may be a direct cathode, which has the filament voltage applied to it. The filament is a material (usually tungsten) through which a current passes to heat the cathode and cause the emission of electrons. Filament voltages are usually 6.3 or 12.6 V ac. The anode (or plate) of the tube is positively charged to attract the electrons emitted from the cathode.

Basic operation of a diode vacuum tube is as follows:

1. Filament voltage is turned on and heating of the cathode begins.
2. Electrons are emitted from the cathode and are in a "space charge" around the cathode. This acts as a reservoir of electrons to be supplied upon demand.

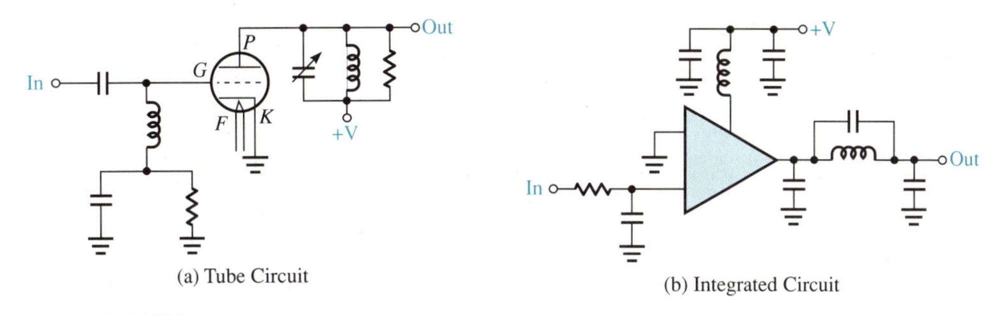

Figure 9.21 Power amplifier.

3. A positive voltage is placed on the plate, which causes a flow of free electrons from the cathode through the open region and to the plate.

This operation creates the same diode action as any solid-state diode.

To obtain the necessary high-power output from commercial AM or FM transmitters, we obviously must have a device that exhibits amplification. This feature is not present in the diode just discussed. To obtain this necessary amplification in the tube, the electrons inside the tube in the open region between the cathode and the plate must be controlled by an input signal. A third element must be introduced to accomplish this—a grid. The **grid**, which is a wire-mesh construction placed between the cathode and plate of the vacuum tube, creates a device called a triode. The **triode** produces the necessary gain for high-power applications. By gain, we mean μ, the change in plate voltage ΔV_p divided by the change in grid voltage $\Delta_g V$. With such a tube, very high gains and high-power outputs are attainable. In Figure 9.21(a) we see where each of the elements are placed in a tube structure. The filament F, cathode K, grid G, and plate P are all shown.

Figure 9.21(b) is a solid-state power amplifier that takes advantage of an integrated circuit. It must be realized that the two power amplifiers shown in Figure 9.21 are not capable of the same power outputs; the tube circuit will develop substantially more power than the integrated circuit amplifier.

Example 9.7

The AM transmitter shown in Figure 9.17 is tested and found to have only an rf carrier at the antenna. When a separate signal is fed into the modulator, there is a normal output from the transmitter. What is the problem?

Solution

Since the modulator is operating properly, as attested to by the injection of an external signal, the problem is the modulation amplifier. However, the operator should also check to be sure that the amplifier has a signal input before replacing the amplifier. It is also possible that no modulating signal is getting to the amplifier.

To show you how all of the pieces we have been discussing fit together and replace the rectangular blocks of Figure 9.17, a representative transmitter is shown in Figure 9.22. The transmitter is discrete except for the modulator and the output power amplifier. It is found in this case that the best arrangement is to have these two components be integrated circuits, thereby enhancing performance, as well as making the entire transmitter much simpler.

Before we move on to the AM receiver, it would be helpful to have some method, other than the ideal oscilloscope displays, used so far to determine if we have truly produced the optimum amplitude-modulated signal. An excellent way to test such circuits is by using the **trapezoidal patterns**, which can be displayed on conventional oscilloscopes and provide much needed information to a system designer.

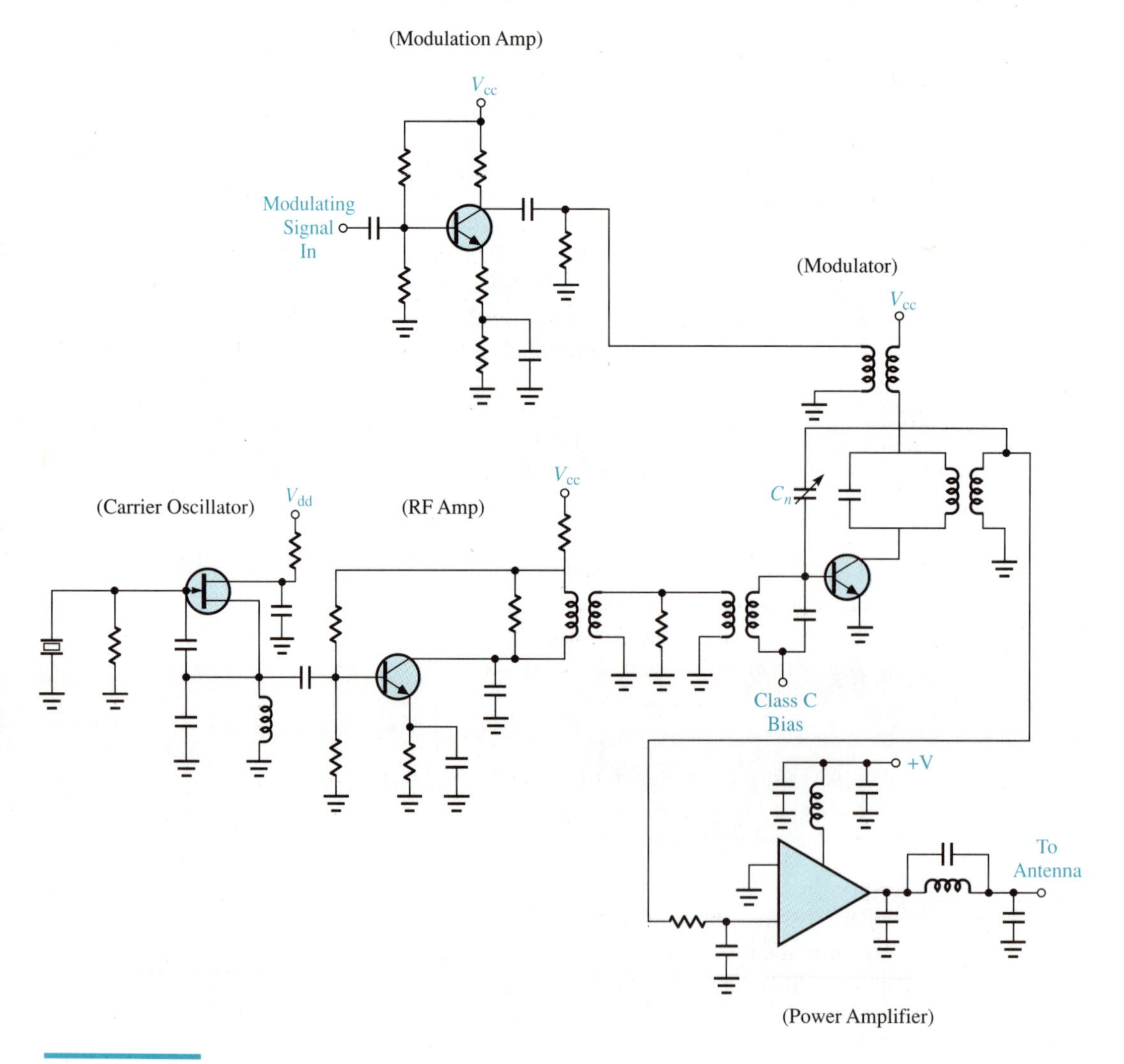

Figure 9.22 Complete transmitter.

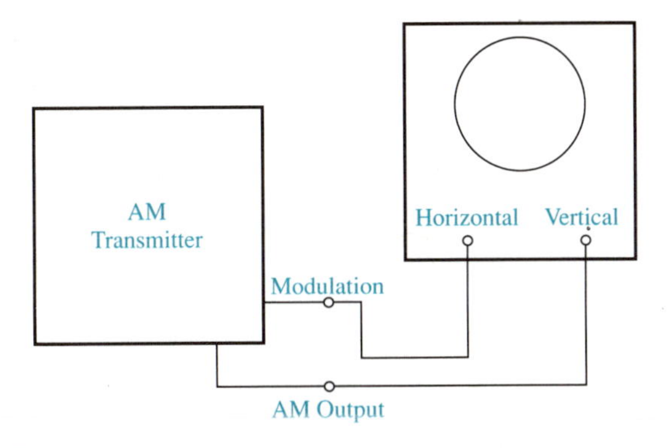

Figure 9.23 Trapezoidal test setup.

The patterns are generated using the test setup shown in Figure 9.23. The output of the transmitter is connected to the *vertical* input on the oscilloscope, and the modulating signal is applied to the *horizontal* input. Care should be taken to ensure that the level of the AM signal coming from the transmitter is sufficiently low that the input of the oscilloscope is not damaged.

With the inputs described above, the trapezoidal pattern shown in Figure 9.24 results. The output follows the rise of the modulating signal and the complete AM

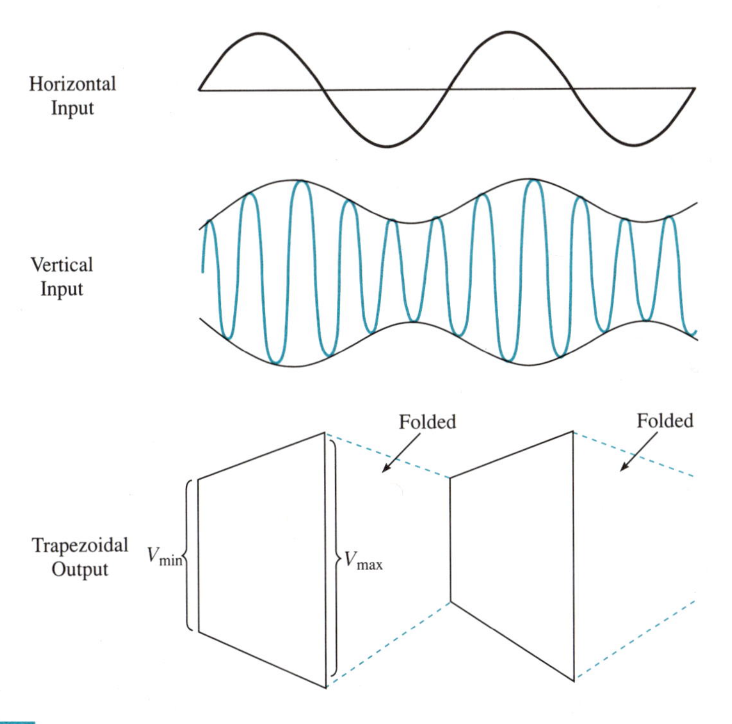

Figure 9.24 Trapezoidal pattern.

envelope to its peak and then folds back on itself to form the trapezoidal pattern. The parameters of interest in this pattern are labeled V_{max} and V_{min} in the figure. Recall that these parameters are very important when determining the modulation index of an AM signal. Using this method of testing, the operator can measure two voltages, make one calculation, and verify if the required modulation index has been achieved.

To understand how this method can be used, consider the patterns shown in Figure 9.25. The first one (part a) shows a case where $V_{max} = 4$ V and $V_{min} = 1$ V. By using Equation (9.8)

$$m = \frac{V_{max} - V_{min}}{V_{max} + V_{min}}$$

we find that $m = 0.60$, or 60%. Figure 9.25(b) is the case for 100% modulation. Once again by using Equation (9.8) and having $V_{min} = 0$ V, we obtain $m = 1$.

Figure 9.25(c) shows a case of overmodulation. The area to the left of the trapezoidal pattern is a null area; that is, nothing is in that area at all. This is the portion of the signal that produces dead spots in the intelligence. The last part of Figure 9.25(d) shows a nonlinear envelope by in which the negative voltage $(-V)$ is larger than the positive portion $(+V)$. This causes a condition called **carrier shift**, which produces an offset time-domain display on an oscilloscope. This is not the optimum type of amplitude modulation. Instead, the symmetrical AM envelopes, such as the diagrams shown throughout this chapter, are the ideal waveforms for good amplitude modulation.

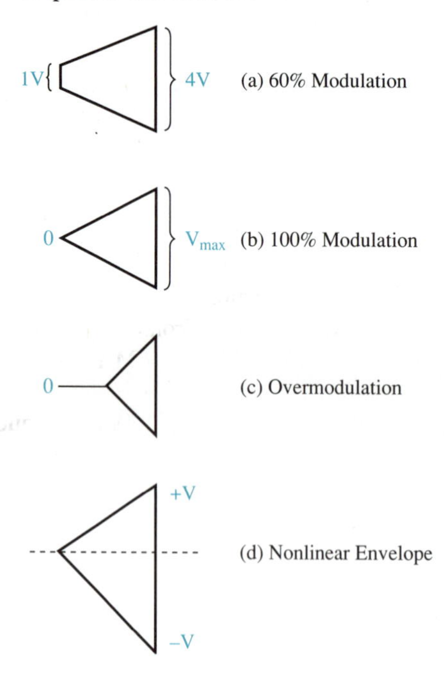

Figure 9.25 Examples of trapezoidal patterns.

Example 9.8

A trapezoidal pattern is presented on an oscilloscope and has the following values taken from it:

$$V_{min} = 0.2 \text{ V}$$

$$V_{max} = 6 \text{ V}$$

What is the modulation index? Is this an acceptable index?

Solution

Using Equation (9.8), we obtain

$$m = \frac{V_{max} - V_{min}}{V_{max} + V_{min}}$$

$$= \frac{6 - 0.2}{6 + 0.2}$$

$$= \frac{5.8}{6.2}$$

$$= .935, \text{ or } 93.5\%$$

This is not an acceptable modulation index because it is so close to 100%.

We have looked at the AM transmitter, described how to generate an amplitude-modulated signal, and explained how to perform tests on it. The next task is to receive this signal and recapture the intelligence so carefully put on the carrier in the transmitter.

9.4 Reception

In Section 9.3 we described how to generate an amplitude-modulated signal; we now explain how to recover the intelligence (modulation) from an amplitude-modulated signal. To aid us, we will examine some representative AM receivers and discuss their demodulation portions.

Figure 9.26 shows a block diagram of a basic **tuned-radio-frequency (TRF) receiver**. This type of receiver is usually used at lower rf frequencies because it requires many rf stages to obtain an acceptable output level, and these stages are difficult to design and are expensive at high rf frequencies. It is, nevertheless, a good first model for an AM receiver. The first block is termed a **selection unit and rf amplifier**, meaning that there is some means of selecting individual frequencies (usually by means of a filtering arrangement) and then amplifying them to an appropriate level. This scheme requires many stages of amplification, and the rf amplifier in this arrangement must have a frequency response that covers the entire band of the receiver input. It must also have very low noise characteristics because it will determine the

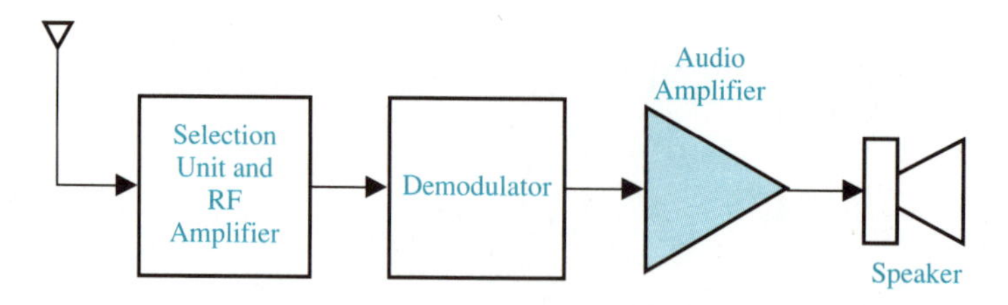

Figure 9.26 Tuned Rf receiver.

overall noise characteristics of the entire receiver. Recall from Chapter 6 that in systems with cascaded stages, such as a receiver, the noise characteristics of the first stage determine the noise characteristics of the entire receiver. This is true if the stage has a reasonable amount of gain along with the low noise figure.

The selection unit, as we have said, is usually made up of a series of filters and switches. If the receiver must operate over a very wide band, there will be many filters with fairly narrow bandwidths. This arrangement can be seen in Figure 9.27. The rf input, which is the AM signal, is applied to the input. The input switch is set for the proper frequency range either manually or automatically by selecting the correct filter to pass the signal on to the rf amplifier. All other signals at the input are attenuated by the skirts of the filter and only the desired one passes. The number and complexity of the filters depends on the particular type of receiver used.

The next block is the **demodulator**. This block can undo everyting that we did in Section 9.3; that is, it removes the modulation from the carrier, eliminates the carrier, and sends the modulation (intelligence) on to an audio (or video) amplifier for further amplification.

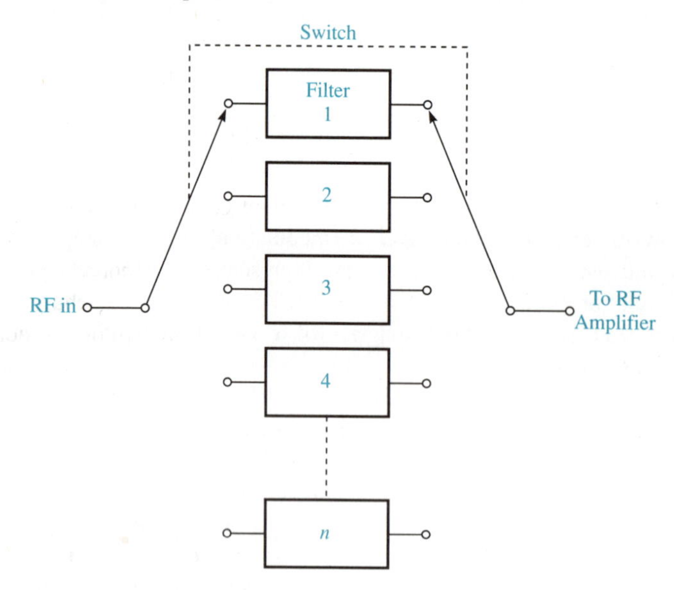

Figure 9.27 Selection unit.

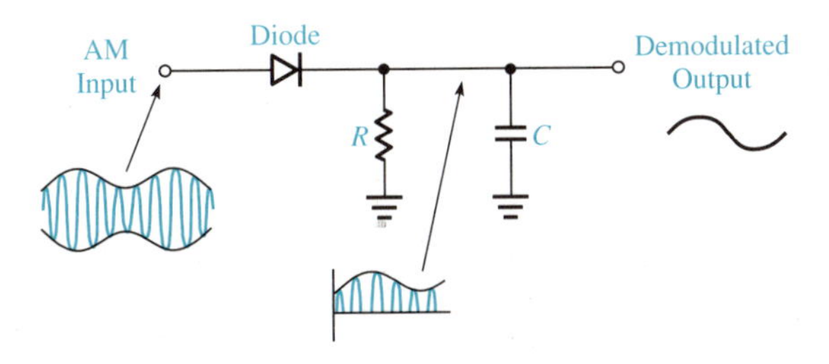

Figure 9.28 Peak amplitude detector.

The most basic type of AM detector, shown in Figure 9.28, is a **peak amplitude detector**. The diode in the circuit conducts whenever the voltage V_{in} exceeds the voltage of the diode V_d. This results in the waveform shown at point *A* in Figure 9.28. Basically, the positive peaks of the carrier vary at the modulating signal rate. The negative peaks of the signal can also be obtained if we were to turn the diode around.

The capacitor following the resistor is used to filter out the rf carrier and to allow the low-frequency modulating signal to pass to the following circuitry. The waveforms shown in Figure 9.28 are not at the same point at the same time. The figure illustrates that the signal shown before the capacitor is what will be there prior to any filtering. That is, without the capacitor, this signal would be the output. The signal termed *demodulated output* is the actual output of the detector with filtering in place. Remember from earlier electronic courses that this process of demodulation is similar to the rectification and filtering schemes used in power supplies. Recall that an ac signal came from the secondary of a transformer; was rectified with a single diode, two diodes, or a bridge rectifier circuit where half-wave or full-wave rectification was achieved; and then the signal was filtered with either a single capacitor, an R–C filter or an L–C filter to produce the required dc output voltage. Similarly, the peak detector has a varying signal input (the AM signal), detects the signal to eliminate the negative (or positive) portion of the signal, and then filters the signal (capacitor) to produce the modulating signal that contains the original intelligence. Thus, you see how similar the two systems are.

The capacitor at the output of the peak detector is important to the fidelity of the modulating signal because the capacitor must charge and discharge at the appropriate times to successfully reproduce the modulating signal. If the capacitor does not charge and discharge with the increasing and decreasing peaks of the signal, the modulating signal will be distorted. Figure 9.29 shows some of the distortion that can occur if the charge and discharge of the capacitor is not proper. Using the signal in Figure 9.29(a) as a reference, we see the types of detector distortion that may be encountered in parts(b) and (c). In Figure 9.29(b) the R–C time is too short—a distortion called **rectifier distortion**. This type of signal is similar to the half-wave rectifier studied in early electronics courses. In Figure 9.29(c) the R–C time is too long—a distortion called **diagonal clipping**. The slope of the output waveform cannot follow the trailing slope of the envelope and thus will *clip off* the information.

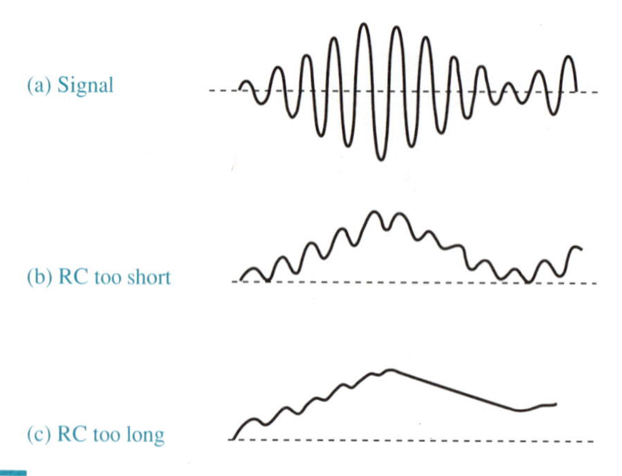

Figure 9.29 Detector distortion.

Integrated circuits can also be used for demodulation schemes. These are circuits that are very reliable, accurate, and easy to assemble and troubleshoot if problems arise. The type of integrated circuit will depend on the particular application and conditions present within a certain system. It is highly recommended that you investigate and determine which, if any, circuit will be best for your particular application.

The **superheterodyne receiver**, shown in Figure 9.30, is often used for amplitude modulation reception. With this receiver, there is a process which takes an rf input signal, mixes it with a local oscillator to produce an intermediate (lower) frequency, and then performs standard demodulation procedures on it. The circuit shown in Figure 9.30 also has a **preselector** at its input, which is similar to the selection unit presented in Figure 9.27. As before, this is very simply a series of filters selected to allow only specific bands of frequencies, or single frequencies, to enter the receiver. Many times the receiver is set to a particular band with a single filter at the input for that band. The preselector shown in Figure 9.30 is connected to the local oscillator block because the preselector is often tunable. As a filter is tuned, an appropriate local oscillator must also be selected. The preselector and the local oscillator must therefore be ganged together to achieve the appropriate intermediate-frequency separation to produce the proper intermediate frequency at the output.

The rf amplifier used in the superheterodyne receiver is the same circuit used for the TRF receiver. It must have a low noise figure and a reasonable gain so that its noise characteristics will determine the noise figure for the entire receiver. This amplifier may use bipolar transistors, field effect transistors, or may be an integrated-circuit amplifier. The amplifier must have its characteristics over the entire frequency range of the receiver, placing many more restrictions on such parameters as gain and noise figure. The type of amplifier is determined by the particular application of the receiver.

The mixer–local oscillator combination was discussed thoroughly in Chapter 7. The mixer is a nonlinear circuit that takes its rf input from the rf amplifier, mixes it

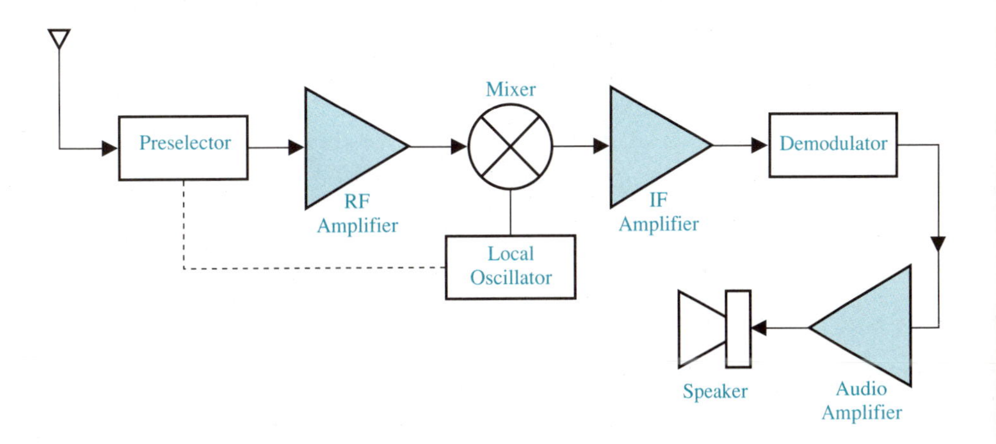

Figure 9.30 Superheterodyne receiver.

with the local-oscillator signal, and produces some intermediate frequency at the output. The level of the local oscillator, along with the frequency, are critical parameters. Recall from the mixer discussion in Chapter 7 that the diodes in the mixer must be driven into a nonlinear region for mixing to occur. This task is assigned to the local oscillator. More specifically, it is assigned to the *level* of the local oscillator. This oscillator generally is required to produce signals in the neighborhood of 0 dBm (1 mW) to +10 dBm (10 mW). This ensures that the diodes are in the nonlinear range and that mixing will result. There are times when this level must be even higher: some mixers require local oscillator levels of +27 dBm (0.5 W) for operation. Generally, however, the 0-dBm to +10-dBm level is more than sufficient. The local oscillator frequency, of course, must be the proper frequency to mix with the rf signal in order to produce the appropriate i-f output.

The intermediate-frequency amplifier has not been addressed thus far. This amplifier is somewhere between the rf amplifier at the input and the audio amplifier at the output—truly an **intermediate-frequency amplifier**. This type of amplifier is generally asked only to amplify a single frequency because the output of the mixer is usually only a single frequency. For a commercial AM receiver, this frequency is 455 kHz. Of course, this single-frequency operation simplifies the design of such a circuit substantially, as you can see with the typical i-f amplifier circuit in Figure 9.31. Figure 9.31(a) is a single-stage bipolar amplifier and Figure 9.31(b) is a cascaded two-stage amplifier with automatic gain-control (AGC) features.

It was stated previously that the i-f for standard commercial AM is 455 kHz. Why is 455 kHz used? What considerations go into choosing an i-f of 455 kHz? These are important questions.

Let us answer the second question first, thereby automatically answering the first question. To have the proper i-f for a superheterodyne receiver, you must minimize interference in the receiver and maximize the selectivity. Interference is any signal, internal or external, that degrades the performance of the receiver. This may

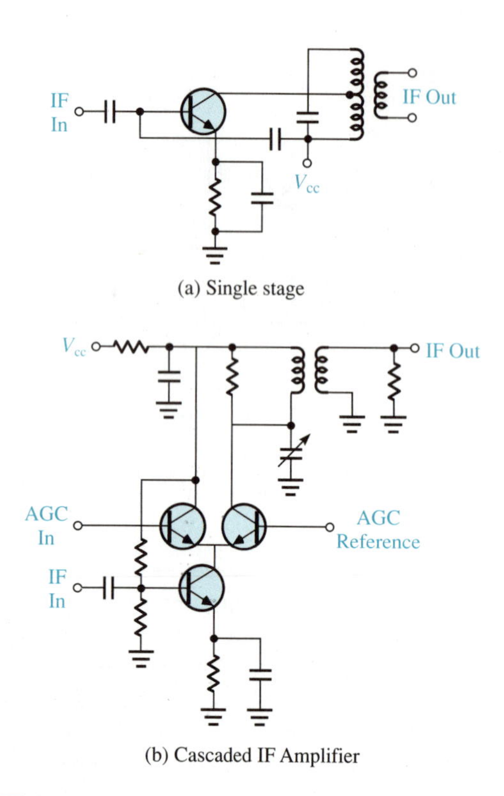

Figure 9.31 IF amplifiers.

be from internal oscillator radiating, power-supply ripple, or demodulator circuit feedback, all of which are types of internal interference. External interference is any signal that falls within the bandwidth of the filters or components used to make up the receiver. All of these forms of interference must be considered when choosing an i-f.

Selectivity is a function of the front end of the receiver. When tuning across a band, the receiver must have the capability of rejecting any channels in the overall receiver band. Adjacent channels, those immediately above or below the desired band of operation, are primarily taken care of with i-f filtering. It is much easier and more efficient to filter these signals here. Filtering them at the rf input puts a high burden on the filter design because the rf input is such a wide bandwidth (550–1600 kHz for commercial AM—nearly a 3:1 bandwidth) and the high Q necessary for a good rejection factor is very difficult to obtain over such a wide bandwidth. Also, since high rejection is used, many poles are necessary for the filters. Many poles create much more attenuation in the filter, which is undesirable. When the filtering is done in the i-f section, these restrictions are not present because the i-f is a single frequency.

Image rejection must also be considered when discussing intermediate frequencies. **Image rejection** results when the intermediate frequency is equal to the rf plus the LO, or high-side injection. The image is exactly one intermediate frequency above the local oscillator, or

$$F_{im} = F_{LO} + F_{if} \tag{9.17}$$

If we consider that the commercial i-f is at 455 kHz and F_{LO} = 1.455 MHz, the image for this case would be 1.900 MHz. Any signal at 1.9 MHz with the rf set to 1.00 MHz would interfere with the receiver operation. Thus, images play an important part in determining what the i-f will be in a receiver.

We mentioned automatic gain control (AGC) above in the i-f amplifier section. The idea behind AGC is that the signal levels coming into the front ends of many communications receivers can vary greatly. To have a constant amplitude at the i-f output so that the signals can be processed, AGC is incorporated into the i-f amplifier or rf amplifier or both. The scheme uses a feedback voltage to adjust the gain of the amplifiers to keep the overall gain of the system constant. If the input signal decreases in amplitude, the AGC circuitry will cause the rf or i-f amplifier to increase its gain to compensate. If the signal increases, the opposite situation occurs. This is a valuable circuit in a communications receiver because of the previously mentioned variability in the levels of many communications signals.

Two terms used with receivers that must be understood are gain and sensitivity. The **gain** of any system is a relative number. There is a certain level at the input and a certain level at the output. The ratio of the two, if the output is greater, is the gain of the system. If, for example, an amplifier had an input of 0.2 V and an output of 2.7 V, the gain would be 13.5, or 22.6 dB (20 log 13.5). All of these numbers, 13.5 and 22.6 dB, are relative numbers and therefore have no units associated with them.

The gains called out in a receiver may be rf gain, i-f gain, or overall receiver gain. The first two are self-explanatory. One is the gain in the rf stages, and the other is the gain of the i-f stages. The third is of the greatest value for receiver design or analysis. This gain is the *net* gain within the receiver from the rf input to the demodulated output. This gain takes into account all gains and losses within the entire receiver. Referring to Figure 9.30, we see how the overall gain of a receiver can be determined.

Example 9.9 The receiver in Figure 9.30 has the following gains and losses:

- ☐ Preselector −3.0 dB
- ☐ Rf Amplifier +35 dB
- ☐ Mixer −6.5 dB
- ☐ I-f Amplifier +30 dB
- ☐ Demodulator −8.0 dB
- ☐ Audio Amplifier +22 dB

What is the overall receiver gain?

Solution

$$\text{Losses} = (-3) + (-6.5) + (-8) = -17.5 \text{ dB}$$

$$\text{Gains} = +35 + 30 + 22 = +87 \text{ dB}$$

$$\text{Overall gain} = +87 - 17.5 = 69.5 \text{ dB}$$

Example 9.10

In the receiver in Example 9.9, there is an input signal level of −47.5 dBm. What power level is present at the input to the audio amplifier?

Solution

The gain to the input of the audio amplifier is the overall gain minus the amplifier gain; that is, 69.5 dB − 22 dB, or 47.5 dB. Therefore, if there is a −47.5-dBm input, the audio amplifier will have a 0-dBm input to it. This is also 1 mW since dBm = 10 log $P/1$ mW.

Sensitivity refers to the weakest (or smallest) signal at the output that can achieve the desired performance of the receiver. Using the circuit of Figure 9.30 and the numbers in Example 9.9, we see such a signal. Recall that the overall gain of this circuit was 69.5 dB. If we required an audio signal level of at least −10 dBm at the output, the sensitivity, or weakest signal, at the input of the receiver is −79.5 dBm. Any signal lower than this level would not satisfy the requirements for operation. Also note that this number is an absolute number as opposed to the gain figures, which were relative numbers.

The demodulator and audio amplifier are the same types as used in the TRF receiver discussed previously. Their main objective is to remove the intelligence from the carrier and amplify it to a suitable level so that the information can be interpreted and used.

Example 9.11

The receiver shown in Figure 9.30 does not have an output. Upon checking points in the circuit, we find that there is no output from the audio amplifier, demodulator, i-f amplifier, or mixer. However, there is an rf input and a local-oscillator input to the mixer, and the mixer has been checked by itself and found to be good. The local-oscillator level has also been checked and found to be proper. What is the trouble?

Solution

Since everything within the receiver has checked out, there is only one conclusion to be drawn. That is that the local oscillator has been pulled off frequency and the i-f that was needed is now some other frequency. A frequency test of the local-oscillator output should verify this.

As stated previously, noise characteristics of a receiver are primarily determined by the noise figure of the first stage of that receiver. The filters within the receiver also help to improve the system noise by eliminating unwanted noise with the narrowest bandwidth filters possible. This bandwidth must be narrow enough to eliminate the unwanted noise but still wide enough to pass all of the intelligence bandwidth so that the information being received can be processed. The best position for all the filtering is in the i-f section of the receiver. This is why a superheterodyne receiver is so valuable and finds so many uses in AM communications systems.

When calculating the signal-to-noise ratio (predetection), the i-f bandwidth is generally used because it is narrow and more representative of what is occurring in the system. Recall the equation used was $N = kTB$, which was the noise. Another method is to obtain a bandwidth-improvement factor which tells how much better the signal-to-noise ratio is in the i-f amplifiers than at the rf input. This improvement factor is a decibel reading and is equal to

$$10 \log \frac{B_{rf}}{B_{if}}$$

Example 9.12

If a receiver has an rf bandwidth of 6.3 MHz and an i-f bandwidth of 250 kHz, what is the predetection signal-to-noise ratio if the system noise figure is 7.3 dB and the input signal-to-noise ratio is 31 dB?

Solution

$$\frac{S}{N_{out}} = \frac{S}{N_{in}} + NF + BI$$

$$BI \text{ (Bandwidth improvement)} = 10 \log \frac{6.3 \times 10^6}{250 \times 10^3}$$

$$BI = 14 \text{ dB}$$

$$\frac{S}{N_{out}} = 31 - 7.3 + 14$$

$$= 37.7 \text{ dB}$$

If multistages are used in an i-f section, the bandwidth of each stage does not remain the same even though each stage has the same characteristics. To calculate how **bandwidth narrowing** occurs, use

$$BW = BW_1\sqrt{2^{1/n} - 1} \qquad (9.18)$$

Where: BW = Total bandwidth (in hertz)
BW_1 = Bandwidth of each stage (in hertz)
n = Number of stages with BW_1

Example 9.13

A superheterodyne receiver has two stages of i-f amplification with a bandwidth of 15 kHz for each. It is found that an additional stage is needed for proper operation. What must the bandwidth of each stage be to maintain the bandwidth that was present with two stages?

Solution

With two stages the bandwidth was

$$BW = 15 \times 10^3\sqrt{2^{1/2} - 1}$$
$$= 10 \times 10^3\sqrt{1.414 - 1}$$
$$= 15 \times 10^3(0.375)$$
$$= 5.63 \text{ kHz}$$

With three stages the bandwidth was

$$5.63 \times 10^3 = X\sqrt{2^{1/3} - 1}$$
$$= X\sqrt{1.2596 - 1}$$
$$= X(0.5095)$$
$$X = 11.049 \text{ kHz}$$

Thus, the bandwidth of each of the three stages must now be 11 kHz instead of the original 15 kHz.

A variation of the superheterodyne receiver discussed above is the double conversion receiver shown in Figure 9.32. This type of receiver makes use of the best of many worlds. For good image-frequency rejection in a superheterodyne receiver, a fairly high i-f frequency is selected. This was the case for the receiver shown in Figure 9.30. If a high gain and very frequency-selective i-f amplifier is to be used, a lower i-f is much more desirable. To get both conditions in one receiver, the double conversion receiver is used. The receiver amplifies the signals from an antenna and sends

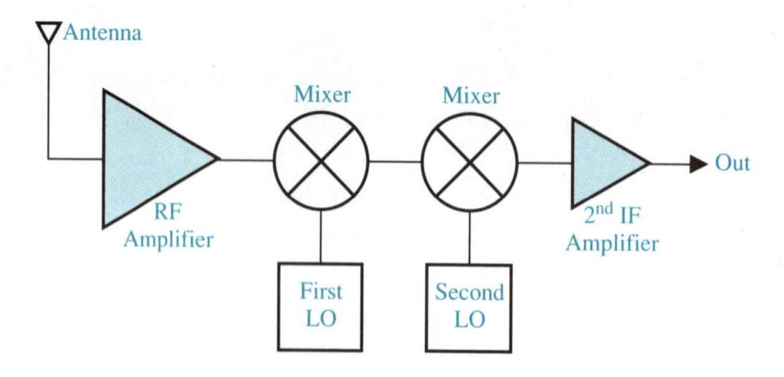

Figure 9.32 Double conversion receiver.

them to a mixer circuit, just as in the previous receiver. This is a fairly high i-f in order to obtain good image-frequency rejection. This i-f is then sent to a second mixer where a lower i-f is obtained allowing a high-gain and frequency-selective i-f amplifier to be used. This type of system is used for many types of communications receivers where an overall high gain is required.

9.5 Summary

This chapter presented a discussion of amplitude modulation. It began with a general discussion of modulation and then presented the theory of amplitude modulation and how it is achieved. Such areas as carriers, sidebands, voltage levels, and modulation index were covered with examples of calculations used to characterize systems for communications use.

The AM transmitter was then presented, showing how the carrier can be amplitude-modulated in order to send the desired intelligence out over the air and to the appropriate receiver.

Finally, the AM receiver was presented. The TRF, superheterodyne, and double conversion receivers were covered and components within each of these receivers was explained.

Questions

9.1 Introduction

1. Name three parameters of the carrier that can be varied to produce modulation.
2. Define **amplitude modulation**.

9.2 Theory

3. Draw a time-domain and frequency-domain picture of a double-sidedband full-carrier signal.
4. Define **interelement capacitance**.
5. Explain why 100% modulation is *not* an ideal value.
6. Define **modulation index**.
7. Define **intelligence bandwidth**.
8. What is the difference between DSB-FC and DSB-SC?
9. Why is a nonlinear circuit needed for a modulator?
10. Explain the mathematical representation of Equation (9.3).
11. Which modulation index would have the smallest value of V_{min}, 50% or 80%? Why?
12. Define **overmodulation**.
13. Explain what happens with Equation (9.9) when the modulation index goes to zero.
14. Which representation is better for AM, time domain or frequency domain?
15. Name four parameters that can be determined directly from a time-domain representation of an AM signal.

16. Name two parameters that can be determined indirectly from a time-domain representation of an AM signal.
17. How can the actual bandwidth of a system be determined from a spectrum-analyzer display of an AM signal?
18. What happens to the modulation index of an AM signal if the carrier level remains constant and the sideband level increases?
19. If we have a difference in carrier and sideband level of 14 dB, can a 50% modulation index be obtained?

9.3 Transmission

20. Draw a block diagram for an AM transmitter and explain each block.
21. Draw a trapezoidal pattern for a nonlinear envelope.
22. What is the term for an envelope that is offset?
23. How do signal levels differ for mixing and modulation processes?
24. Describe class-C operation.
25. What is the output tank circuit of an rf amplifier tuned to? Why?
26. Why would a crystal-controlled oscillator be used for some AM transmitters?
27. Name two functions of an amplifier which is at the output of an oscillator.
28. What is a prime consideration when analyzing power amplifiers?
29. Define **thermionic emission**.
30. Explain the operation of a triode vacuum tube.
31. Why are vacuum tubes used for high-power applications in some communications systems?
32. What are **trapezoidal patterns**?
33. If a trapezoidal pattern is observed on a scope and suddenly the value of V_{min} decreases by $\frac{1}{2}$, what has happened?

9.4 Reception

34. Draw a peak detector circuit and explain its operation.
35. What is the detector distortion called when the R-C time constant is too short?
36. What is the detector distortion called when the R-C time constant is too long?
37. Why is double detection used in some AM receivers?
38. What is a disadvantage of a TRF receiver?
39. Describe a selection unit for a receiver.
40. Compare a power-supply circuit to a peak detector circuit.
41. Describe a superheterodyne receiver.
42. Why is a superheterodyne receiver used for so many communications applications?
43. What is a preselector?
44. Describe AGC in a receiver.
45. What factors must be considered when choosing an i-f for a receiver?
46. Define **selectivity**.
47. Define **sensitivity**.
48. How is a receiver's overall gain determined?
49. How are S/N ratio and filter-bandwidth related?
50. If the number of stages in an i-f section increases, what happens to the bandwidth-improvement factor?

Problems

9.2 Theory

1. An AM signal is shown on the oscilloscope with $V_{max} = 5.2$ V with $V_{min} = 0.8$ V. What is the modulation index?
2. A modulation index of 67% is needed. The V_{max} is known to be 3.1 V. What does V_{min} need to be?
3. The modulation index for four signals within a spectrum are known to be $m_1 = 40\%$, $m_2 = 30\%$, $m_3 = 20\%$, and $m_4 = 32\%$. Find the total modulation index for the system.
4. The modulation index for a system cannot exceed 80%. There are two signals within the spectrum. One has a modulation index of 40%. What must the other signal have for a modulation index?
5. A system has a carrier of 5 MHz and is modulated with a 2.5-kHz signal. What is the bandwidth of this system?
6. We have a spectrum-analyzer display with the carrier at 0 dBm (10 MHz) and the sidebands at −16 dBm (10.01 and 9.99 MHz). What is the percentage modulation of this system?
7. With a peak carrier voltage of 42 V and a 61% percentage modulation, what is the power in each of the AM spectral components if an effective resistance of 47 Ω is available?
8. A spectrum-analyzer display has sidebands that are 14 dB, 16 dB, and 18 dB down from the carrier. What is the total modulation index?

9.3 Transmission

9. A trapezoidal pattern has a V_{max} of 5.1 V and a V_{min} of 3.1 V. What is the modulation index?
10. If we need a modulation index for an AM broadcast station of 87% and the trapezoidal pattern has a maximum of 2.2 V, what should the V_{min} be?
11. An AM transmitter has a rated power output of 25 W. When tests are run, it will develop no more than 20 W. All circuits up to the power amplifier have checked out as operational. The power amplifier has four output stages and is only drawing 6 A of current instead of the rated 8 A. What is the problem?
12. A transistor power amplifier operates with +24 V at 1.6 A, has an input power of 0.5 W and a gain of 6 dB. What is the efficiency and how much of the input power is wasted in heat?
13. A power-amplifier circuit is required to have at least 50% efficiency. If the dc input is +15 V at 1.2 A, and 1.3 W is available at the input, what is the gain of the stage?

9.4 Reception

14. The peak detector of an AM receiver has an output on a frequency counter that fluctuates between many random frequencies. It should read only 1 kHz. What is the problem?
15. The detector output waveform appears as shown in Figure 9.29(b). What has happened to the detector?
16. The filter capacitor in the peak detector is found to be open. What effect does this have on the output waveform?

17. An AM receiver operates from 500 kHz to 1.2 MHz and an i-f of 450 kHz. If the rf input is tuned to 1.0 MHz and the LO is high-side injected, what is the image frequency? Would a signal at 1.450 MHz be a problem? Why?
18. A superheterodyne receiver has the following characteristics:

☐ Preselector	−2.5 dB
☐ Rf Amplifier	+28 dB
☐ Mixer	−4.0 dB
☐ I-f Amplifier	+18 dB
☐ Demodulator	−6.2 dB
☐ Audio Amplifier	+15 dB

What is the receiver overall gain?
19. If the receiver in Problem 18 has an input signal of −36 dBm, what is the level present at the input to the mixer?
20. The receiver in Problem 18 uses an i-f amplifier that can have no more than 0 dBm at its input. If the receiver input level is at −30 dBm, is this condition met?
21. The receiver in Problem 18 must have an overall gain of 60 dB to operate properly; any change must take place in the i-f amplifier. What must the gain of the i-f amplifier be to satisfy this condition?
22. A superheterodyne receiver has three stages of i-f amplification. Each stage has 12 kHz of bandwidth. It is found that only two stages are needed to have the system operate. What must the bandwidth of each stage be to maintain the bandwidth that was present with three stages?

10

Outline

Objectives

- To introduce the student to the concepts of angle modulation. Angle modulation consists of frequency modulation and phase modulation. The main emphasis of this chapter is on frequency modulation with references made to *phase modulation* where applicable
- To look at examples of transmitters and receivers used to relay FM information
- To present FM stereo systems for a familiar example of FM information

Key Terms

frequency modulation
angle modulation
instantaneous phase deviation
instantaneous frequency deviation
frequency deviation
Bessel functions
preemphasis
FCC
FM stereo
discriminator
phase modulation
instantaneous phase
instantaneous frequency
deviation sensitivity
modulation index
information bandwidth
varactor diode
FDM
deemphasis

Frequency Modulation

10.1 Introduction

In Chapter 9 we noted that three parameters of the carrier in a communications system can be varied to produce a modulated signal: amplitude, frequency, and phase. We discussed the effects of amplitude variation in Chapter 9, and we will now concentrate on frequency and phase. These two parameters can not really be separated since any time you change the frequency of a signal you may also change the phase of it. The reverse is also true; a change in phase may result in a change in frequency. Therefore, they will be treated simultaneously. However, we will focus on frequency modulation, making reference to phase modulation where appropriate.

The general term used for frequency and phase modulation is **angle modulation**, which occurs when the phase angle θ of a sinusoidal wave is varied with respect to time. This variation in phase angle produces either a frequency or phase modulation on the carrier. Although the two schemes are virtually the same, there are some subtle differences that will be explained as they arise.

10.2 Theory

As stated above, angle modulation results whenever a phase angle θ of a sinusoidal wave is varied with respect to time. Mathematically, this statement is

$$m(t) = V_c \cos [\omega_c(t) + \theta(t)] \tag{10.1}$$

Where: $m(t)$ = Angle modulated carrier
V_c = Peak carrier amplitude
ω_c = Carrier frequency, $2\pi F_c$
$\theta(t)$ = Angle modulation (in radians)

Equation (10.1) shows that there is some rf carrier $2\pi F_c$ and a method of varying the phase angle of the carrier that is indicated by $\theta(t)$. This quantity, indicated as a time-varying parameter, is the key to producing an angle-modulated signal. In amplitude modulation the amplitude varies at a specific rate, in angle modulation the phase angle varies as a function of time. Recall that frequency is equal to 1/time, and you will see how a frequency or phase-modulated signal can result.

The major difference between FM (frequency modulation), and PM (phase modulation) is which parameter (frequency or phase) is varied directly by the modulating signal. This indicates that both FM and PM must occur whenever either form of angle modulation is performed. When there is direct FM, there will be indirect PM. Similarly, when there is direct PM, there will be indirect FM. In **direct FM** the frequency of the carrier is directly varied by the modulating signal. With this typical representation of an FM system the objective of the modulating signal is to change only the frequency of the carrier wave. This is a direct process and thus termed **direct** FM. Similarly, **indirect FM** is the result of deviating the phase of the carrier wave. This, as we will see shortly, is a form of phase modulation. There is no direct change in frequency by the modulating signal as in direct FM, but there will be a slight change in frequency because of the phase deviation, resulting in **independent FM**. We will consider the distinctions between the FM and PM before we proceed into the actual process of modulation and demodulation.

Consider once again what we are referring to when we call out a specific frequency. Figure 10.1 shows a single-frequency signal with a certain maximum amplitude and within a certain time frame T. One cycle encompasses 2π radians or 360°. The frequency of this signal is equal to $1/T$. The phase relationships of the signal are defined for a specific time frame within the 0 to T range.

Figure 10.2 now places two full cycles within that same time frame T. The time for one cycle is now $1/2T$, rather than $1/T$. This means that the frequency is now twice what it was in Figure 10.1; the frequency has changed, or deviated from the original signal. We have maintained the same time reference T and varied the signal within that time frame, basically producing a form of frequency modulation. Also,

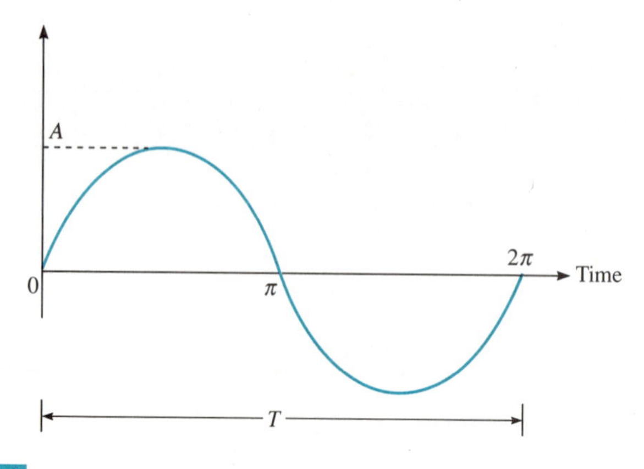

Figure 10.1 Single frequency signal.

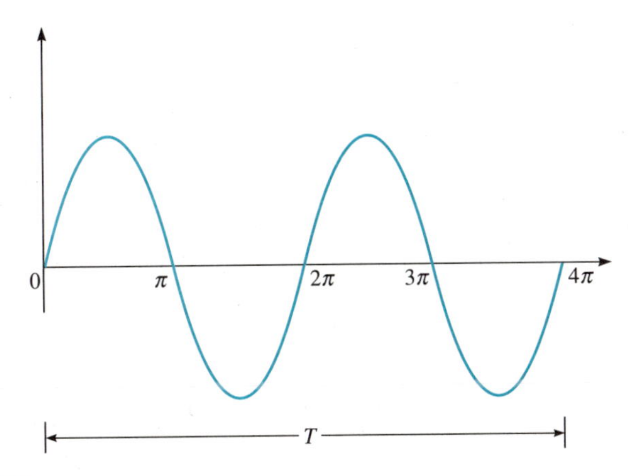

Figure 10.2 Signal at twice the frequency.

note that in Figure 10.1 there was a phase relationship as a function of time. At $T/2$, for example, the phase relationship said that the signal had progressed π radians, or 180°, as a function of time. In Figure 10.2, however, the same time frame $T/2$ says that the phase relationship is now 2π radians, or 360°, again a change from the original signal. This time it is a phase change. Thus, the statement that FM and PM must occur whenever angle modulation is performed holds true.

To further clarify the relationship of phase and frequency, let us now look at four terms vital to the understanding of these concepts: instantaneous phase, instantaneous phase deviation, instantaneous frequency, and instantaneous frequency deviation.

Instantaneous phase is the exact *phase* of the carrier at a given instant of time. Expressed mathematically, it is

$$\text{Instantaneous phase} = \omega_c t + \theta(t) \tag{10.2}$$

Where: $\omega_c = 2\pi F_c$ (Carrier frequency)
$\theta(t)$ = Phase change as a function of time

It is important to understand instantaneous phase so that a solid reference can be obtained at a specific time.

Instantaneous phase deviation is the instantaneous *change* in phase of the carrier at a given instant of time, shown in Equation (10.2) as $\theta(t)$. This term indicates how much the phase of the carrier is changing with respect to the reference, or instantaneous, phase.

Instantaneous frequency is the exact *frequency* of the carrier at a given instant of time. It is formally defined as being the first time derivative of the instantaneous phase. Expressed mathematically, it is

$$\begin{aligned} \text{Instantaneous frequency} &= \frac{d}{dt}[\omega_c t + \theta(t)] \\ &= \omega_c + \theta'(t) \end{aligned} \tag{10.3}$$

(Since this text does not require calculus as a prerequisite, this expression is presented only as an illustration of how the term can be expressed mathematically.)

Instantaneous frequency, like instantaneous phase, is important in establishing a reference frequency with which to compare the system in future time frames.

Finally, **instantaneous frequency deviation** is the instantaneous *change* in frequency of the carrier. This is once again a first-time derivative, but this time it is of the instantaneous phase deviation $\theta(t)$. Thus, it is designated as $\theta(t)$ in Equation (10.3).

You can see from the definitions of the four terms above that a common thread ties each of them together—*time*. The two instantaneous terms (frequency and phase) were each defined at a specific instant of time and set up as a reference. Then, at some other instant of time, the variations in either frequency or phase could be presented and classified as its *deviation*.

To further define these terms and clarify their relationship to phase and frequency modulation, **deviation sensitivities** are given for each type of modulation. For phase modulation, we will use K_p, and for frequency modulation we will use K_f. These sensitivities are actually the input–output transfer functions of the modulator; that is, how the modulator reacts to the specific type of modulation. These sensitivities are expressed in the following units:

- ☐ K_p in radians per volt
- ☐ K_f in hertz per volt
- ☐ or in $2\pi F$ radians per volt

To summarize the information presented to this point, we can say that if we modulate a carrier with a signal that is V_m cos mt, the phase- and frequency-modulated signals will be

$$\text{Phase modulation} = V_c\omega_c t + K_p V_m \cos \omega_m t \tag{10.4}$$

and

$$\text{Frequency modulation} = V_c \cos\left[\omega_c t + \left(\frac{K_f V_m}{\omega_m}\right)(\sin \omega_m t)\right] \tag{10.5}$$

You can see how the two modulations are similar, but still have their own distinctions.

The terms covered thus far all use the term *instantaneous* and refer to an exact time when an event takes place. What is needed to work with angle-modulation systems is a much simpler term—frequency deviation. **Frequency deviation** can be defined as the change in frequency of the carrier when it is acted upon by a modulating signal. This is referred to as F and is typically given as a peak frequency shift in hertz. Frequency deviation is proportional to the amplitude of the modulating signal V_m and the rate at which the frequency changes is proportional to the modulating frequency F_m. The mathematical expression for frequency deviation is

$$F = K_f V_m \tag{10.6}$$

or

$$F = \frac{K_f V_m}{2\pi} \text{ (in radians)} \tag{10.7}$$

Where: K_f = Deviation sensitivity
V_m = Peak modulating signal amplitude

Example 10.1

A frequency-modulated system is found to have a deviation sensitivity K_f of 2 kHz per V. If the modulating signal is $5 \cos 3 \times 10^2 t$, what is the frequency deviation?

Solution

Using Equation (10.6) and noting that the amplitude of the modulating signal is 5 V, we obtain

$$\begin{aligned} F &= K_f V_m \\ &= 2 \text{ kHz per V } (5 \text{ V}) \\ &= 10 \text{ kHz} \end{aligned}$$

We can now utilize all of the terms presented thus far to introduce a term that is similar to one discussed in Chapter 9 on amplitude modulation—modulation index. As in the case of the amplitude modulation, the **modulation index** is important for the modulation system because it measures how well the modulation signal is actually modulating the carrier.

The modulation index for a sinusoidal FM signal is expressed as

$$m = \frac{\Delta F}{F_m} \tag{10.8}$$

Where: ΔF = Frequency deviation
F_m = Modulating frequency

You can reason out this expression if you remember what frequency modulation is and what the terms in Equation (10.8) are saying. Remember that FM is the variation (deviation) of a carrier frequency by means of a modulating signal (F_m). Equation (10.8) shows that the modulation index (percentage achieved) of the carrier is the ratio of the frequency deviation (ΔF) and the modulating frequency (F_m). If the deviation is the same as the modulating frequency, the carrier is 100% modulated. If the system is not this efficient with regards to the modulation process, the percentage of modulation will be something less than 100%. If the system has a problem and a modulation signal produces no frequency deviation, then there is a modulation index of 0, or simply a carrier present.

If we relate the frequency deviations in Equation (10.8) to deviation sensitivity K_f, we have

$$m = \frac{K_f V_m}{F_m} \tag{10.9}$$

For phase modulation,

$$m = K_p V_m \tag{10.10}$$

Example 10.2

Given: A 2.5-MHz VCO with a measured sensitivity K of 4.2 kHz per V is modulated with a 3-V peak, 5-kHz sinusoid.

Determine:
(a) The maximum frequency deviation
(b) Peak phase deviation
(c) Modulation index for $V_m = 3 \cos [2\pi(7 \text{ kHz})t)]$

Solution

(a) Maximum frequency deviation is given as KV_m, so

$$\Delta F = KV_m$$
$$= 4.2 \text{ kHz per V } (3 \text{ V})$$
$$= 12.6 \text{ kHz}$$

(b) Peak phase deviation is modulation index, therefore

$$m = \frac{\Delta F}{F_m}$$
$$= \frac{12.6 \text{ kHz}}{5 \text{ kHz}}$$
$$= 2.52 \text{ radians, or } 2.5$$

(c) From section (a) the deviation is found to be 12.6 kHz. (This holds true since the amplitude of the signal is still 3 V.) The modulating frequency, however, is now 7 kHz.

$$m = \frac{\Delta F}{F_m}$$
$$= \frac{12.6 \text{ kHz}}{7 \text{ kHz}}$$
$$= 1.8$$

Modulation index can be used to determine whether a system is wide-band or narrow-band FM. As a rule of thumb, if the modulation index is greater that 0.25 it is wide-band FM, and if a modulation index is less than 0.25 it is narrow-band FM.

The percentage of modulation for FM systems is defined as the ratio of the frequency deviation produced to the maximum frequency deviation allowed. That is

$$\text{Percentage modulation} = \frac{\text{Deviation (actual)}}{\text{Deviation (allowed)}} \tag{10.11}$$

To illustrate this equation, the FCC allows FM stations to deviate ±75 kHz. If a system deviates ±60 kHz, the percentage of deviation is 60 kHz/75 kHz, or 80%. Thus, each station must keep a close watch on the characteristics of its stations to ensure that the maximum limits are not exceeded.

Example 10.3

A 3.2-MHz source has a sensitivity of 1.3 kHz per V and is modulated with a 0.5-V 2.7-kHz signal. Is this a wide-band or narrow-band system?

Solution

To determine whether it is a wide- or narrow-band system, the modulation index must be known. We first find F (Equation 10.6) and then the modulation index (Equation 10.8).

$$\begin{aligned}\Delta F &= K_f V_m \\ &= 1.3 \text{ kHz per V } (0.5 \text{ V}) \\ &= 0.65 \text{ kHz, or } 650 \text{ Hertz}\end{aligned}$$

$$\begin{aligned}m &= \frac{\Delta F}{F_m} \\ &= \frac{0.65 \text{ kHz}}{2.7 \text{ kHz}} \\ &= 0.24\end{aligned}$$

This modulation index indicates that this is a narrow-band FM system. Recall that any modulation index less than 0.25 was narrow-band.

Having defined our terms, it is now time to examine the FM (or PM) signals more precisely. They will be presented in both the time and frequency domains.

Figure 10.3 shows a time-domain presentation of an FM signal. The components shown are the rf carrier, the modulating signal, and the frequency-modulated signal. The first two are inputs to the modulator; the third one is the result. You can see that the modulated output is a constant-amplitude signal, which is at the rf carrier frequency until the modulating signal is impressed on it. This constant amplitude is one of the greatest assets of FM. You have undoubtedly noticed that when traveling in an automobile, an AM radio station is lost when you go under power lines. An FM station on the same radio will suffer no ill effects from the power lines because the modulation is not a function of amplitude, but of frequency. This change in carrier frequency is proportional to the modulating-signal frequency. We present cases for a single-

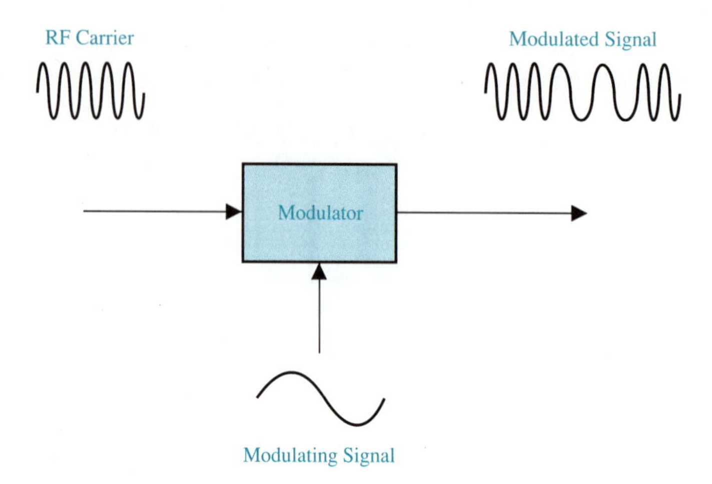

Figure 10.3 Frequency modulation (time domain).

frequency-modulating signal. In actuality, the modulating signal may be an audio or video signal that is much more complex than the single frequency shown here.

The frequency-domain presentation for an FM signal, shown in Figure 10.4, illustrates that there are many more components to the spectrum for FM than there were for an amplitude-modulated signal. This presentation has a series of components, each separated from the carrier by an amount equal to the modulating frequency. It is actually much easier to see frequency modulation in the frequency domain because any change in the signal results in a change in the frequency presentation on the analyzer. The time-domain presentation does not always clearly show what is happening, but the frequency domain presentation is always an absolute picture.

The important part of Figure 10.4 is the amplitude of the frequency components in the display. Whereas the amplitude of the AM sidebands were rather easily calculated, the amplitudes of the sideband components of an FM signal are very com-

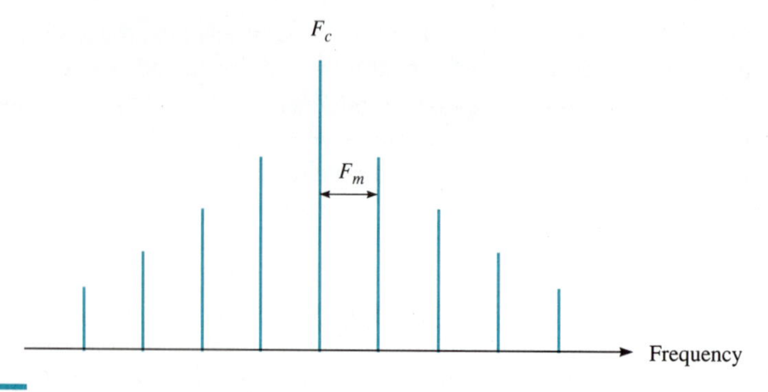

Figure 10.4 Frequency modulation (frequency domain).

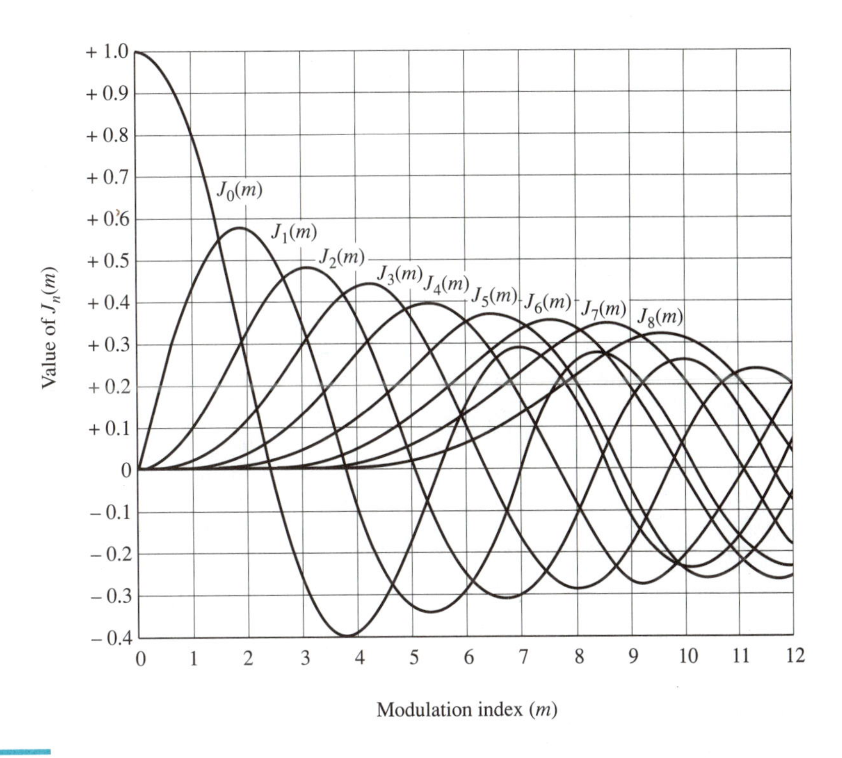

Figure 10.5 Bessel functions.

plicated. They are, instead, tabulated in table form. These tabulations are called **Bessel functions**, or **Bessel functions of the first kind**, named for the German astronomer and mathematician Friedrich Wilhelm Bessel (1784–1846). Figure 10.5 shows the functions with the modulation index along the x axis and the value of the function along the y axis. Table 10.1 shows the tabular results of Figure 10.5. In the table, the J_0 term is the carrier with the other terms, $J_1(m)$, $J_2(m)$, and so on, as the sidebands associated with the carrier. Note that each component has various points where it goes to zero. The values for the carrier occur at modulation indices of 2.40, 5.52, 8.65, 11.79, 14.93, 18.07, and for carrier zeros beyond 6 the modulation index is 18.07 + $\pi(n - 6)$. These are carrier **null points**—that is, where the carrier is at zero. These points, up to the modulation index of 11.79, can be verified in Figure 10.5 and in Table 10.1. From Table 10.1 you can see that when $m_f = 2.4$, for example, that J_0 (carrier level) is 0. Also, if we interpolate numbers, we see that at $m_f = 5.52$ and 8.65, the carrier also is 0.

Similarly, if we investigate the harmonics; J_1, J_2, J_3, . . . , we see that at $m_f = 0$, all of the sideband components are 0. This is understandable because there is no modulation, only a carrier present at this time. Also, Figure 10.5 verifies that all of the Bessel curves begin at 0 except for J_0, the carrier. If we follow sideband J_1, we see that the curves show that this sideband will go to 0 at $m_f = 3.7$, 7, and 10.2. This can also be verified by Table 10.1. Thus, you can see that the level of the

Table 10.1: Bessel Functions of the First Kind, $J_n(m)$

m_f	J_0	J_1	J_2	J_3	J_4	J_5	J_6	J_7	J_8	J_9	J_{10}	J_{11}	J_{12}	J_{13}	J_{14}
0.00	1.00	—	—	—	—	—	—	—	—	—	—	—	—	—	—
0.25	0.98	0.12	—	—	—	—	—	—	—	—	—	—	—	—	—
0.5	0.94	0.24	0.03	—	—	—	—	—	—	—	—	—	—	—	—
1.0	0.77	0.44	0.11	0.02	—	—	—	—	—	—	—	—	—	—	—
1.5	0.51	0.56	0.23	0.06	0.01	—	—	—	—	—	—	—	—	—	—
2.0	0.22	0.58	0.35	0.13	0.03	—	—	—	—	—	—	—	—	—	—
2.4	0	0.52	0.43	0.20	0.06	0.02	—	—	—	—	—	—	—	—	—
2.5	−0.05	0.50	0.45	0.22	0.07	0.02	0.01	—	—	—	—	—	—	—	—
3.0	−0.26	0.34	0.49	0.31	0.13	0.04	0.01	—	—	—	—	—	—	—	—
4.0	−0.40	−0.07	0.36	0.43	0.28	0.13	0.05	0.02	—	—	—	—	—	—	—
5.0	−0.18	−0.33	0.05	0.36	0.39	0.26	0.13	0.05	0.02	—	—	—	—	—	—
6.0	0.15	−0.28	−0.24	0.11	0.36	0.36	0.25	0.13	0.06	0.02	—	—	—	—	—
7.0	0.30	0.00	−0.30	−0.17	0.16	0.35	0.34	0.23	0.13	0.06	0.02	—	—	—	—
8.0	0.17	0.23	−0.11	−0.29	−0.10	0.19	0.34	0.32	0.22	0.13	0.06	0.03	—	—	—
9.0	−0.09	0.25	0.14	−0.18	−0.27	−0.06	0.20	0.33	0.31	0.21	0.12	0.06	0.03	0.01	—
10.0	−0.25	0.05	0.25	0.06	−0.22	−0.23	−0.01	0.22	0.32	0.29	0.21	0.12	0.06	0.03	0.01

spectral content by the FM signals can be predicted by knowing the modulation index and referring to Table 10.1 or Figure 10.5.

The following are other interesting points about Figure 10.5 and Table 10.1:

- For a modulation index of less than 0.25 (narrow-band FM), the carrier drops by less than 2% and only one set of sidebands is present. These are the first-order sidebands $J_1(m_f)$.
- For the modulation index greater than 0.25, additional sidebands become very prominent.
- All sets of sidebands are 0 for no modulation, whereas the carrier J_0 starts at 1.0, as previously stated.
- The carrier and sidebands reach a positive (+) value, then go through a null (as the carrier does for the modulation index of 2.4), and then go negative (−).

To see how the table can be used to find specific values, consider Example 10.4.

Example 10.4

Suppose we have an FM system with a modulation index of 1.5. From the table we find 1.5 on the left side of the table and read across to obtain the sidebands that go with this modulation index. It should be pointed out

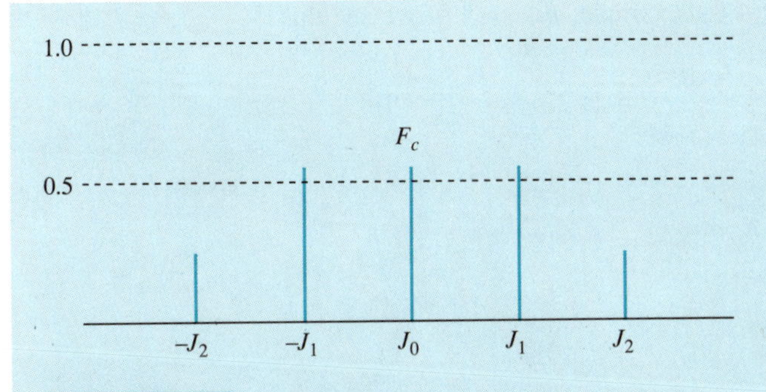

Figure 10.6 Example: $M_f = 1.5$.

here that the values shown in Table 10.1 are for the carrier and one side of the spectrum. To make the picture complete, there must also be a mirror image of the sidebands on the left of the carrier.

Having found the modulation index of 1.5, we now read the carrier and sideband levels. In terms of the carrier or sideband, and the Bessel functions, they are

□ Carrier (J_0)	0.51
□ First-order sideband (J_1)	0.56
□ Second-order sideband (J_2)	0.23
□ Third-order sideband (J_3)	0.06
□ Fourth-order sideband (J_4)	0.01

In most cases, values of 0.06 and lower would be neglected, and we would only consider the sidebands of the first- and second-order components. This is shown in Figure 10.6 with the amplitudes of all the components indicated.

We stated previously that an FM signal is a constant-amplitude signal, a fact which you can see when you observe the modulated signal on an oscilloscope. The Bessel functions can be used to prove that the amplitude is constant (as shown in Example 10.5). This can be accomplished and shown, as the display on the oscilloscope shows, that as the modulation index increases the amplitude will remain constant. This is true only if all of the power in the sidebands is accounted for in the calculations. To do this we consider the following relationship, which is the relative power in the signal:

$$\frac{P}{P_{\text{mod}}} = [J_0(m_f)]^2 + 2[J_n(m_f)^2] \tag{10.12}$$

If we take, for example, $m_f = 2.0$, we see that

$$\frac{P}{P_{\text{mod}}} = [(.22)^2] + 2[(.58)^2 + (.35)^2 + (.13)^2 + (.03)^2]$$

$$= 0.0484 + 2(.3364 + .1225 + .0169 + .0009)$$

$$= 0.0484 + 2(.4767)$$

$$= 0.0484 + 0.9534$$

$$= 1.0018$$

This result says that all of the sideband components are accounted for and that there is a 0.18% error.

Example 10.5

What percentage of error is present if the modulation index is 1.5?

Solution

For $m_f = 1.5$, $J_0 = .51$, $J_1 = .56$, $J_2 = .23$, $J_3 = .06$, $J_4 = .01$

$$\frac{P}{P_{\text{mod}}} = [J_0(m_f)]^2 + 2[J_1^2 + J_2^2 + J_3^2 + J_4^2]$$

$$= [.51^2] + 2[(.56)^2 + (.23)^2 + (.06)^2 + (.01)^2]$$

$$= .26 + 2[.31 + .05 + .0036 + .0001]$$

$$= .26 + .72$$

$$= .98, \text{ or } 0.02\% \text{ error}$$

Thus, you can see how the Bessel functions may be used to show that the ratio of power to modulated power remains virtually constant over a varying modulation index for FM systems.

As in the case for amplitude modulation, or any other modulation scheme, it is important to know the bandwidth of the system. As we have stated many times, the total power in an angle-modulated (FM or PM) system is constant as long as all of the sidebands are accounted for, whether there be one pair or a dozen pairs of sidebands. This statement implies that the system has an infinite bandwidth. Quite obviously, however, this cannot be the case. To make the system more realistic we consider that the signal is basically accounted for if the ''significant'' sidebands are considered. The significant sidebands are defined as those whose Bessel functions in Table 10.1 are >0.01. The 99% bandwidth determined by including all of the significant sidebands is the **information bandwidth** for a given modulation index. As a general rule, the bandwidth is

$$\text{Bandwidth} = 2(F_m + F_c) \tag{10.13}$$

or

$$\text{Bandwidth} = 2F_m(1 + m_f) \quad (10.14)$$

Example 10.6

A modulation frequency of 5 kHz is used to frequency modulate a carrier and results in a modulation index of 2.0. What is the information bandwidth?

Solution

Using Equation (10.14), we obtain

$$\begin{aligned}\text{Bandwidth} &= 2F_m(1 + m_f)\\ &= 2(5\text{ kHz})(1 + 2)\\ &= 10\text{ kHz }(3)\\ &= 30\text{ kHz}\end{aligned}$$

Example 10.6 helps to support the previous statement about which sidebands are significant. It was stated previously that Bessel function values of 0.06 and lower are neglected. In the P/P_{mod} example we showed that for a modulation index of 2.0 there was $J_0 = .22$, $J_1 = .58$, $J_2 = .35$, $J_3 = .13$, and $J_4 = .03$. If we had the modulating frequency of 5 kHz, as presented in Example 10.6, the spectrum shown in Figure 10.7 would appear. There is 5 kHz between each sideband, resulting in a bandwidth of 40 kHz, as shown in the lower portion of the figure. If we neglect any components less than 0.06, we see that only J_0, J_1, J_2, and J_3 are involved. This results in a 30-kHz information bandwidth, which is exactly the number calculated in Example 10.6. This confirms that if we neglect any sideband with Bessel amplitudes

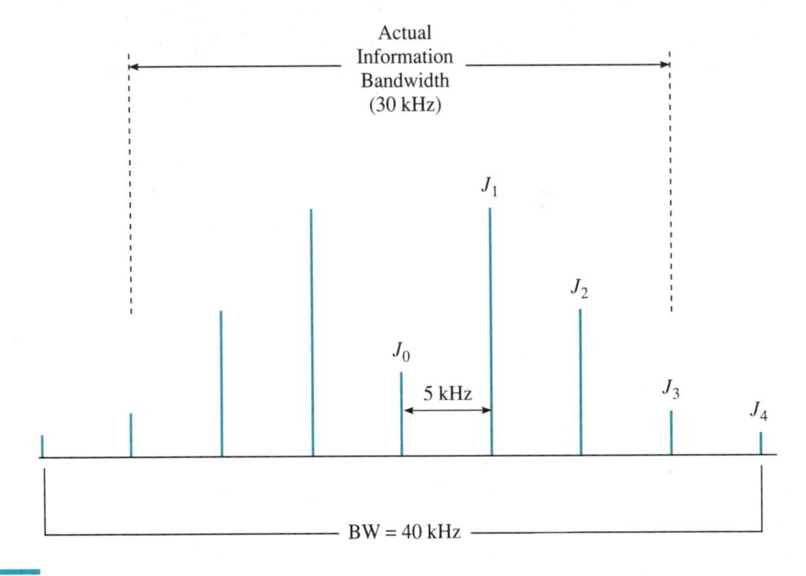

Figure 10.7 Information bandwidth example.

<0.06, we obtain the information bandwidth of the system, which is what will be the most useful.

To show a typical bandwidth for an FM system, we will look at the bandwidth allocations for a commercial FM-broadcast station (88–108 MHz). Each station is allocated a 150-kHz channel plus a 25-kHz guardband at both the upper and lower edges of the allocated frequency. This results in a total channel of 200 kHz for each station, as shown in Figure 10.8. Also, only alternate channels are assigned within a particular geographical area to ensure a minimum possibility of interference between stations. Thus, the bandwidth of an FM system must be taken into consideration when designing and working with FM signals within an area.

To help you relate the information presented thus far, we will present a simple and practical example—a common signal-generator output. If the generator is set to a particular frequency with its output displayed on a spectrum analyzer, a single signal will be present. This signal appears to be solid and at one frequency. If the analyzer display is expanded so that the signal is spread out to display a few kilohertz per centimeter, you can see that the signal, rather than being stable, fluctuates around the required frequency. Recall that when a carrier frequency is changed from its normal value, frequency-modulated signals result. This is what happens with the signal-generator output. It has an FM component called **residual FM**, which indicates how much ($\pm$ hertz) the signal varies around the required output frequency. This parameter is present in every generator and is one that should be considered when using or specifying a generator.

Before proceeding into the transmission and reception of angle-modulated signals, we'd like to focus on phase modulation for a short period. We have mentioned some of its features previously, but now will specifically look at PM and compare it with the now familiar FM.

Phase modulation is used for many satellite applications and for deep-space missions because of its excellent noise properties. These properties are also present in FM applications, but unlike FM, phase modulation may be obtained with simple circuits driven from a frequency-stable crystal-controlled carrier oscillator. The fre-

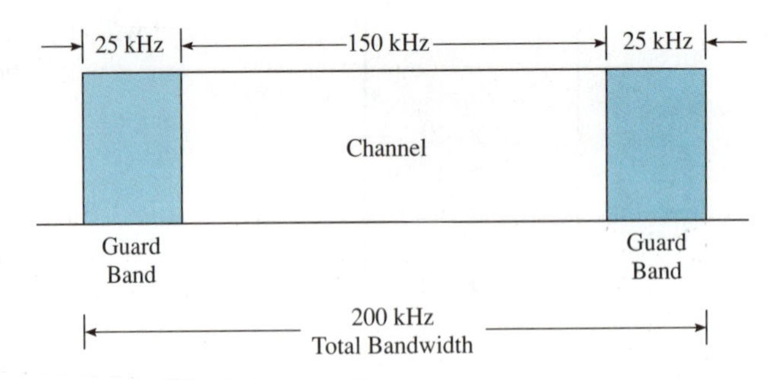

Figure 10.8 Broadcast channel allocations.

quency of a VCO is intentionally made highly variable to produce high deviations and a high modulation index. Consequently, the average carrier frequency tends to drift, resulting in the phase-modulation scheme being set up. Recall that Equation (10.4) described phase modulation as

$$V_c = A \cos [\omega_c t + K_p V_m(t)]$$

The phase variation θ is proportional to $V_m(t)$. For a linear phase modulation, $\theta = K_p V_m(t)$, where K_p is the modulation sensitivity described earlier. The modulation index m_p is equal to $\Delta\theta$ (peak), since $\Delta\theta$ is the peak phase deviation. So, for the narrow-band case, where the peak phase deviation (or modulation index) is less than 0.25,

$$V_p = A \cos (\omega_c t) - A[(m_p \cos (\omega_m t)] \sin (\omega_c t) \quad (10.15)$$

We have said many times that PM and FM are very similar to one another. If this is true, how do you really tell the two apart? There is only one real way that someone can tell PM from FM and that is to put them on the spectrum analyzer. When the signal is on the analyzer, the operator can increase or decrease the modulation in the system. If the modulation is PM, the modulation index will remain constant. If the modulation is FM, the modulation will change according to the change that was originally made. This is about the only definitive way of telling the two apart.

Having discussed both FM and PM, we should now look at the relative advantages of each, so you may know where to look for specific modulation methods. One large disadvantage of phase modulation, quoted for years in textbooks and by engineers, was that it required the use of a phase-locked loop for demodulation. This was considered a large expense for communications receiver manufacturers, and is probably the one real reason why we have FM radios today instead of PM radios. This argument doesn't apply today with current technology because PLL's are much more inexpensive than only a few years ago.

Phase modulation has better noise performance than FM because the PM signal is generated after the carrier oscillator has developed a very stable crystal-controlled frequency. This means that carrier drift is generally reduced with PM.

One advantage of FM over PM is that the FM VCO can directly produce high-index FM, which is not possible for PM without using multiplier circuits. Another advantage is that with FM any long-term change (slow telemetry changes, for example) can be detected, which is not possible with PM because the PLL has only a short-term phase memory. So each type of modulation has its advantages. It is up to the individual to match the advantages to the applications.

We have presented the concept of angle modulation and the terms important to understanding it. With this background, we now proceed to show how a carrier is frequency- (or phase-) modulated, transmitted to a receiver, and how the intelligence is removed from the modulated wave.

10.3 Transmission

As we did for amplitude modulation, we will now discuss the generation and transmission of angle-modulated (FM and PM) signals. That is, the angle modulation process.

Figure 10.9 shows a block diagram for a typical FM transmitter. The oscillator is an rf oscillator (similar to those discussed in Chapter 7), which generates the carrier for the modulation system. This oscillator may be an R-C, L-C, or crystal oscillator, depending on the parameters of the particular FM transmitter. The parameters may require that the oscillator be variable, in which case either the R-C oscillator or the Colpitts L-C oscillator would be appropriate. If the parameters require a very precise frequency output, the crystal oscillator would then be used. The signal thus generated is then sent to the modulator circuit as shown.

At the same time, a modulating signal is placed at the input to an amplifier (audio amplifier, for example). This amplifier is the same type you studied in many earlier electronics courses, an example of which is shown in Figure 10.10. You can see in the figure that a single transistor uses a voltage-divider type of bias (R_1 and R_2) and has coupling capacitors at the input and output of the stage (C_1 and C_3, respectively). The emitter resistor R_E is bypassed by a capacitor C_2, and the combination provides stabilization of the circuit over a wide range of temperatures.

Before continuing through the block diagram of Figure 10.9, we want to point out that an FM transmitter must be a low-noise type of system. Let us see how this quality is achieved.

To understand the differences between the effects of noise on FM as opposed to AM, consider what happens when a random noise enters an AM receiver. Any noise causes a change in the amplitude of a signal. This is a serious problem with AM because the intelligence (information) is contained in the amplitude variations of the signal. In the case of a frequency-modulated signal, the noise still produces an amplitude change, which is usually a rapid spike in the response. This spike, however, does not affect the information because an FM signal has its information contained in frequency changes, not amplitude changes. Although the spike does not affect the

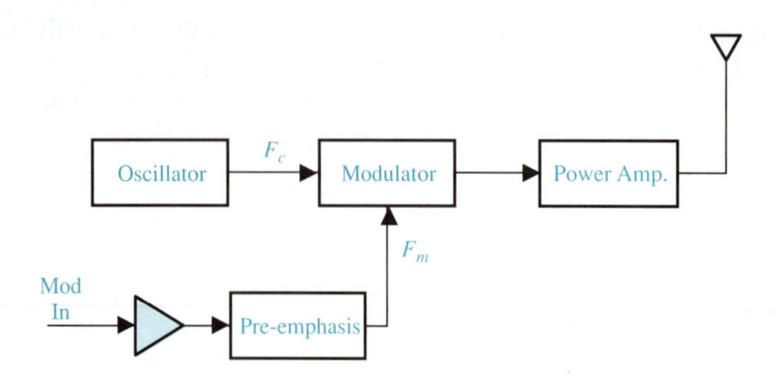

Figure 10.9 FM transmitter.

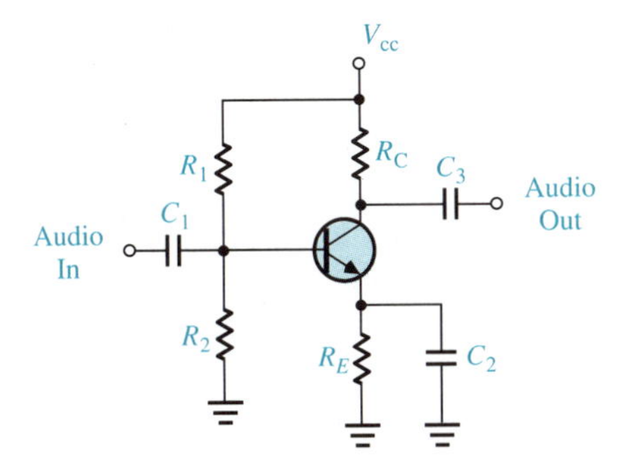

Figure 10.10 Audio amplifier.

signal with regards to amplitude, it will create a small phase shift in the signal, which cannot be removed from the system. The amount of shift (or deviation) occurring can be calculated as follows:

$$\delta = \phi F \tag{10.16}$$

Where: δ = Frequency deviation (in hertz)
ϕ = Phase shift (in radians)
F = Intelligence signal frequency (in hertz)

This is a case where a direct change in phase (PM) by the noise signal causes an indirect frequency deviation (FM). To calculate the amount of phase shift that results from noise, we need to know the signal-to-noise ratio (S/N) and proceed as follows:

$$\phi = \sin^{-1}\frac{N}{S} \tag{10.17}$$

Where ϕ is expressed in degrees (radians = degrees/57.3).

(Note that Equation 10.17 shows N/S, that is, noise-to-signal ratio, the reciprocal of S/N)

To see how noise can affect an FM system, consider the following example.

Example 10.7

Determine the worst-case output S/N for a broadcast FM with an intelligence signal frequency of 6.2 kHz and an input S/N = 2.6:1.

Solution

For an input S/N = 2.6:1,

$$\phi = \sin^{-1}\frac{N}{S}$$

$$= \sin^{-1}\frac{1}{2.6}$$

$$= 22.62° = 0.39 \text{ radians}$$

So,

$$\delta = \phi F$$
$$= 0.39 \times 6.2 \times 10^3$$
$$= 2.418 \text{ kHz}$$

Recall that the maximum deviation for a broadcast FM receiver is 75 kHz, so the worst-case S/N is

$$\frac{S}{N} = \frac{75 \text{ kHz}}{2.418 \text{ kHz}}$$
$$= 31.10 \text{ (14.92 dB)}$$

If we are not dealing with broadcast FM receivers, the maximum deviation will not be 75 kHz. You should be aware of what the restrictions are for the system you are using.

Example 10.8

If the problem in Example 10.7 were the same, except the maximum deviation was reduced to 25 kHz, what would be the worst-case S/N?

Solution

$$\phi = 0.39$$
$$\delta = 2.418 \text{ kHz}$$

So,

$$\frac{S}{N} = \frac{25 \text{ kHz}}{2.418 \text{ kHz}}$$
$$= 10.34 \text{(10.14 dB)}$$

Noise performance can be further improved by decreasing the maximum deviation to approximately 10 kHz (This is for narrow-band FM operation). This also reduces the maximum intelligence signal frequency that can be used. You must, therefore, make some trade-off in your system to get the best performance.

The output of the audio amplifier in Figure 10.9 is now sent to what is called a **preemphasis circuit**, which is used in FM transmitters to boost the amplitude of the high-frequency modulating signals prior to the modulation process. It is usually a high-pass filter that enhances the higher frequencies and levels out the modulating signal level. We have thus defined a preemphasis circuit, but what does it look like, and how does it work? We will answer these questions to help give you a good idea of what the circuit is and to enable you to recognize it when you come upon it in a system.

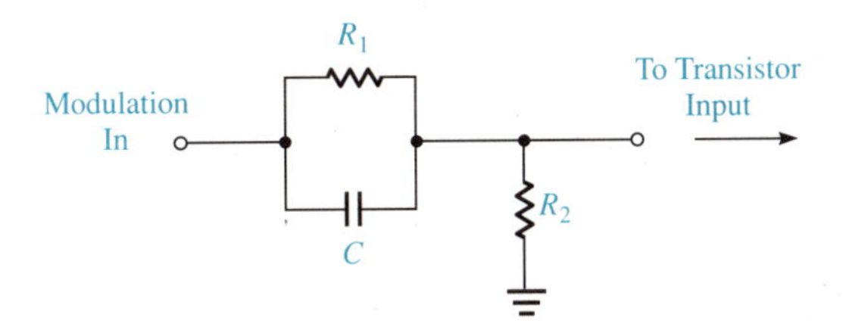

Figure 10.11 Preemphasis circuit.

To answer the first question, the preemphasis circuit increases the gain of the system at the higher frequencies. This is necessary because the ability of an FM system to suppress noise at high frequencies is greatly reduced, and there is therefore much worse noise performance as the frequencies increase. To overcome this discrepancy, the gain at high frequencies is increased proportionally to effectively increase the signal-to-noise ratio and improve the noise performance of the entire system. (Signal-to-noise ratio was discussed in Chapter 6.)

The circuit itself looks much like the one shown in Figure 10.11. It is a high-pass circuit, consisting of a parallel R-C combination (R_1, C) and a resistor to ground R_2. This type of circuit arrangement results in two resonant frequencies that need to be calculated, as follows:

$$F_1 = \frac{1}{2\pi R_1 C} \tag{10.18}$$

and

$$F_2 = \frac{R_1 + R_2}{2\pi R_1 R_2 C} \tag{10.19}$$

Where: F_1 = Lower cutoff frequency
F_2 = Upper cutoff frequency

The frequencies presented in Equations (10.18) and (10.19) are illustrated in Figure 10.12. For a properly operating preemphasis circuit the frequencies are $F_1 = 2.12$ kHz and $F_2 = 30$ kHz.

The operation of the circuit is as follows:

1. The modulating signal, usually an audio signal, is applied to the input of the circuit.
2. The time constant τ, which is the product of R_1C in Figure 10.11, is set to equal 75 μs (in Europe this time constant is 150 μs).
3. A constant amplitude input signal will result in a constant increase in the output voltage for frequencies above 2.12 kHz, as shown in Figure 10.12.
4. This increase produces a larger carrier deviation and consequently a larger m_f, which results in a preemphasis of the higher audio frequencies.

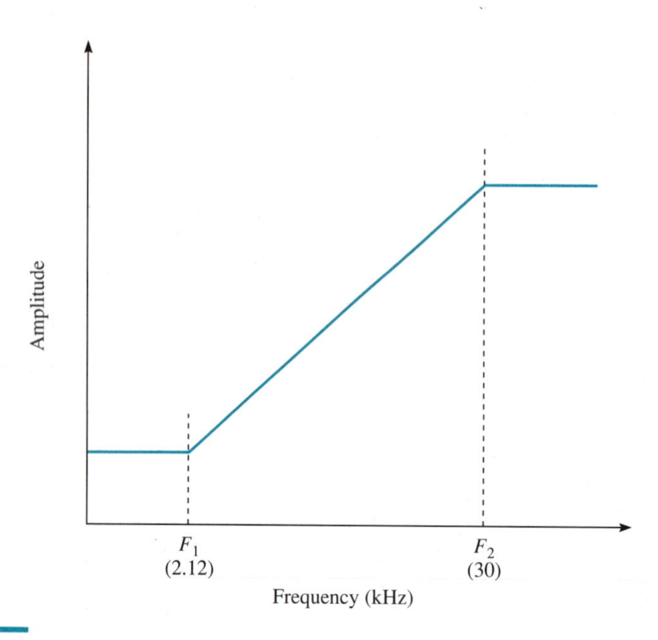

Figure 10.12 Preemphasis frequency response.

5. There is an upper limit placed on the system of 30 kHz so that the increase will not be too great over a wide band of frequencies and possibly cause oscillations at some higher frequencies.

Example 10.9

If the resistor R_2 was a potentiometer and was set to the proper value of 5.6 kΩ and then accidentally bumped so that this now read 75 kΩ, what would be the result? ($C = 1$ nf)

Solution

The new value of F_2 would now need to be calculated by using Equation (10.19). F_1 is not affected because R_2 is only in the equation for F_2.

$$F_2 = \frac{R_1 + R_2}{2\pi R_1 R_2 C}$$

$$= \frac{75 \times 10^3 + 75 \times 10^3}{2\pi(75 \times 10^3)(75 \times 10^3)(1 \times 10^{-9})}$$

$$= \frac{150 \times 10^3}{35.325}$$

$$= 4.236 \times 10^3$$

$$= 4.236 \text{ kHz}$$

The result is no preemphasis because the first cutoff is 2.12 kHz and the second would be at 4.236 kHz as shown in Figure 10.13.

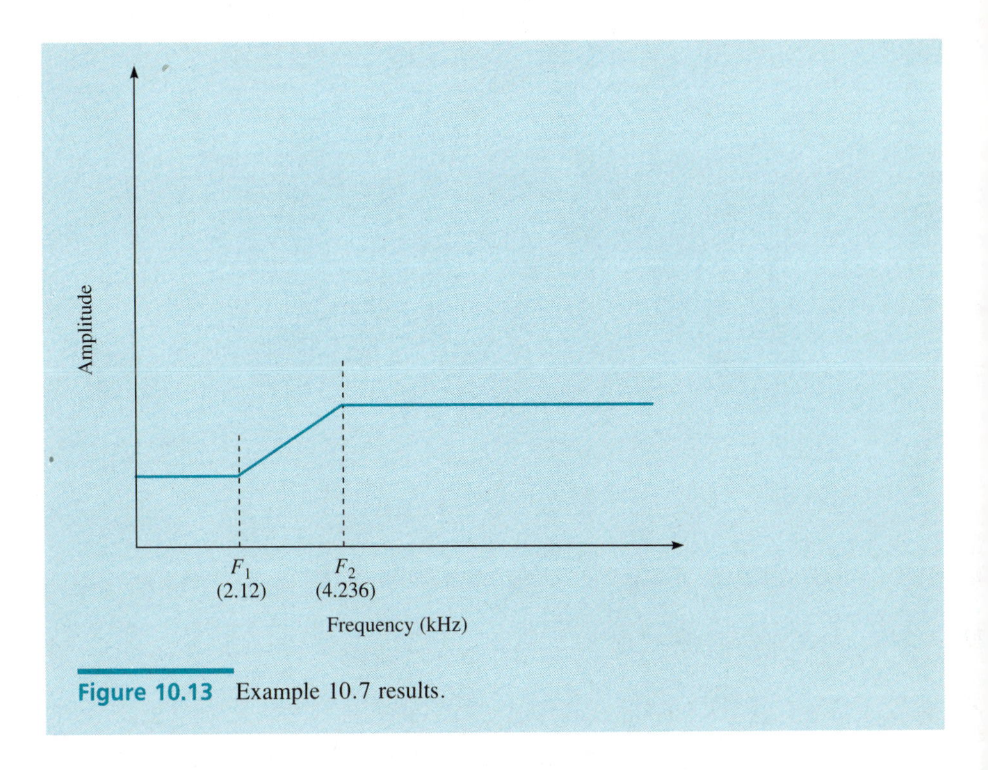

Figure 10.13 Example 10.7 results.

A variation in the preemphasis (and deemphasis in the receiver) is the **Dolby system**, which is used when the 75-μs time constant referred to previously for preemphasis is reduced at the transmitter to 25 μs. This increases the cutoff frequencies significantly and, thus, emphasizes them even more than the preemphasis circuit does.

The Dolby system is considered a dynamic as opposed to a passive system for the conventional preemphasis circuitry, meaning that the amount of preemphasis will vary depending on the level (or loudness) of the signal. Preemphasis in previous discussions did not vary with level. This means that there are varying degrees of high-frequency boost depending on the signal level. Typical examples of boost may be as much as 10 dB for a signal from 1 to 10 kHz at an input level of -40 dB, as opposed to virtually no boost for the same frequency range with an input of -5 dB or 0 dB. Thus, there is a dynamic reaction to noise level with the Dolby system.

The compensated level from the preemphasis circuit is now sent to the modulator circuit, which is the heart of the FM transmitter.

There are a number of ways that the carrier and modulating signal can be combined to form a frequency-modulated (or phase-modulated) system. Generally, there is one component that is a vital part of that combining process—the **varactor diode** (also called a **varicap**). This is a two-element device that varies the capacitive reactance of the diode as the bias voltage is changed. The diode is actually inserted into a circuit as a variable component, and rather than a physical variation, the reactance is varied with a voltage. This variable reactance is provided by the junction capacitance C_j, which varies as the bias across the junction is changed. A typical

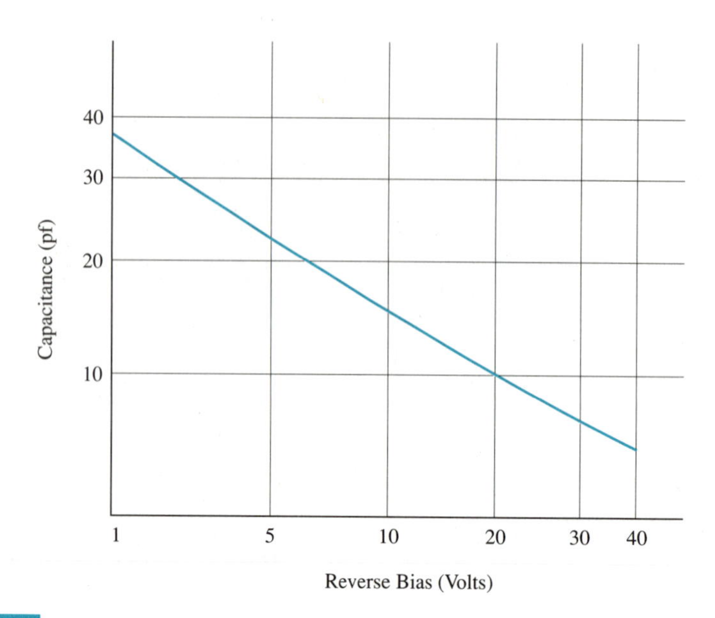

Figure 10.14 Varactor diode curve.

curve of bias voltage versus capacitance is shown in Figure 10.14. In contrast to how many of the semiconductor devices are biased, the varactor diode works when it is *reverse* biased. Taking these characteristics into consideration, we conclude that the varactor diode is a nonlinear device. This nonlinearity can produce three different effects. One effect is the modulation of signals through variations in the reactance of the diode. The second is the generation of harmonics of the input signal. The third effect is the mixing action resulting when two signals are applied. The effect of interest to us is the modulation effect.

The active element of a varactor diode consists of a semiconductor wafer containing a junction. An equivalent circuit for the diode is shown in Figure 10.15(a), with the schematic symbol shown in Figure 10.15(b). The parameters are defined as follows:

- C_j is junction capacitance, a function of the applied voltage.
- R_j is junction resistance, in parallel with C_j and also a function of the applied voltage.
- R_s is series resistance, it may be a function of the bias. Includes the resistance of the semiconductor on either side of the junction through which the conduction current passed and the resistance of the electrical contact of the wafer.

Typical values for these parameters are: $C_j(0)$, the junction capacitance at 0 V bias, ranging from 2 to 5.5 pF; $R_j(0)$, usually greater than 10 MΩ (megohm); and $R_s(0)$, ranging from 0.45 to 0.7 Ω.

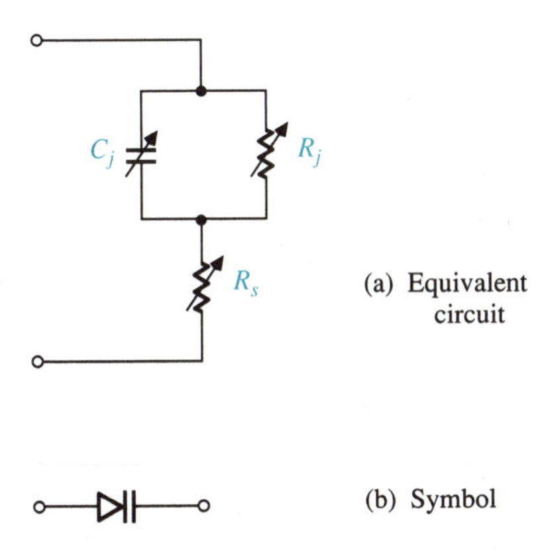

Figure 10.15 Varactor diode.

Having established the basic function and design of a varactor diode, let us now look at some circuits that use such a device. A basic varactor diode modulator, seen in Figure 10.16, shows the rf carrier applied to the primary of a transformer. The signal is then coupled through the transformer to a tank circuit made up of some value of inductance (L) and the varactor diode. (C_1 is an rf bypass capacitor.) The modulating signal is also applied to this tank and places a bias on the varactor diode, thereby varying the capacitance of the diode and, thus, the resonant frequency of the tank circuit. This changes the rf carrier's frequency at the rate of the modulating signal. Figure 10.16 is probably one of the most basic types of circuits illustrating the functioning of the varactor diode and how it achieves a frequency-modulation scheme. This type of circuit is also used in many basic electronics circuit courses to introduce a varactor diode. You may recognize it from one of these courses, except that the modulation signal was then simply a dc voltage that biased the diode.

Another type of modulator used for FM (or PM) signals is the **reactance modulator**, shown in Figure 10.17. This is a basic amplifier circuit that has its operating point changed by the modulating signal, causing a nonlinear operation and producing the modulation at the output of the device. Recall from introductory transistor courses that there is an operating, or Q, point set so that a linear response is achieved for the

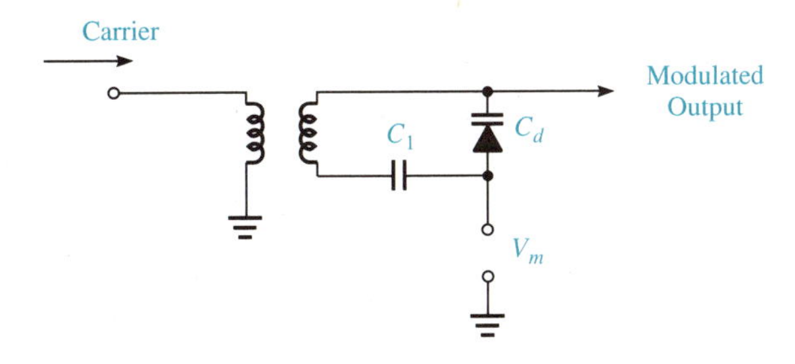

Figure 10.16 Varactor diode modulator.

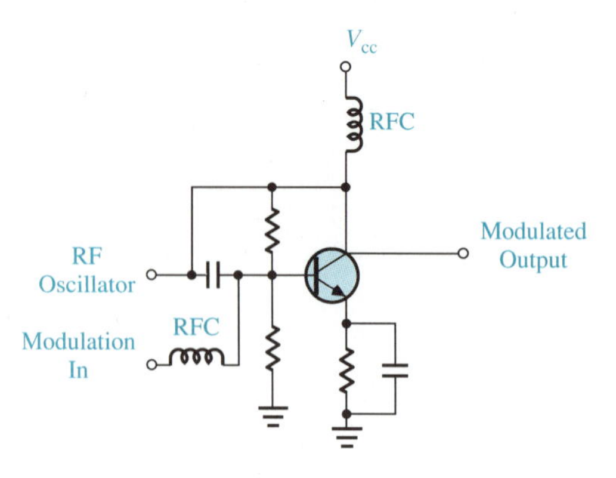

Figure 10.17 Reactance modulator.

circuit. This is where an input to a device produces a corresponding output for that same device. Figure 10.18(a) shows a linear response for a system. Recall from previous discussions that when the system is linear, there is simply a direct relationship between input and output of a device.

In contrast to the response of Figure 10.18(a), is the response shown in Figure 10.18(b), which is the response for a circuit driven into a nonlinear region. The circuit may be driven into this region simply by overdriving the input (too much power), or it may be driven there by changing the operating, or Q, point on the transistor. The latter is what is happening in the reactance modulator of Figure 10.17. The modulating signal causes the Q point of the transistor to change with the signal and drive the system into a nonlinear operation. Once the device is nonlinear, there will be a mixing of the two signals (carrier and modulating signal), creating the final FM signal. Thus, the reactance modulator takes advantage of a very basic theory of semiconductor devices to produce an efficient and useful FM modulator.

The circuit of Figure 10.17, although shown using a bipolar junction transistor, may also be used with FET's.

The modulators discussed above are all used for direct FM, that is, angle modulation in which the frequency of the carrier is varied directly by the modulating signal. As mentioned before, however, an indirect FM can also be used, wherein the frequency of the carrier is indirectly changed by the modulating signal. Generally, when we speak of indirect FM we mean a signal that directly modulates a carrier with PM and indirectly results in FM. The circuit usually used varies the phase of a crystal oscillator output. From previous discussions you should understand how varying the phase of a signal also varies the frequency. Thus, a phase-modulated crystal oscillator will produce indirect frequency modulation.

One very important area where frequency modulation is used is commercial FM stereo. This concept was first approved by the FCC in 1961 and is very common today. As a matter of fact, you would probably be hard pressed to find a commercial FM station today that was not in stereo.

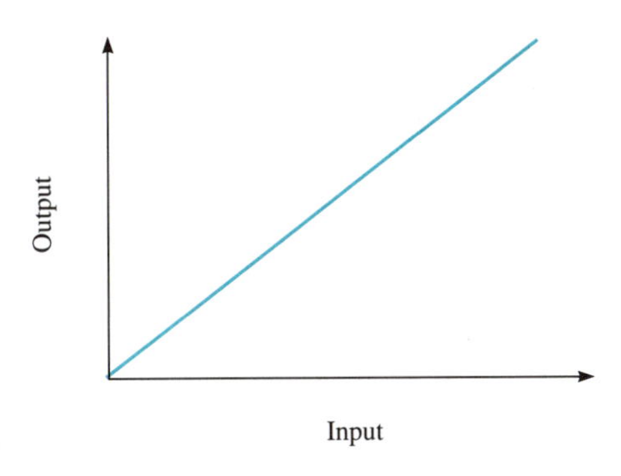

(a) Linear Operation

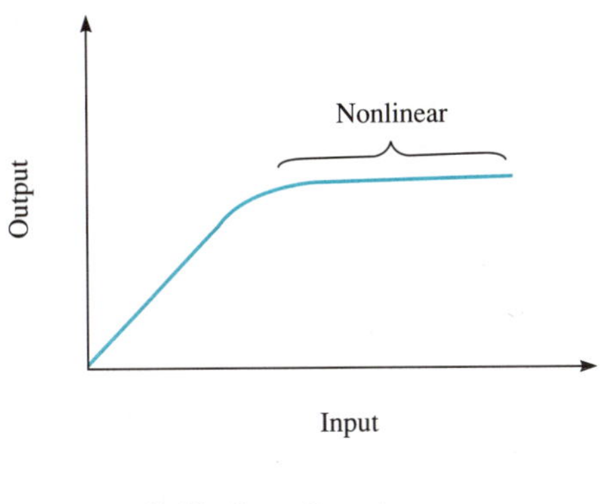

(b) Nonlinear Operation

Figure 10.18 Linear and nonlinear operation.

The major difference between ''plain'' FM and stereo FM, of course, is that there are two channels, left and right, broadcast for stereo FM as opposed to a single channel for conventional FM. Beyond that basic observation is the spectrum used to accomplish this. Stereo FM supplies twice as much information in a modulated system than previously sent. This amount of information is to be placed in the same 200-kHz bandwidth described in Figure 10.8. **Frequency-division multiplexing (FDM)** is used to accomplish this. In FDM different frequencies are transmitted simultaneously in the same medium, and many sources that may have originally occupied the same frequency band are transmitted over a single transmission medium. It is an analog multiplexing, which means that all of the information entering the system is analog and remains analog throughout. There is no analog-to-digital or digital-to-analog conversion to perform in this type of multiplexing. The most common form of FDM is one which you probably never associated with it—commercial AM and FM bands.

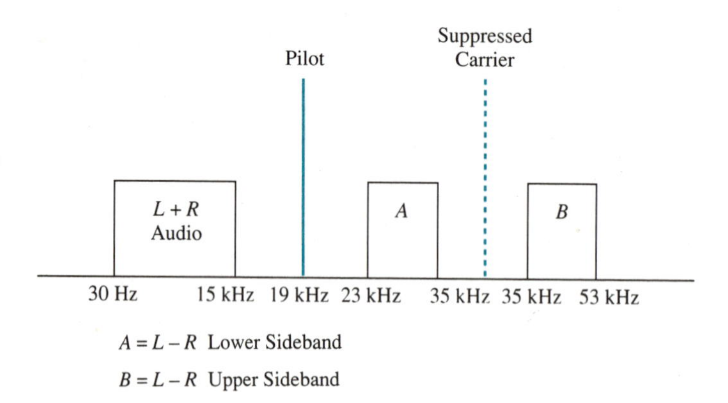

Figure 10.19 FM stereo spectrum.

With this arrangement each "station" is allowed a certain carrier frequency with a bandwidth within the standard broadcast bands of 550 to 1600 kHz for AM, and 88 to 108 MHz for FM. There is a frequency allocation for each station as described earlier for each band. Thus, both channels of the FM stereo system can be contained within the allotted 200-kHz bandwidth. The spectrum for FM stereo is shown in Figure 10.19, where L is the left channel, R is the right channel, $L - R$ is the left minus the right, and $L + R$ is the left plus the right channel. To see where these signals are generated, refer to Figure 10.20, which is a block diagram for an FM stereo transmitter.

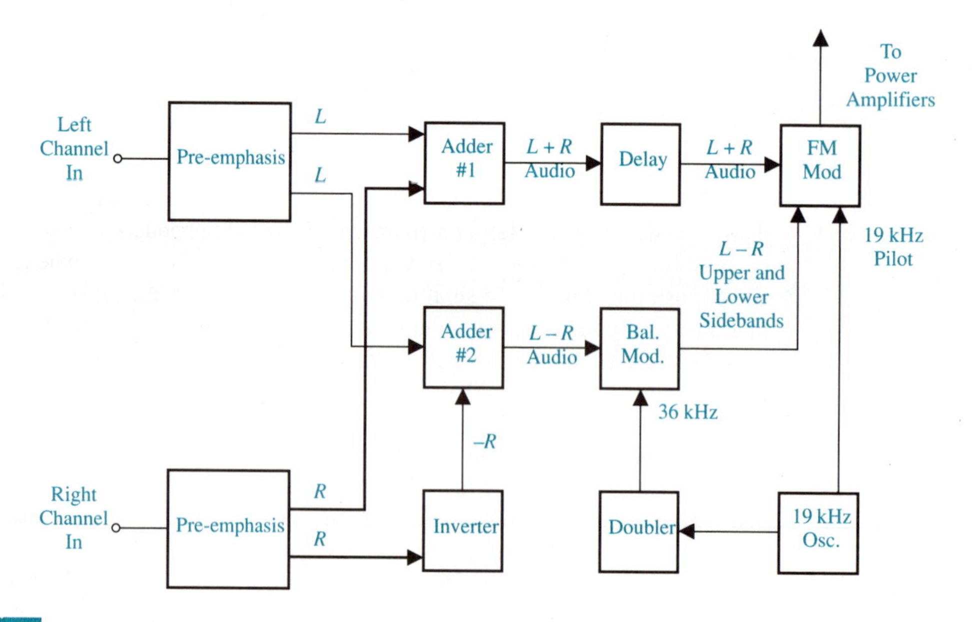

Figure 10.20 FM stereo transmitter.

You can see how each channel (left and right) is inserted at the input to the transmitter. Each channel has its own individual preemphasis circuit, which is the same type as described earlier. The left channel is then sent to two separate adder circuits where the left and right channels are summed together. The right-channel preemphasis output is sent to two different circuits: one to an adder just as the left channel was (adder #1), and the other to an inverter, resulting in an output of $-R$ to the adder circuit (adder #2). This combination of events results in an output from adder #1 of the $L + R$ (left plus right) audio and from adder #2 the $L - R$ (left minus right) audio.

The $L + R$ output is now delayed slightly so that both the $L + R$ and $L - R$ signals are applied to the FM modulator in the same phase. This is necessary because the $L - R$ signal will encounter a natural delay when sent through the balanced modulator. The $L - R$ signal, as mentioned, is sent to the balanced modulator and is mixed with a 38-kHz signal from a frequency doubler. The 38 kHz is obtained from a 19-kHz master oscillator, which is shown in Figure 10.19 in the frequency spectrum.

There are three signals applied to the FM modulator: the $L + R$ audio from the delay network, the upper and lower $L - R$ sidebands from the balanced modulator, and the 19-kHz oscillator signal. These signals make up the spectrum shown in Figure 10.19, which is now sent to the appropriate power amplifier to be transmitted into the air and received by home and automobile receivers.

It should be obvious that frequency stability is of vital importance for all of the systems discussed thus far. Stability for commercial FM stations, for example, is the assigned frequency ± 2 kHz with frequency stability mathematically expressed as $\Delta F_c/F_c$, in parts per million (ppm). As an example, a frequency of 100 MHz that varies ± 2 kHz will exhibit a stability of 2×10^{-5} or 20 ppm. To maintain this frequency stability many systems use crystal oscillators. Other systems that do not use crystal oscillators will sometimes use ovens to maintain a constant temperature for the oscillator and thus a constant frequency, or will use a technique called **automatic frequency control (AFC)**. Chapter 8 on phase-locked loops introduced the concept of AFC, a feedback system that compares the frequency at a specific output point with that of a reference frequency in a comparator. Any deviation beyond a certain limit (the stability limit) will change a VCO frequency to bring the system within the prescribed limits. This is the circuit that is switched in whenever you push the button on your FM receiver marked AFC.

We have presented the FM transmitter and shown how it differs from the AM transmitter. We also described the components necessary for providing good quality, reliable frequency-modulated signals. Now we will study FM reception.

10.4 Reception

A receiver for FM or PM looks very similar to that of the AM signal shown in Chapter 9 because the superheterodyne scheme is widely used for both systems. A block diagram for an angle-modulated receiver is shown in Figure 10.21. The front end of

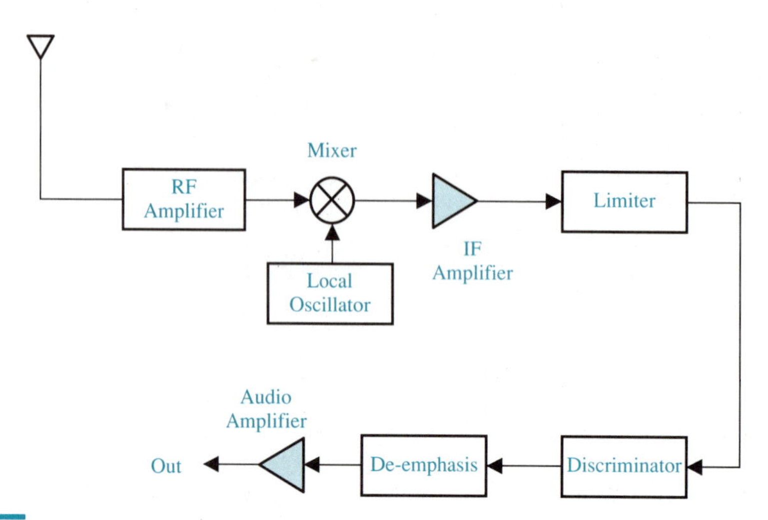

Figure 10.21 FM (PM) receiver.

the diagram should be familiar because it is basically the same as that for the AM receiver. The rf amplifier (the same type as described in Chapters 7 and 9) is used to amplify the input signal and to establish a noise figure for the entire receiver. The mixer–local oscillator combination mixes the radio frequency down to an intermediate frequency (10.7 MHz for commercial FM) that is much easier to work with. Recall from previous discussions that nonlinearity causes the mixer circuit to perform the mixing operations. The next block is a limiter circuit that ensures that the FM signal is always at the same level and that no amplitude modulation is superimposed on the angle-modulated signal. This is vital to the receiver because the FM receiver, or any FM system, must have a signal that varies only in frequency (or phase), not amplitude. The limiter circuit guarantees this. It is basically an overdriven i-f amplifier that essentially limits its output as soon as a specific value of input is reached. Recall from previous discussions how an amplifier can be driven into a compression region where a higher input level will not produce any more output. In that example we were actually discussing a limiter.

Another means of limiting is a diode that conducts at a certain level and absorbs energy rather than passing it on to another circuit. This is a common method used at the front end of a receiver to protect it from high-level input signals. The limiter in an angle-modulated receiver is designed to eliminate any amplitude variations and allow the system to have only frequency or phase variations.

If the signal progressing through the receiver of Figure 10.21 was checked before and after the limiter, we would notice two interesting things. First, and obviously, any amplitude variations prior to the limiter would not be present at the output. Second, the noise measured before and after the limiter would be considerably different. The action of the limiter performs a *quieting* effect on the signal; that is, it results in much less noise appearing at the output of the limiter than was present at the input. This quieting is related to **receiver sensitivity**, which is defined as the input

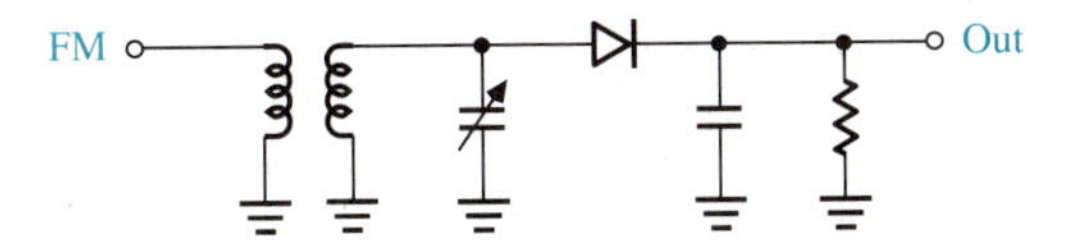

Figure 10.22 Slope detector.

level necessary to produce a certain amount of quieting (usually 30 dB). Thus, there is much more to a limiter than simply a circuit that squeezes an input level down to a specific amplitude. This circuit can either make the receiver perform admirably or fail all together.

Whereas the modulator was the heart of the angle-modulated transmitter, the **discriminator** is the heart of the angle-modulated receiver. This is where the intelligence placed on the rf carrier in the transmitter by the modulator is removed and sent on to be processed. The circuits in this block are frequency-dependent and produce an output voltage proportional to the instantaneous frequency at the input. Some of the types of discriminators are:

- ☐ Slope detector
- ☐ Foster-Seeley discriminator
- ☐ Ratio detector
- ☐ Quadrature detector

The slope detector, Foster-Seeley discriminator, and the ratio detector are **tuned-frequency discriminators** that convert FM to AM and then demodulate it with a peak detector. These discriminators revert to the process of amplitude modulation for a demodulation process because, as discussed in Chapter 9, the process of demodulating an AM signal by means of a peak detector is basic and efficient. Therefore, the most common types of FM discriminators translate the FM signal to AM and proceed as in Chapter 9 with the peak detector.

The **slope detector** is shown in Figure 10.22. The tank circuit shown in Figure 10.23 is tuned to F_0 where the output is maximum. The i-f center frequency F_c is usually set so it falls on the most linear portion of the response curve. As the frequency

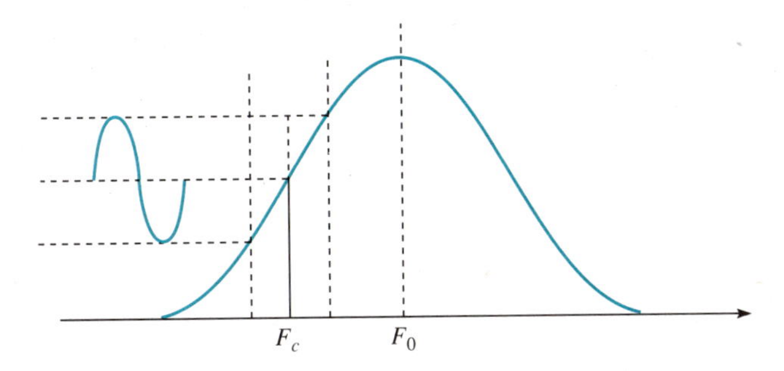

Figure 10.23 V_{out} vs. frequency curve.

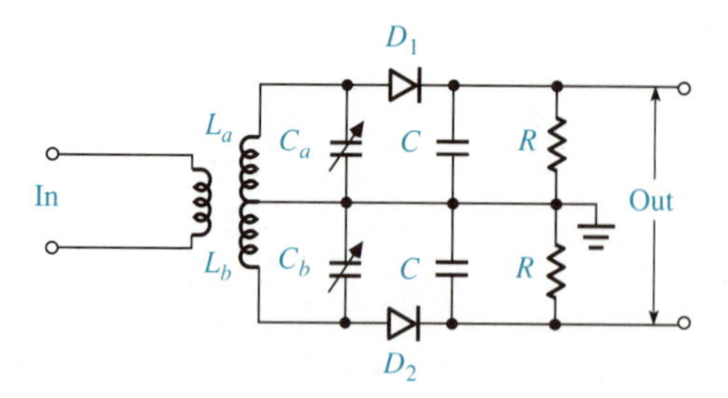

Figure 10.24 Balanced slope detector.

increases the output voltage increases, and as the frequency decreases, the output voltage decreases. In this way, the circuit converts the frequency variations to amplitude variations. If there is any amount of frequency variation in the circuit, the response is usually nonlinear, and for this reason this circuit is not used for many demodulation applications.

A modification of the slope detector described above is the balanced slope detector shown in Figure 10.24. The **balanced slope detector** is actually two slope detectors connected in parallel and fed 180° out of phase. The components L_a, C_a, L_b, and C_b perform the FM-to-AM conversion that was explained in the single slope detector. The balanced peak detector consists of D_1, C_1, R_1, D_2, C_2, and R_2.

At the intermediate frequency of the receiver, the outputs of the detectors are equal and out of phase. They will cancel and $V_{out} = 0$ V. If the signal is below the resonant frequency, the output from the lower tank circuit is greater and V_{out} is negative. Similarly, if the frequency is above the resonant frequency, the output from the upper tank is greater and V_{out} is positive. Figure 10.25 shows the output curve for

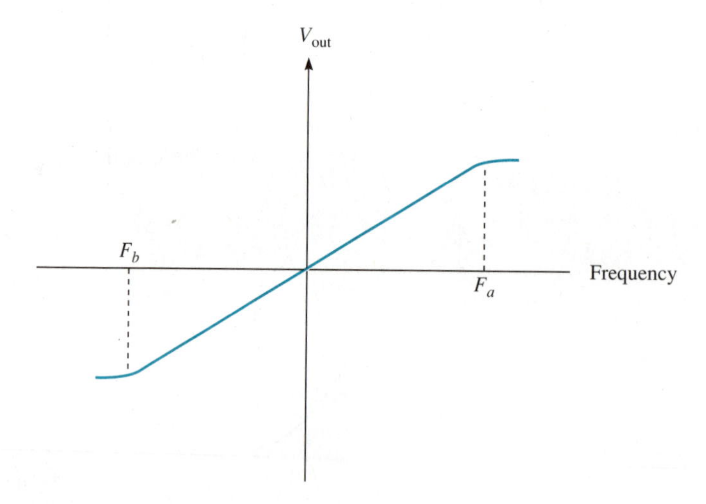

Figure 10.25 Balanced slope detector response curve.

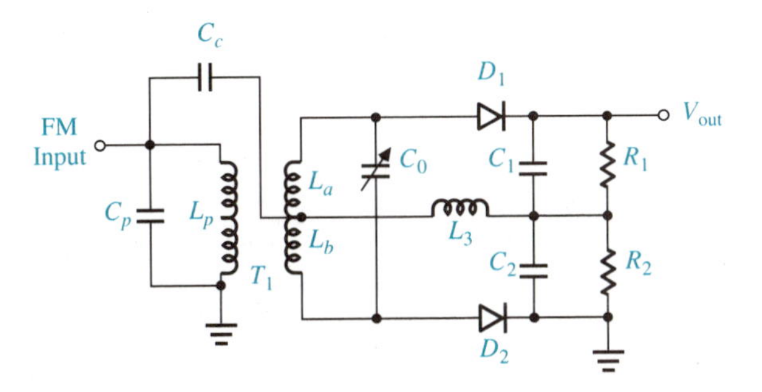

Figure 10.26 Foster-Seeley discriminator.

these conditions. As can be seen in the figure, the more the frequency varies from the intermediate frequency, the higher the amplitude of V_{out}. Thus, there is the frequency-to-amplitude conversion necessary for operation.

The **Foster-Seeley discriminator** is very similar to the balanced slope detector. In the circuit shown in Figure 10.26 notice the differences from the detectors covered previously. There is a coupling capacitor between the primary and the secondary of the transformer; there is an adjustable capacitor from D_1 to D_2; and there is an inductor in the center line of the circuit. The coupling capacitor C_c between the primary and secondary of the transformer is a short circuit to the carrier frequency so that the carrier can appear directly across the center line inductor L_4, which is an rf choke (RFC). The capacitor from D_1 to D_2 adjusts the tank circuit consisting of the combination of L_a, and L_b, and C_2. This is a high-Q resonant circuit. When the circuit is at resonance, the voltages across the secondary are 180° out of phase, canceling everything out. With $F_{in} > F_0$, the upper portion of the circuit is dominant and the diode D_1 produces the output. Similarly, when $F_{in} < F_0$, the lower portion of the circuit is dominant and D_2 produces the output. Thus, for the Foster-Seeley discriminator, the output voltage is directly proportional to the magnitude and direction of the frequency deviation. Once again, it involves converting FM to AM and then demodulating with a diode.

The **ratio detector**, shown in Figure 10.27(a), is a circuit that is basically immune to amplitude variations of the input signal. The circuit looks similar to the

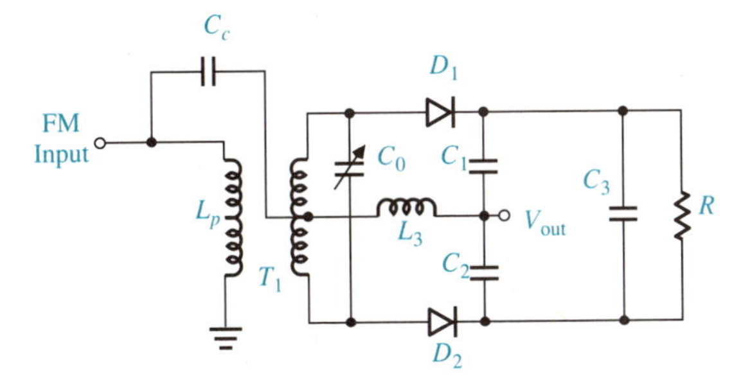

Figure 10.27 Ratio detector.

Foster-Seeley discriminator, but is much simpler in that the output is taken to ground, whereas the discriminator output was taken between two points.

The voltage-versus-frequency curve for the ratio detector is shown in Figure 10.27(b). This curve has the same basic shape as the other curves but is labeled as having an average positive voltage and maximum positive voltage. No negative voltages are indicated. One other change is that at resonance, V_{out} does *not* equal 0 V but is equal to one-half the voltage across the secondary winding of the transformer.

Because of its amplitude characteristics, the ratio detector is chosen many times for FM demodulation. The discriminator is sometimes chosen over the ratio detector because the discriminator has a more linear voltage-versus-frequency curve. The choice of detectors is dependent on the particular application and performance requirements.

The final demodulator is the **quadrature FM demodulator** shown in Figure 10.28. The circuit extracts information from the i-f waveform by multiplying two quadrature signals. (The term *quadrature* simply means 90° out of phase.) The circuit shown has three parts:

1. 90° Phase shifter
2. Tuned circuit
3. Product detector

The 90° phase shifter produces a signal in quadrature with the received i-f; the tuned circuit converts frequency variations to phase variations; and the product detector multiplies the received i-f by the phase-shifted signal. The 90° shift is produced by the combination of C_i, L_0, R_0, and C_0. This network has values that will produce the 90° shift at the appropriate frequency. The tank circuit, itself consisting of L_0, R_0, and C_0 is tuned to the i-f center frequency and produces an additional phase shift proportional to any frequency deviation. When the circuit is at resonance the tank circuit is purely resistive because the reactance of L_0 and C_0 cancel out. Any deviation from resonance produces a positive or negative phase shift so that the product-detector output voltage is proportional to the phase difference between the two input signals. The R and C component at the output make up a filter for the second harmonic of the

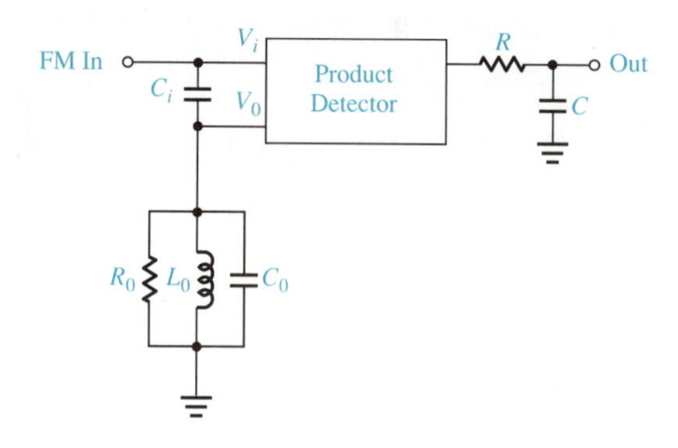

Figure 10.28 Quadrature FM demodulator.

input, which is produced in the product detector. With this filtered out, the final output is proportional to $V_0 \sin \phi$, where $\phi = \tan^{-1} pQ$ (p is the fractional frequency deviation, $2\pi F/F_0$, and Q is the quality factor of the tank). This combination of events produces a very efficient and useful way of demodulating FM or PM signals.

Returning to Figure 10.21, we examine the **deemphasis circuit**, which undoes the preemphasis discussed in Section 10.4. Because the preemphasis circuit uses a high-pass filter to enhance the high frequencies, the deemphasis circuit uses a low-pass circuit to attenuate these same high frequencies as shown in Figure 10.29(a). This is a simple R-C circuit like that discussed in Chapter 7. The formula for the deemphasis circuit is the same as that for F_1 of the preemphasis circuit; that is, $F = 1/2\pi RC$. The cutoff frequency must be the same as F_1 for the preemphasis circuit, 2.12 kHz. Thus, the values for R and C must be multiplied together for result in 75 μs. So, if we use 75 kΩ and 1.0 nF, the cutoff frequency will be 2.12 kHz with the response shown in Figure 10.29(b).

The final stage shown in Figure 10.21 is the audio amplifier. This is the same type of amplifier used in the transmitter, in AM systems, and anywhere that audio must be amplified. The amount of gain will depend on the level of the signal coming from the deemphasis circuit, and the level of signal is needed to drive the following circuits associated with the receiver.

Thus, the receiver is a fairly straightforward system that incorporates many of the circuits discussed in Chapters 7 and 9. These are, however, representative circuits, and the exact value of components used in such circuits will depend on the amount of gain and other parameters governed by the system being used.

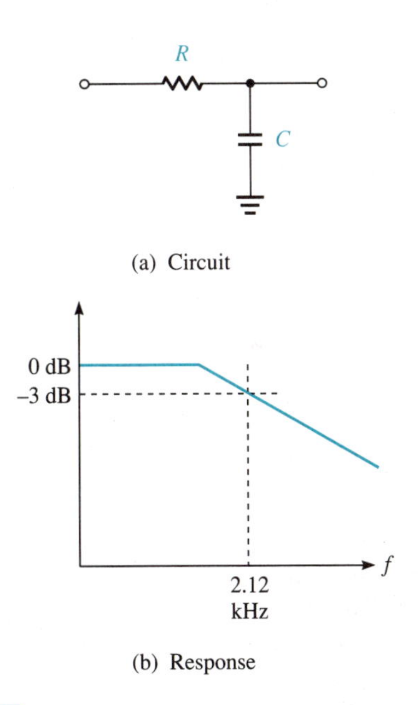

Figure 10.29 Deemphasis circuitry and response.

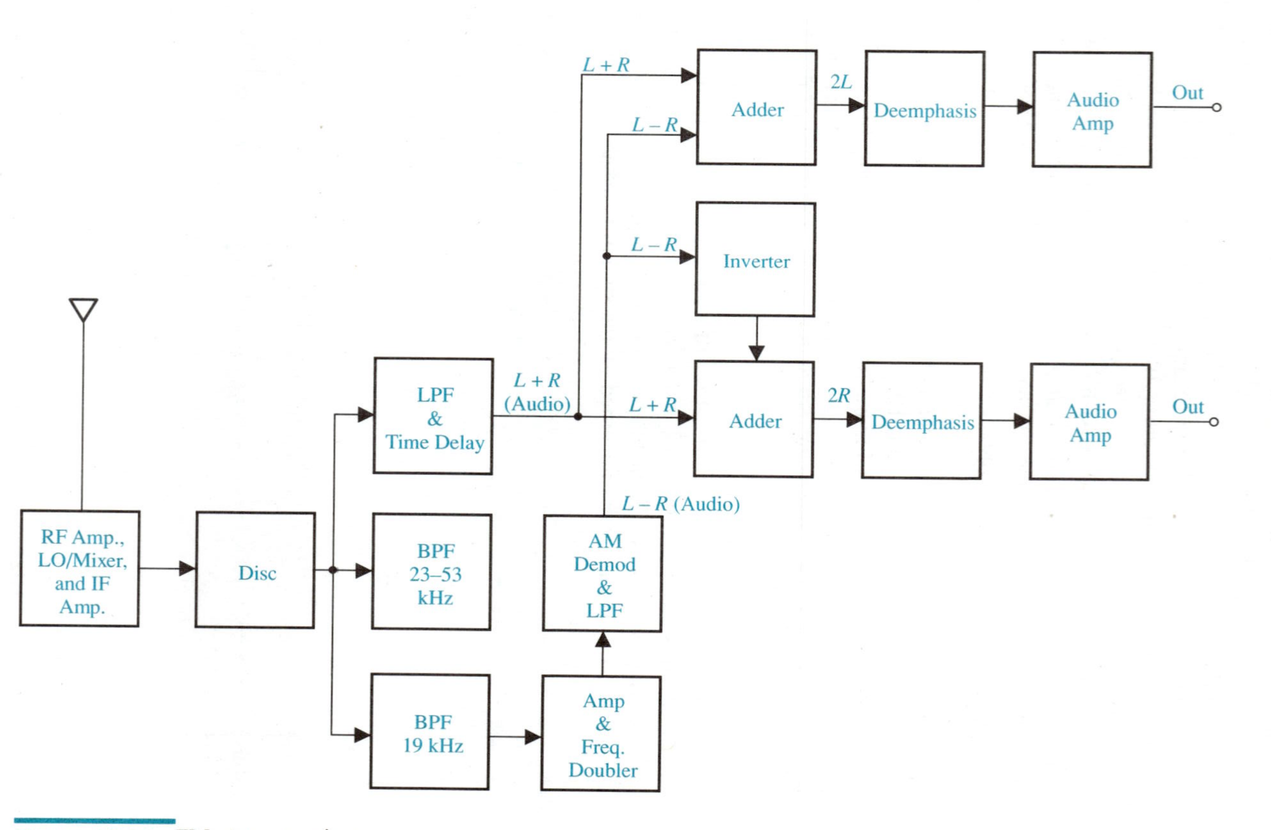

Figure 10.30 FM stereo receiver.

Before leaving the topic of FM receivers, we should look at how we are going to receive the FM stereo signal we developed in Section 10.3. Figure 10.30 is a block diagram of an FM stereo receiver. The input from the antenna is sent to an rf amplifier (as described earlier in the transmitter section), a local oscillator–mixer combination, and an i-f amplifier much the same as any other superheterodyne receiver. As in other FM receivers, a discriminator is the next block. If we were dealing with conventional monophonic FM, we would add an audio amplifier at this point, place a speaker at the amplifier output, and enjoy the programming. This is not the case, however, and we must add considerable circuitry to the receiver.

The output of the discriminator is now sent to three locations within the receiver. The first is to a low-pass filter and time-delay circuit that places everything back into phase as far as the $L - R$ and $L + R$ channels. The second route is a bandpass filter that selects the 23–53 kHz band of the $L - R$ channel. The third route is a bandpass filter that passes only the 19-kHz pilot signal. The output of this filter is sent to an amplifier and frequency doubler to create a 38-kHz signal that is applied to an amplitude modulator. The other signal applied to this modulator is the output of the 23–53 kHz bandpass filter. This output is now sent to a low-pass filter, and the $L - R$ audio is then sent to an adder and inverter as shown. The $L + R$ channel is also sent to adders as in the reverse of the FM transmitter. The result is a signal in one channel that is $2L$ and in the other that is $2R$. Each of these signals is then sent to its own deemphasis circuit and audio amplifiers. The result is a high quality FM stereo signal.

Thus, the FM receiver has now been discussed and explained so that the student may know how the FM communications system will do the tasks it is asked to perform. The theory of FM stereo should make the student much more aware of what is occurring when that "STEREO" light appears on the car radio panel.

10.5 Summary

In this chapter we presented the concepts and theory of angle modulation, including both frequency and phase modulation. Examples of applications of angle modulation were also included.

The FM transmitter was then explained using a block diagram. The ideas of FM modulators and preemphasis were presented, and the section concluded with a discussion of FM stereo transmitters.

Finally, the FM receiver was discussed, expanding the conventional superheterodyne concepts to fit the FM applications. The FM stereo receiver was also presented.

Questions

10.1 Introduction

1. Define **angle modulation**.
2. What is the difference between FM and PM?

10.2 Theory

3. What is the exact phase of the carrier called?
4. What is the exact frequency of the carrier called?
5. Why is the term **deviation** so important in angle modulation?
6. What is **deviation sensitivity**?
7. What determines narrow-band FM? Wide-band FM?
8. Draw a time-domain and frequency-domain picture of FM.
9. Define **informational bandwidth**.
10. Compare frequency and phase.
11. What is direct FM?
12. What is indirect FM?
13. Define **instantaneous phase**.
14. Define **instantaneous phase deviation**.
15. Define **instantaneous frequency**.
16. Define **instantaneous frequency deviation**.
17. What are the units for phase and frequency-deviation sensitivities?
18. What is **frequency deviation**?
19. Define modulation index for FM.
20. What are **Bessel functions**?
21. Explain the importance of having Bessel functions go to zero.
22. Why is an FM signal a constant-amplitude signal?
23. What is the difference between the informational bandwidth and the total bandwidth?
24. Define **residual FM**.
25. Name two applications of PM.
26. How is FM better than PM?
27. How is PM better than FM?

10.3 Transmission

28. What is a **preemphasis** circuit and why is it used?
29. Why is a varactor diode used for angle-modulation systems?
30. What type of oscillator would be used in an FM transmitter if the frequency were variable?
31. Why doesn't noise affect FM?
32. If the phase shift increases in an FM system, what happens to the frequency deviation?
33. Describe narrow-band FM in terms of maximum deviation.
34. How are the preemphasis frequencies calculated?
35. What is the time constant for a good preemphasis circuit?
36. Explain the operation of a preemphasis circuit.
37. What is a Dolby system?
38. How does the Dolby system differ from conventional preemphasis?
39. Explain the operation of a varactor diode.
40. How is a varactor diode biased?
41. Explain C_j in a varactor diode.
42. Explain R_j in a varactor diode.
43. Explain R_s in a varactor diode.
44. What does $C_j(0)$ mean for a varactor diode?
45. How is a varactor diode used in an FM modulator?
46. In the reactance modulator, how is the Q point varied?
47. Compare a reactance modulator and an overdriven amplifier.
48. How is phase modulation usually accomplished?

49. Describe the spectrum for FM stereo.
50. Describe frequency-division multiplexing (FDM).
51. How is FDM used for FM stereo?
52. How is the 38-kHz signal obtained in an FM stereo system?
53. Why is frequency stability so important in an FM system?
54. What is AFC?
55. Describe an AFC system.

10.4 Reception

56. Name three types of frequency demodulation circuits.
57. What is the intermediate frequency for a commercial FM receiver?
58. What is the primary purpose of a limiter?
59. What devices can be used to make a limiter?
60. What is the condition of the signal before and after the limiter?
61. Define **quieting**.
62. How is quieting related to sensitivity?
63. What is a **discriminator**?
64. Why is the practice of converting FM to AM used in FM demodulators?
65. How does the balanced slope detector differ from the standard slope detector?
66. What features make the Foster-Seeley discriminator a common FM demodulator?
67. Why is the ratio detector immune from amplitude variations of the input signal?
68. Name and describe three parts of the quadrature detector.
69. Define **quadrature**.
70. What is the purpose of the R-C circuit at the output of the quadrature detector?
71. Describe the operation of the deemphasis circuit.
72. Describe the basic operation of an FM stereo receiver.

Problems

10.2 Theory

1. A 5-MHz VCO with a sensitivity of 5 kHz per V is modulated with a 4-V peak, 3-kHz sinusoid. Find the maximum frequency deviation, peak phase deviation, and the modulation index for $V_m = 10 \cos [2\pi(3 \text{ kHz})t]$.
2. In Problem 1, what happens to the modulation index if we change VCO's to a 5-MHz oscillator with $K = 3$ kHz per V?
3. Find the percentage modulation if the deviation for an FM station is measured as ± 50 kHz.
4. If the maximum percentage of modulation allowed is 87% for a particular FM station, what is the maximum actual deviation this station can have?
5. Find the relative power in an FM signal with a modulation index of 2.4, 6.0, and 8.0.
6. Plot the percentage of error for the results of Problem 5.
7. If there is an FM station at 101.5 MHz, what would be the nearest frequency that another station could be placed above and below this station?

10.3 Transmission

8. Would a preemphasis circuit still provide preemphasis if $R_2 = 42$ kΩ?
9. If the operating point of a transistor in an FM modulator was kept so that linear operation was still in effect, what effect would this have on the circuit?

10. If the delay circuit was not put in the FM stereo circuit, what effect would this have?
11. Determine the worst-case S/N for a broadcast FM with an intelligence signal frequency of 4 kHz and an input $S/N = 1.3:1$.
12. If the worst-case S/N for an FM station was 12.6 and the intelligence signal frequency was 5.2 kHz, what is the input S/N?
13. If the system in Problem 11 had an intelligence signal frequency of 12 kHz and the maximum deviation was reduced to 12 kHz, what would be the worst-case output S/N?

10.4 Reception

14. If the slope detector was set so that F_c was not on the linear portion of the response curve, what would be the results?
15. The deemphasis circuit built for an FM receiver has only 82-kΩ resistors available. What is the effect of this value on the circuit?
16. If the 82-kΩ resistor in Problem 15 was used, what would the value of C need to be to result in proper deemphasis of the signal?

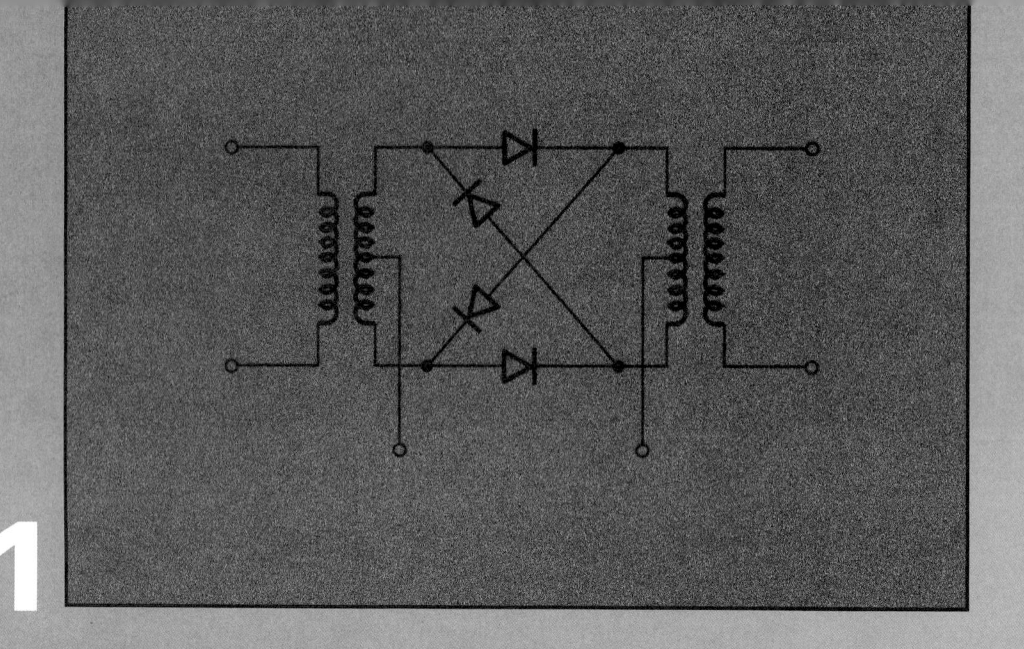

11

Outline

Objectives

- To introduce the advantages of using only one sideband in the transmission of electromagnetic energy
- To present SSB (single-sideband) theory and highlight the balanced modulator that makes such systems possible
- To cover both filter and phase-shift transmission methods, and their particular advantages in SSB systems
- To examine the means of receiving SSB signals

Key Terms

sideband
double-sideband, suppressed carrier (DSB-SC)
single-sideband, reduced carrier (SSB-RC)
vestigial sideband
product modulator
resonator
coherent SSB receiver
noncoherent SSB receiver
double-sideband, full carrier (DSB-FC)
single-sideband, suppressed carrier (SSB-SC)
independent sideband (ISB)
balanced modulator
transducer

Sideband Systems

11.1 Introduction

Having examined all of the features of AM and FM in the previous two chapters, you may wonder why any other type of communication system needs to be considered. Both of these schemes are excellent for many types of communications systems, but, like all systems, they have their limitations.

The one limitation that stands out is in the amplitude-modulation systems. You will recall that in Example 9.5 we calculated the necessary power in the carrier as 8.506 W and that in each of the two sidebands as 0.715 W. This means that of the total power necessary for the signal to be transmitted, over 85% was needed to supply a carrier for the intelligence. There could, therefore, be a considerable savings in power if the carrier was run at a very reduced level.

Also, recall that the AM signal had two sidebands, upper and lower, and the intelligence to be transmitted was exactly the same in both of them. Referring once again to our example, we can reduce the power requirements by another 7% if we eliminate one sideband. More importantly, however, by eliminating one sideband we can cut the bandwidth needed in half. This means that more channels can be placed in the space originally occupied by a double-sideband AM signal.

The sideband systems are based on the reduction in the carrier and elimination of one sideband. We will look at a variety of these systems and show how they can be used in communications systems.

Example 11.1

Determine the percentage of power savings when the carrier is suppressed in an AM signal modulated to 80%.

Solution

From Chapter 9, we use

$$\text{Power savings} = \frac{P_c}{P_c[1 + (m^2/2)]}$$

(Consider P_c to be unity). Thus,

$$\text{Power savings} = \frac{1}{1 + (0.8^2/2)}$$

$$= \frac{1}{1.32}$$

$$= 75.8\%$$

11.2 Theory

To set the stage for the discussions on sideband systems, we will use the conventional **double-sideband, full carrier, amplitude-modulation (DSB-FC)** theory as a base. DSB-FC, represented in Figure 11.1, shows a full-level carrier with an associated upper and lower sideband. The total bandwidth for this system, the width from point *A* to point *B* in the figure, is equivalent to twice the modulation frequency of the system. If we have a carrier of 10 MHz and a modulating frequency of 10 kHz, for example, point *A* would be 9.999 MHz and point *B* would be 10.001 MHz. The bandwidth of this system would be 10.001 − 9.999, or 2 kHz, which is two times the modulating frequency.

Another point to mention in Figure 11.1 is that the carrier amplitude is large relative to the amplitude of the sidebands. Recall from our discussions in the introduction that 85% of the total power in our example was contained in the carrier alone. The remaining 15% was in the sidebands, and thus, the intelligence of the signal. This is shown in the figure and is present on a spectrum analyzer when DSB-FC signals are displayed.

We have said that the carrier power is a significant amount of the total power needed to transmit intelligence in a communications system operating in a double-

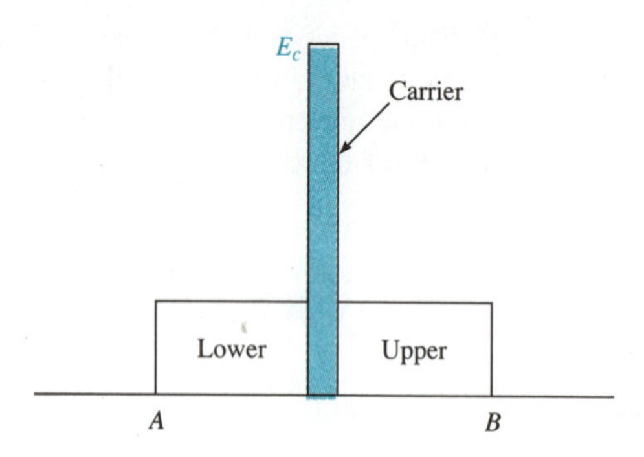

Figure 11.1 Double-sideband full carrier (DSB-FC).

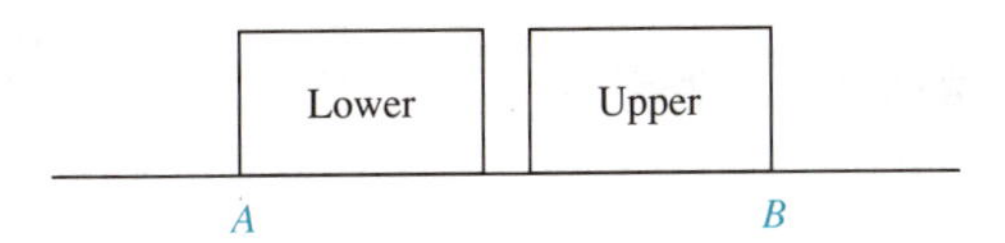

Figure 11.2 Double-sideband suppressed carrier (DSB-SC).

sideband, full-carrier mode. It only seems natural to take the carrier and reduce its amplitude to conserve the power in the system and make the entire system more efficient. This can be done by *suppressing* the carrier, meaning that we produce **double-sideband, suppressed carrier (DSB-SC) signals** (as shown in Fig. 11.2). Figure 11.2 shows the same spectrum as in Figure 11.1, but the carrier has been removed. It should be obvious that this type system results in a considerable savings in power and that the bandwidth of this system is exactly the same as for the DSB-FC system, $2F_m$. This scheme, therefore provides a large savings in power but no reduction in bandwidth.

Next we will consider the **single-sideband, suppressed carrier (SSB-SC) system** shown in Figure 11.3. This figure is actually Figure 11.2 with the lower sideband removed. In addition to the drastic power savings with this system, there is also a conservation of bandwidth. The figure shows that only half of the band is needed for this type of transmission. Thus, the total bandwidth for this case is only F_m, not $2F_m$.

Example 11.2

If we have a system using DSB-FC AM with a carrier of 2 MHz and a modulating signal of 15 kHz, what will be the total bandwidth of this system if it is converted to SSB-SC?

Solution

For SSB-SC, the total bandwidth is equal to F_m. Therefore, the bandwidth is 15 kHz.

Despite all the advantages of the systems shown in Figures 11.2 and 11.3, there are some drawbacks that must be considered. It is very nice to reduce the power consumption of a system, especially if we have a space application where power is at a premium. Remember, however, that when the carrier power is reduced, the range of the entire system is also reduced. Thus, if a specific range for the communications system is needed, the carrier must be present in some form. The usual form is the **single-sideband, reduced carrier (SSB-RC)**. This arrangement, shown in Figure 11.4, represents the best of both worlds. It reduces the level of the carrier (usually the carrier is only 10% as large as the original DSB-FC carrier), and only uses one

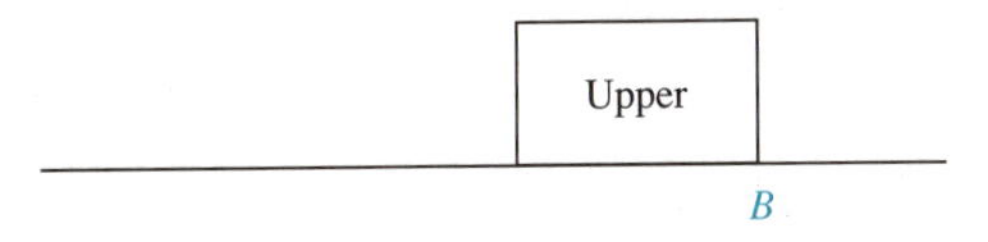

Figure 11.3 Single-sideband suppressed carrier (SSB-SC).

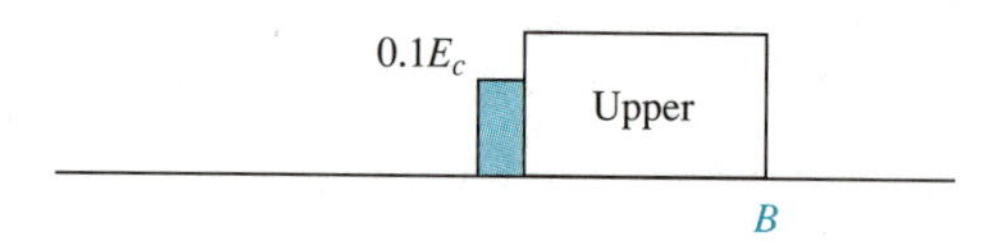

Figure 11.4 Single-sideband reduced carrier (SSB-RC).

sideband (the upper sideband), which reduces the bandwidth. This system will, obviously, reduce the range of the system, but it will not reduce it as far as the suppressed-carrier system does. The SSB-RC scheme is used when there is a moderate range of transmission and many channels are needed within a specific transmission band. To understand this last statement, consider that for a double-sideband system, a certain number of bands can be placed within a specified frequency allocation. If we are allocated 1 to 2 MHz over which we can broadcast, and the allowed double-sideband bandwidth is 5 kHz, a total of 200 channels are available for use. If, on the other hand we used a single-sideband system, the bandwidth would be cut to 2.5 kHz for each channel and 400 channels could possibly be used. Thus, if a band is beginning to fill up, the conversion from double to single sideband may be a viable alternative. The SSB-RC systems will have less power consumption (because of the reduced carrier) and will have the same amount of information as the double-sideband system because all the intelligence is contained in both sidebands. There is no loss of information (intelligence) when one sideband is removed.

To understand further the comparison between DSB-FC and SSB signals, refer to Table 11.1.

The first parameter in the table relates the facts we have been explaining previously. That is, 85% of the total power is in the carrier of the signal. Thus, a factor of 1.2, as opposed to only 0.2, is presented for relative power.

The second parameter is the system bandwidth. Recall from Chapter 9 that the bandwidth is equal to twice the modulation frequency ($2F_m$). Also, the bandwidth of an SSB signal is only F_m. This is expressed in the table.

The next three parameters are related to one another: demodulated signal, noise, and S/N ratio. If we let the amplitude of an AM DSB-FC demodulated signal bc 1, the demodulated signal will be 0.7 (or an rms value of the full AM signal). Similarly, if the noise for an AM signal is 0.1 V per kHz, the SSB noise will be 0.07 V per kHz. These parameters will result in the third parameter, the S/N ratio.

Table 11.1

Parameter	DSB-FC	SSB
Power	1.2	0.2
Bandwidth	$2F_m$	F_m
Demodulated signal	1.0 V	0.7 V
Noise	0.1 V per kHz	0.07 V per kHz
S/N ratio (20 log signal/noise)	20 dB	20 dB

At first glance you may think there is a mistake in Table 11.1 because the S/N ratio remains the same for both modulation systems. Recall that in previous chapters, particularly in Chapter 6, we stressed that when a system bandwidth is reduced the noise characteristics in that system are reduced. But here is a case where the bandwidth is cut in half and nothing happens to the S/N ratio. Let us examine what actually happened to both the signal and noise in Table 11.1. In the case for DSB-FC AM the S/N ratio is

$$\begin{aligned}\frac{S}{N} &= 20 \log \frac{\text{signal}}{\text{noise}} \\ &= 20 \log \frac{1.0}{0.1} \\ &= 20(1) \\ &= 20 \text{ dB}\end{aligned}$$

For the SSB case

$$\begin{aligned}\frac{S}{N} &= 20 \log \frac{\text{signal}}{\text{noise}} \\ &= 20 \log \frac{0.7}{0.07} \\ &= 20 \log(1) \\ &= 20 \text{ dB}\end{aligned}$$

So you see that the signal-to-noise *ratio* has remained the same because both the signal and noise levels have been reduced proportionally. Thus, the table shows some valuable information that can be used to compare double-sideband and single-sideband systems.

A variation of the single-sideband scheme, which does not reduce the bandwidth of the system but increases the amount of information that can be transmitted is the **independent-sideband (ISB) system**. With this arrangement a single carrier (which is at a reduced level) is modulated by two independent, modulating signals. This can be considered as a double-sideband, reduced-carrier system that contains two single-sideband systems. The two sidebands are completely independent of one another and are shown (Fig. 11.5) in the same drawing because they are symmetrical about the common carrier. The two signals are designated as MOD 1 and MOD 2 to distinguish them from each other. A very popular application of ISB is AM stereo. With ISB,

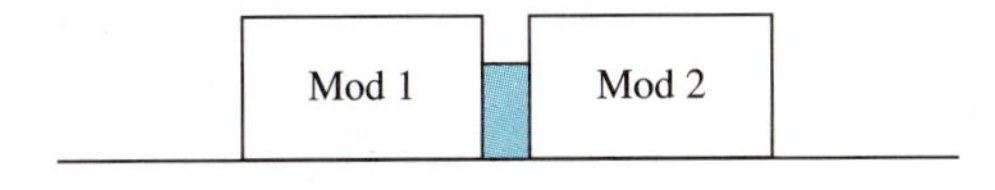

Figure 11.5 Independent sideband (ISB).

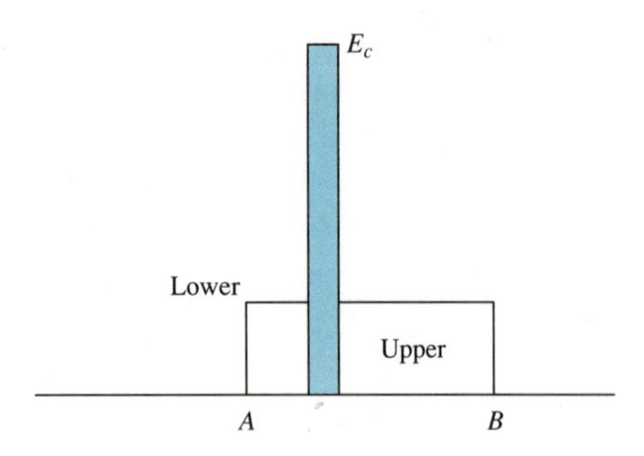

Figure 11.6 Vestigial sideband.

the left channel is transmitted on the lower sideband, and the right channel is transmitted on the upper sideband. This is very convenient because the carrier is common, and the two channels ride along symmetrically about this carrier to form the complete stereo signal. It can be seen how ISB can conserve both transmission power (by running with a reduced carrier level) and bandwidth (by transmitting two information sources within the same frequency spectrum used for a single DSB signal).

Another form of single-sideband system, which is used in television transmission, is the **vestigial sideband**. This arrangement is shown in Figure 11.6 and will be covered in more detail in Chapter 12. This system consists of a full carrier and two sidebands; however, it is different from the conventional DSB-FC system previously covered because the upper sideband is a full sideband, whereas the lower one is only a partial band. This type of scheme causes somewhat of a reduction in bandwidth because two full sidebands are not required or used. The lower sideband is smaller due to the filtering that takes place in the transmitter. This filtering leaves only a trace, or vestige, of the entire sideband that was generated. There must be some lower sideband left because it is not possible to filter out the entire lower sideband without affecting the phase and amplitude of the upper sideband. This is undesirable.

When the vestigial sideband scheme is used, some unique properties are involved. The low frequencies transmitted benefit from a DSB-FC system and have the benefit of 100% modulation. The upper frequencies are transmitted single-sideband, and are thus limited to a 50% modulation. With this arrangement, the lower frequencies are emphasized much more in the demodulator and produce larger signals than the high frequencies.

Example 11.3

A carrier of 250 MHz is modulated with a 1-kHz tone. Show the frequency spectrum for this system for:

(a) AM
(b) DSB-SC
(c) SSB-SC

Solution

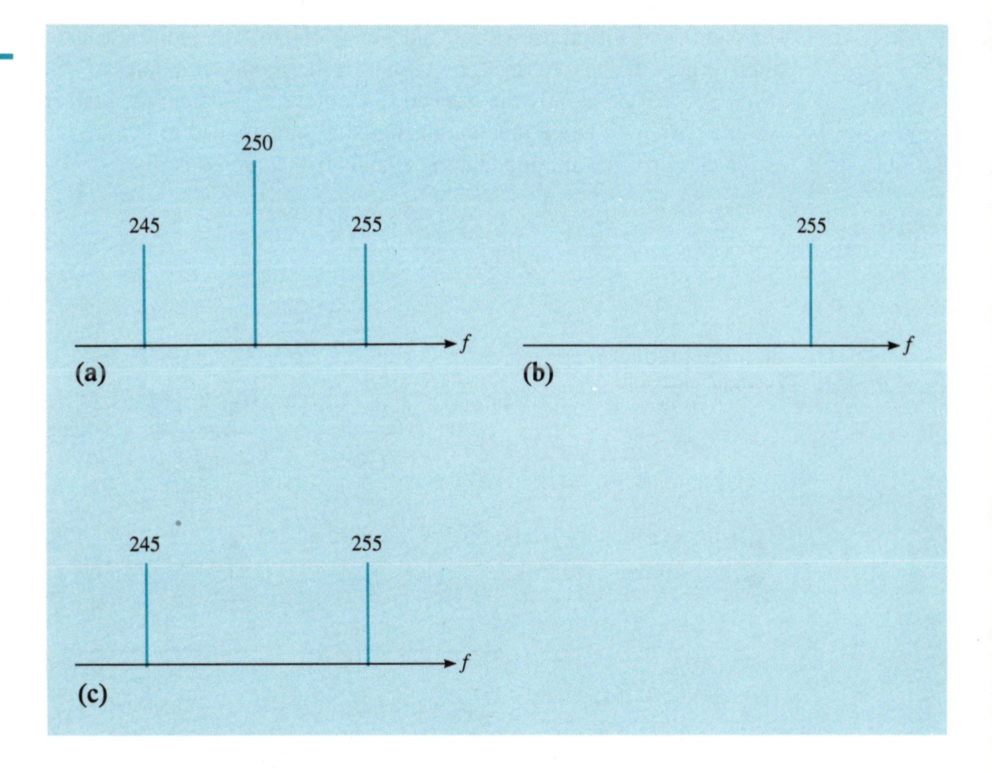

11.3 Balanced Modulator

To understand the transmission methods for sideband systems that will be covered in later sections, it is important to understand the operation of a valuable component—the balanced modulator. The **balanced modulator** is more commonly called a **product modulator**. With this terminology it is easier to comprehend what is trying to be accomplished when a balanced modulator is placed in a circuit.

Figure 11.7 is a basic block diagram of a product (balanced) modulator showing its inputs and its output. One of the inputs is the carrier signal; the other is the modulating signal. The output is the product of the two and is only the sum (upper

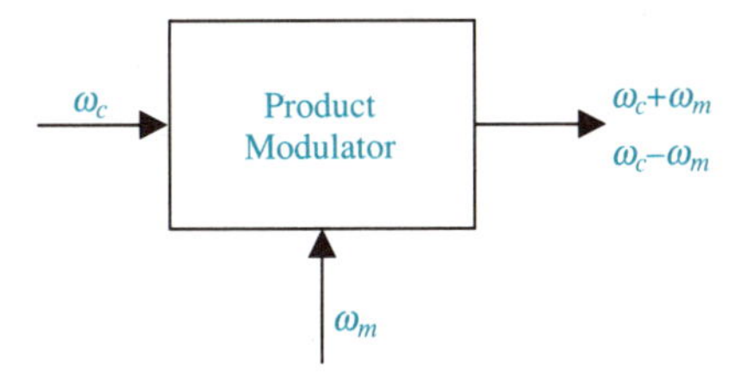

Figure 11.7 Basic block for product modulator.

sideband) and difference (lower sideband) of the two input signals. There is a distinctive lack of the carrier at the output of the product modulator.

To verify that only the sidebands are present, consider the mathematics involved when a carrier F_c and a modulating signal F_m are applied to the modulator. The result is the carrier times the modulating signal. If the carrier is given as

$$F_c = \sin(\omega_c)t$$

and the modulating signal is

$$F_m = \sin(\omega_m)t$$

then the product is

$$\text{Output} = \underbrace{[\sin(\omega_c)t]}_{\text{Carrier}} \times \underbrace{[\sin(\omega_m)t]}_{\text{Modulating signal}}$$

By using elementary trigonometry the product is

$$\text{Output} = \underbrace{\frac{1}{2}[\cos(\omega_c - \omega_m)t]}_{\text{Lower sideband}} - \underbrace{\frac{1}{2}[\cos(\omega_c + \omega_m)t]}_{\text{Upper sideband}} \quad (11.1)$$

It is clear from the mathematics that only the sidebands, $(\omega_c - \omega_m)$ and $(\omega_c + \omega_m)$ are left after the multiplication process. The carrier has completely disappeared.

Having presented the mathematics of the product modulator, it is time to look at some of the circuits that can produce these results.

The first circuit is the balanced ring modulator, shown in Figure 11.8. This circuit should look somewhat familiar because it is similar to the balanced mixer circuits presented in Chapter 7. The circuit consists of two transformers and a series of four diodes. If the circuit were redrawn you could see that the four diodes form a bridge exactly like the circuits of the Wien bridge oscillator covered in Chapter 7. The output is impressed across the primary winding of the output transformer. This winding is across the bridge the same as it was for the Wien bridge circuit. The only variation from the Wien bridge circuit is that both the secondary of the input trans-

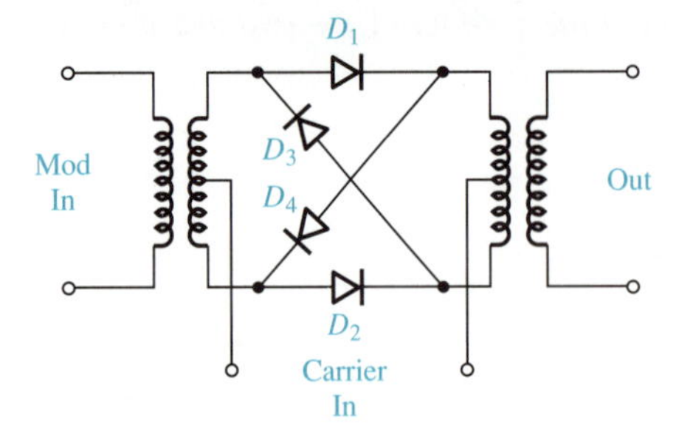

Figure 11.8 Balanced ring modulator.

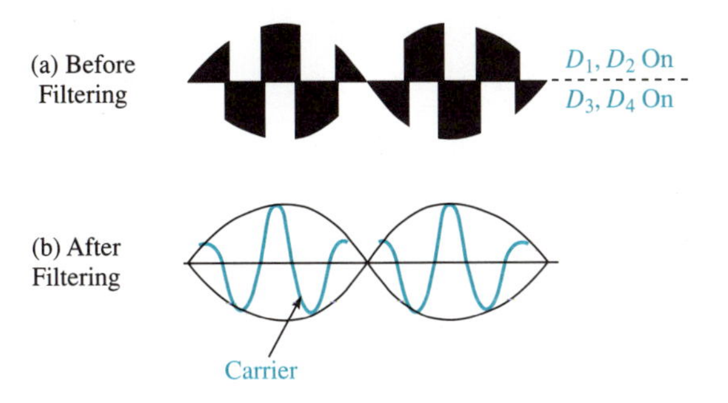

Figure 11.9 Modulation waveforms.

former and the primary of the output transformer are tapped to allow the carrier signal to combine with the modulation. The modulating action takes place by periodically turning diodes D_1 and D_2 on, while keeping diodes D_3 and D_4 off. On the reverse cycle, diodes D_3 and D_4 are on, while D_1 and D_2 are off. This action produces a chopped waveform (Fig. 11.9a) at the output with the final filtered output being the one shown in Figure 11.9(b).

It is important to notice the periodicity of the waveform in Figure 11.9(a). The complete cycle is divided into ten sections. During the first half of the cycle (the positive half), three segments are turned on for the positive swing because D_1 and D_2 are on mostly during the positive half of the input cycle. The remaining two segments are on with D_3 and D_4. Similarly, during the second half (the negative half), three segments are turned on for the negative swing because D_3 and D_4 are on mostly during the negative half of the input cycle. The remaining two segments are on with D_1 and D_2. This periodicity and switching of diodes creates the modulation process and creates the products of the carrier and modulating signal.

One requirement must be fulfilled when applying the carrier and modulating signals to the balanced modulator: the carrier must be much larger in amplitude than the modulating signal. The difference in level, sometimes 6 to 7 times larger than the modulating signal, ensures that the carrier controls the on–off characteristics of the four diodes being used. Note that this difference in level resembles that in the local oscillator in mixer circuits being at a higher level so it will drive the diodes into a nonlinear region and produce mixing. Similarly in the modulator, the multiplication process cannot take place if the diodes are not alternately switched on and off.

Notice also that the output of the balanced modulator takes the shape of the modulating signal. This verifies that there is no carrier at the output, only the sidebands, which are the modulating signal.

Another type of modulator uses FET's. The FET's are the nonlinear devices used in place of the diodes in the previous modulator circuits. Figure 11.10 shows the configuration for such a modulator. The transistors are placed in a **push–pull** arrangement, which aids in the efficiency of the circuit and allows each of the transistors to conduct on only one-half of the input cycle. In the conventional push–pull amplifier, an input signal is applied to each of the transistors. As mentioned above, each of the transistors conducts on half of the input cycle, this is class-B operation.

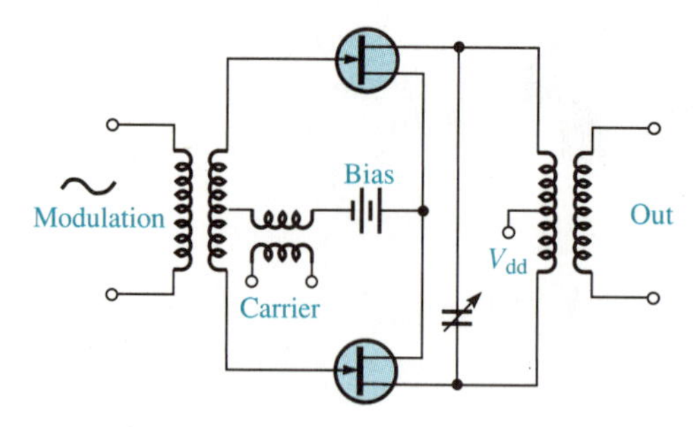

Figure 11.10 FET modulator (push-pull).

In the FET balanced modulator, the main difference between the amplifier and the modulator is that there are two input signals to the modulator instead of only one. The two signals are the carrier and the modulating signal.

The FET balanced modulator, like the balanced ring modulator, is a product detector, and therefore, all the mathematics for the ring modulator apply to the FET modulator. As such, the carrier is eliminated and the upper and lower sidebands are present at the output. This combination is present because of the nonlinear characteristics of the FET's. The variable capacitor shown in the output circuit is there to complete a tuned circuit. This tuned circuit is necessary to select the desired output frequencies from all of the other frequency components generated in the nonlinear multiplication process.

An integrated form used as a modulator (and a demodulator) is the LM1496/1596 balanced modulator–demodulator. The full data sheets for this device are shown in Figure 9.15 (Chapter 9). A typical circuit layout using the chip for single-sideband operation is shown in Figure 11.11. The pin numbers listed in the

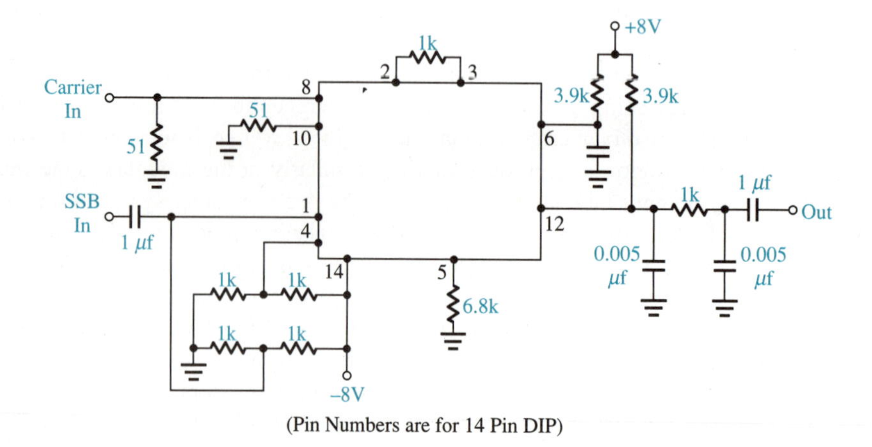

Figure 11.11 Integrated circuit product detector.

figure are for the 14-pin dual in-line package (DIP), and should not be confused with those for the 10-pin, metal-can package. Be aware of which package is being used and construct your circuits accordingly. The circuit shown in Figure 11.11 is set up as a demodulator (detector); to use the circuit as a modulator you would apply the carrier as shown (pin 8) and place the modulation at pin 1. The SSB output would then be taken at the output (pin 12). This circuit is convenient and highly reliable, and is therefore used for many communications applications and in a variety of laboratory experiments to teach the fundamentals of sideband systems.

11.4 Transmission Methods

The transmission of sideband signals involves different methods from those discussed for the conventional DSB-FC AM systems in Chapter 9. We will cover two methods commonly used for sideband transmission: the filter method and the phase-shift method.

11.4.1 Filter Method

As the name implies, the **filter method** relies heavily on filters for its proper operation. We discussed filters in detail in Chapter 7 and will only make general statements about them here. Figure 11.12 shows a detailed block diagram of a possible SSB transmitter using the filter method. In the figure a series of three balanced modulators suppress the carrier, and various filters throughout the system suppress the unwanted sidebands. Also notice that the first balanced-modulator section generates a DSB-SC signal. Following the bandpass filter for this first modulator, the remaining circuits are then SSB-RC signals, which is the desired output of the transmitter.

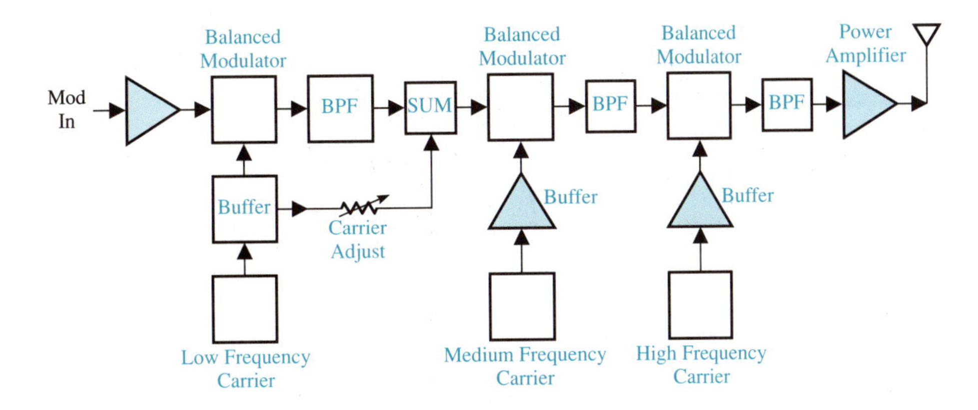

Figure 11.12 Filter method.

The filters, as can be seen in Figure 11.12, play a vital part in generating the proper signal in this method of transmission. For this application, L-C and R-C filters discussed in Chapter 7 do not have a high enough Q to fulfill the operational requirements for this type of transmitter. When we refer to high Q, we mean the narrowness of the filter response and the swiftness of the ''skirts'' falling off once we are out of the passband of the filter. This fall off of the skirts, or roll off of the response, is necessary for SSB operation because adjacent channels, which are relatively close together, must be rejected. This rapid roll off is a function of the filter Q, which should be in the thousands (1000–20,000) in order to provide the required signal rejection. Also, all of the required signals must be passed through a flat passband of the filters. These two characteristics are not generally associated with an L-C or R-C filter, whose Q's are more in the hundreds and generally have more passband ripple than can be tolerated in SSB applications. Thus, for SSB applications, crystal, ceramic, and mechanical filters are used. The crystal filter uses the properties of crystals described in detail in Chapter 7. One arrangement used for a filter circuit (shown in Fig. 11.13) has two sets of matched crystals connected between tuned input and output filters: this is called a **crystal-lattice filter**. One set of crystals is series connected (X_1 and X_2), and the other is parallel connected (X_3 and X_4). Crystals X_1 and X_2 are cut to operate at the lower cutoff freqeuncy F_L in Figure 11.13(b) and crystals X_3 and X_4 are cut to operate at the high cutoff frequency, shown as F_H in the figure. The input and output filters (which are transformers with capacitors across them to form a tuned L-C circuit) are tuned to the center of the desired passband of the overall filter, F_C in Figure 11.13(b). This combination tends to spread out the differences between the series and parallel crystal frequencies and form the entire band shape.

When the reactances of the arms are equal and have the same sign, the input

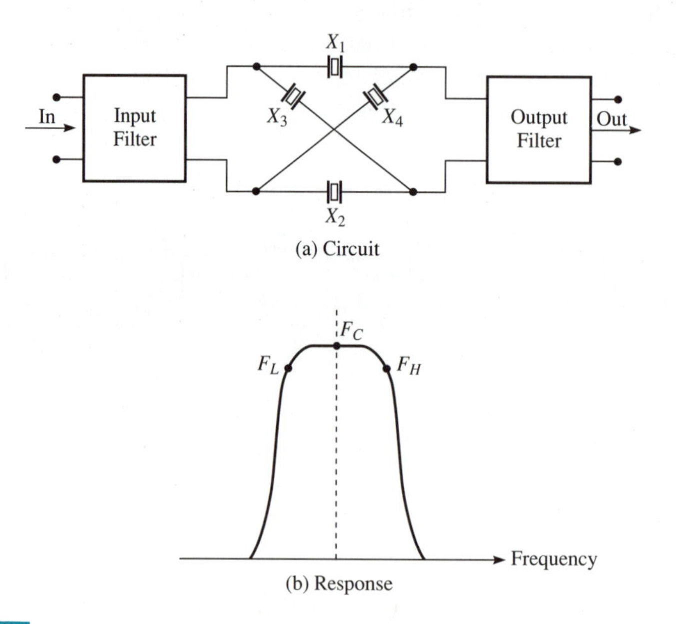

Figure 11.13 Crystal lattice filter.

signal cancels (this is the skirts). When the reactances are equal and opposite, the maximum signal is transmitted through the filter (this is the passband). The value of Q for this type of filter can range into the 100,000 region with insertion losses ranging from 1.5 to 3 dB.

Ceramic filters are like crystal filters in that they make use of the piezoelectric effect. The material most commonly used for such a filter is lead zirconate titanate, which is used for ceramic phonograph cartridges. The Q for a ceramic filter, in the range of 2000, is not as high as that for a crystal filter, but ceramic filters far outperform conventional L-C filters.

Although the ceramic filter is less expensive, smaller, and more rugged than the crystal filter, it has more loss, 2 to 4 dB, than the crystal filter. Ceramic filters, aside from being used for SSB applications, are sometimes used as replacements for i-f transformers in superheterodyne receivers.

The mechanical filter is exactly what the name implies, a mechanical structure that provides a filtering action. The filter consists of four basic elements:

1. Input transducer
2. Series of mechanical resonator discs
3. Coupling rod
4. Output transducer

The **input transducer** converts the electrical signal into mechanical energy by changing the varying signal to a corresponding vibration. This mechanical energy causes the **mechanical resonator** discs, which are tuned to a specific frequency, to resonate at that frequency. The bandpass and general filter characteristics depend on the size and number of these discs. The **coupling rods** serve to connect the resonator discs together much the same way a coupling capacitor does in an L-C filter. The **output transducer** converts the mechanical energy within the filter back to an electrical signal by taking the vibrations and converting them to a varying electrical signal, to be sent to other parts of the system. The value of Q is in the order of 10,000 for mechanical filters. The bandwidth of such a filter may be as narrow as 500 Hz or as wide as 35 kHz. Thus, this type of filter is a good choice as the filter method of SSB transmission.

A simpler method of transmission using filters is shown in Figure 11.14. This would be ideal for a single frequency, or very narrow band system because the band-

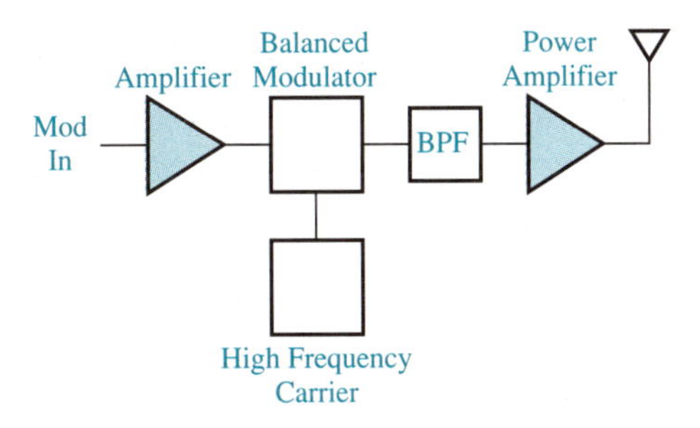

Figure 11.14 Simpler single frequency filter method.

pass filter would have to be a variable filter if the bandwidth was of any size at all. Thus, for single-frequency or narrow-band operations this is a much simpler method, and it will probably have a much higher degree of reliability associated with it.

Example 11.4

For the single-frequency filter method shown in Figure 11.14, the modulation input is 2 kHz and the high-frequency carrier is 100 kHz. If the bandpass filter will pass 101–104 kHz, what is the single frequency that the power amplifier must amplify?

Solution

Inputs to the balanced modulator are 2 kHz and 100 kHz. The output is one sideband at 98 kHz and one at 102 kHz (according to Equation 11.1). They are also $\frac{1}{2}$ amplitude. Since the filter passed 101–104 kHz, the output frequency is only 102 kHz.

11.4.2 Phase-Shift Method

The filter method has many advantages and is an excellent method for generating SSB signals. There are, however, three areas that present problems for certain applications:

1. High-Q filters are bulky and expensive.
2. The filter method may cause difficulty in switching from one sideband to another.
3. Intermediate balanced modulators are needed to generate the desired transmitted frequency.

The phase-shift method can overcome the disadvantages that occur with the filter method.

Figure 11.15 is a block diagram of the phase-shift method. Notice right away that there are no filters in the circuit. This is because the undesired sideband is canceled in the output of the modulator, which eliminates the need for any sharp filtering.

The mathematical arrangement of the circuit shows us the following:

Input to balanced modulator A:

$$\sin(\omega_m)t \qquad \text{and} \qquad \sin(\omega_c)t$$

Input to balanced modulator B:

$$\cos(\omega_m)t \qquad \text{and} \qquad \cos(\omega_c)t$$

output of balanced modulator A:

$$\frac{1}{2}[\cos(\omega_c - \omega_m)]t - \frac{1}{2}[\cos(\omega_c + \omega_m)]t \tag{11.2}$$

Output of balanced modulator B:

$$\frac{1}{2}[\cos(\omega_c - \omega_m)]t + \frac{1}{2}[\cos(\omega_c + \omega_m)]t \tag{11.3}$$

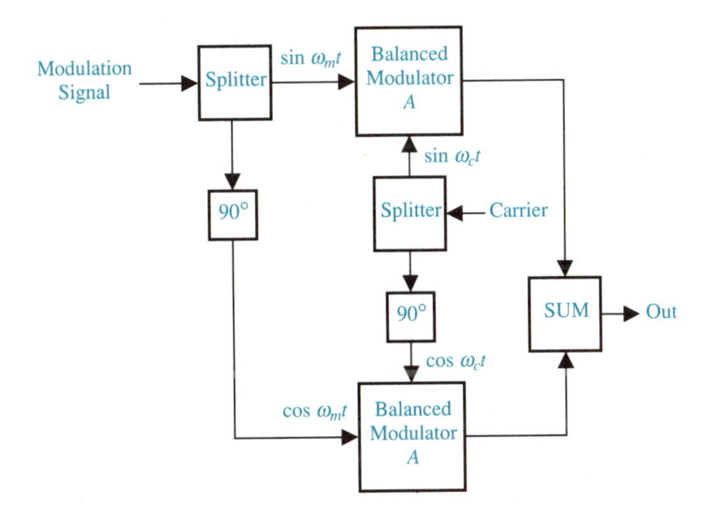

Figure 11.15 Phase shift method.

When these relationships are added together the following output is obtained:

$$\begin{array}{r} \frac{1}{2}[\cos(\omega_c - \omega_m)]t - \frac{1}{2}[\cos(\omega_c + \omega_m)]t \\ \frac{1}{2}[\cos(\omega_c - \omega_m)]t + \frac{1}{2}[\cos(\omega_c + \omega_m)]t \\ \hline \cos(\omega_c - \omega_m)t + 0 \end{array} \qquad (11.4)$$

Thus, the lower sideband is the only one generated in this case.

Having detailed the mathematics of the phase-shift method, let us look at the basic operation of the transmitter, using the block diagram in Figure 11.15. The modulation singal sin $(\omega_m)t$ is applied to the signal splitter (divider) and proceeds in one direction to balanced modulator A and in the other direction to a 90° phase shift. This phase shift may be a length of transmission line or a capacitor that produces the required phase characteristics. This operation results in a sine term at modulator A and a cosine term at modulator B of the same frequency—the modulating frequency.

The second input to the circuit is the carrier, which is similarly applied to a splitter with the same results as the modulation signal splitter—a sine input to modulator A and a cosine to modulator B. We now have one modulator with two sine inputs (one at the modulating frequency and one at the carrier frequency) and the other modulator with two cosine inputs (also at the modulating and carrier frequencies). The outputs of these modulators are expressed mathematically in Equations (11.2) and (11.3). The two outputs are now added together in the output summer, resulting in Equation (11.4), the lower sideband.

It should be apparent that this is a much simpler system, and therefore less expensive than the previous filter method. The required output is produced without a single filter. Two strategically placed phase-shift networks that utilize balanced modulator theory make for a reliable and efficient system.

11.5 Reception Methods

Having presented the theory of sideband systems and the methods for transmitting such a signal, it is now time to examine the reception of such a signal. We will look at two basic types of receivers: noncoherent and coherent. Figure 11.16 is a block diagram of a **noncoherent SSB receiver**. It is termed *noncoherent* because the rf local oscillator and beat-frequency oscillator are not synchronized to the transmitter local oscillators.

The **beat-frequency oscillator (BFO)** is an oscillator that is "beat" against the i-f frequency to produce the original information-signal spectrum. BFO's are commonly used to find the frequency of a signal by beating (combining) the unknown frequency against the known BFO signal, changing the BFO frequency until the tone you hear as a result of the beating process goes away. You then read the frequency of the BFO. Recall from early circuits courses that when you use Lissajous patterns on an oscilloscope you place one signal on the horizontal input and another on the vertical input and adjust one frequency until a circle appears on the display. The circle appears when the two signals are at the same frequency, having been "beat" together until there was zero difference.

In a sideband receiver we are not after a complete **zero beat**, but only a difference equal to the original information signal. To accomplish this we must know that information is supplied to the transmitter in the noncoherent receiver, because the oscillators (LO and BFO) are not synchronized to the transmitter LO. Regardless of the end result, the beating process is still the same.

The circuit in Figure 11.16 shows the received input signal going through an rf amplifier to improve or preserve gain and noise characteristics of the incoming signal. The preselector is in parenthesis in the diagram because it is an optional component of the system. Recall that a preselector is a series of filters that can be switched at the input to the receiver to improve the selectivity of the overall system. The exact number and type of filter used will depend on the receiver requirements and application.

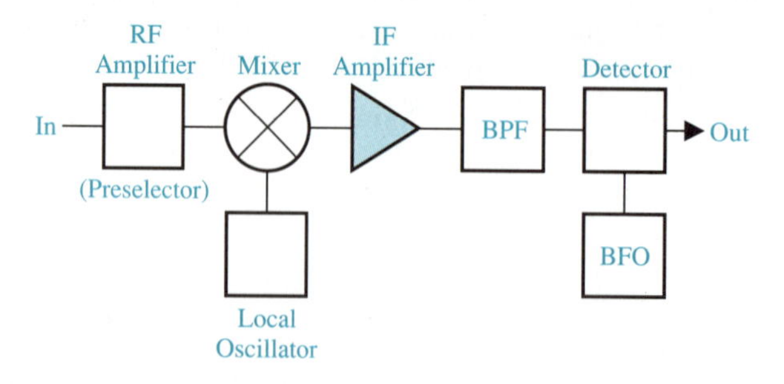

Figure 11.16 Noncoherent SSB receiver.

The output of the rf amplifier is sent to a conventional mixer that converts the signal to an i-f, which is determined by the characteristics of the receiver. Such areas as image frequencies, interfering signals, and bandwidth considerations must all go into the characterization of an intermediate frequency for this receiver. These requirements will also determine what frequency the local oscillator must be designed for. The mixer characteristics will also dictate many items of this receiver, such as the rf, i-f, and LO frequencies, as well as determining what the output power levels and stability of the local oscillator must be. The i-f output of the mixer is the SSB i-f difference signal.

The i-f amplifier and bandpass filter combination perform what can be called an amplification and clean-up operation. The amplifier raises the level of the signal and the filter ensures that only the i-f (with a narrow bandwidth) proceeds to the detector for further processing. The second detector–BFO section actually recovers the information sent from the transmitter. The output of the detector is

$$\text{Output} = ([\cos(\omega_c - \omega_m)]t) \times [\sin(\omega_b)t]$$

Where: ω_b = BFO frequency

Using the fact that $\omega_b = \omega_c$ and trigonometry identities for products of sines and cosines of different frequencies, we get an output of

$$\text{Output} = \frac{1}{2}[\sin(2\omega_c + \omega_m)t] + \frac{1}{2}[\sin(\omega_m)t]$$

(The term $\sin 2\omega_c$ is filtered out)

Thus, the output is

$$\text{Output} = \sin(\omega_m)t$$

which is what we are after, since the desired output of the receiver is the intelligence originally placed on the carrier at the transmitter $\sin(\omega_m)t$.

Example 11.5

For the receiver shown in Figure 11.16, the rf received signal is 45 to 45.005 MHz, the LO is 30 MHz, an i-f of 15 to 15.005 MHz and a BFO of 15 MHz. Find the demodulated first intermediate-frequency band and the demodulated information spectrum.

Solution

$$\text{Output spectrum of rf mixer} = F_{\text{rf}} - F_{\text{LO}}$$

$$F = (45 - 45.005) - 30$$

$$= 15 \text{ to } 15.005 \text{ MHz}$$

$$\text{Demodulated information spectrum} = F_{\text{i-f}} - F_b$$

Therefore,

$$(15 - 15.005) - 15 = 0 - 5 \text{ kHz}$$

Example 11.6

If the receiver in Example 11.5 had a 0- to 6-kHz filter on the output, and the BFO drifted, what would be the frequency of the BFO for the response to be out of the filter passband?

Solution

The spectrum is $F_{\text{i-f}} - F_b = 15 - 15.005$ MHz. For the system to be out of the passband of the filter, this must be greater than 0 to 6 kHz. Since the BFO frequency cannot be a negative number, the BFO has to decrease in frequency, thus

$$6 \text{ kHz} = 15.005 - F_b$$

$$0.006 \text{ MHz} = 15.005 - F_b$$

$$F_b = 15.005 - 0.006$$

$$= 14.999 \text{ MHz}$$

A coherent SSB receiver is shown in block diagram in Figure 11.17. With this arrangement the carrier and BFO frequencies are synchronized with the transmission carrier oscillator. This is where the term *coherent* applies. The operation of **a coherent SSB receiver** is the same as the noncoherent system, with the exception that a phase-locked loop is used for tracking in the carrier recovery and synthesizer block shown in Figure 11.17.

The carrier-recovery circuit, which is the main portion of the coherent receiver, is a narrow-band, phase-locked loop with a tracking operation that follows and removes the suppressed carrier from the SSB-RC spectrum received at the antenna. The PLL then uses the recovered pilot to regenerate carrier frequencies, which are coherent, in the synthesizer portion of the circuit. The task of the synthesizer, therefore, is to generate the proper frequencies for the rf, LO, and BFO circuitry. Any changes in the transmission carrier frequency is compensated for in the receiver by the recovery circuitry, thus eliminating any frequency offset errors in the system, which can be a distinct problem in a noncoherent receiver.

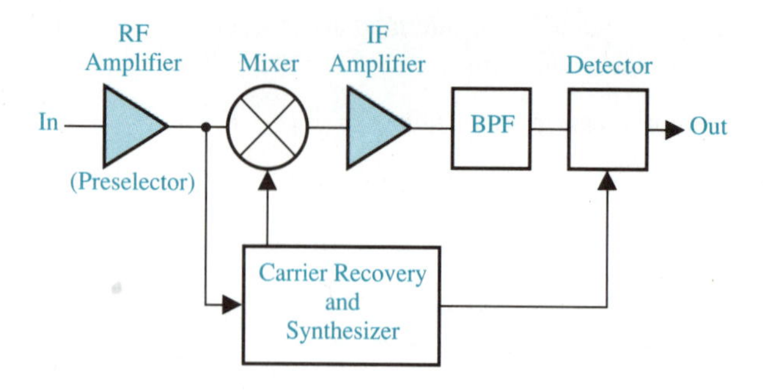

Figure 11.17 Coherent SSB receiver.

Before leaving the topic of sideband systems, we should look at some of the basic methods used in measuring such a signal. Ratings on SSB transmitters are in peak envelope power (PEP) and peak envelop voltage (PEV). It is important to understand that for a single-frequency modulating signal the modulated output signal for SSB is not an envelope as you would expect with the standard AM, but a continuous single-frequency tone. Therefore, for test purposes a two-frequency test signal must be used with equal amplitudes for the test signals. The time inverval between maximum peaks of such a modulated signal is $T = 1/(F_1 - F_2)$.

Two and only two tones are used to measure SSB performance for the following reasons:

1. As previously stated, one tone will produce a continuous single-frequency tone that does not produce intermodulation.
2. A single-frequency output is not a typical type of conversation.
3. More than two tones make for an impractical analysis procedure.
4. Two tones of equal amplitude place a more demanding requirement on the transmitter than usually occurs under normal operating conditions.

If we call the rms voltage of the two-tone test tones e_1 and e_2, the total voltage developed across a particular load will obviously be $e_1^2 + e_2^2$. To find PEP

$$\text{PEP} = \frac{e_1^2 + e_2^2}{R} \text{ (in watts)}$$

Because we have set the measurement so that $e_1 = e_2$,

$$\text{PEP} = \frac{(2e)^2}{R} = \frac{4e^2}{R} \text{ (in watts)} \tag{11.5}$$

When you consider that the average power dissipated in the load is the sum of the two tones,

$$P_{\text{ave}} = \frac{e_1^2}{R} + \frac{e_2^2}{R} = \frac{2e^2}{R} \text{ (in watts)} \tag{11.6}$$

and since

$$\text{PEP} = \frac{4e^2}{R}$$

$$P_{\text{ave}} = \frac{\text{PEP}}{2} \tag{11.7}$$

Example 11.7

Given: A two-tone test signal of 2.1 and 4.2 kHz and a carrier frequency of 150 kHz

Find: **(a)** The output frequency spectrum
(b) For $e_1 = e_2 = 5$ V, $R = 75\ \Omega$, determine PEP and average output power.

Solution

(a) The spectrum contains two upper side frequencies,

$$F_1 = 150 \text{ kHz} + 2.1 \text{ kHz} = 152.1 \text{ kHz}$$

$$F_2 = 150 \text{ kHz} + 4.2 \text{ kHz} = 154.2 \text{ kHz}$$

(b) Using Equation (11.5), we obtain

$$\text{PEP} = \frac{4(e^2)}{R} = \frac{4(5)^2}{75} = \frac{100}{75} = 1.33 \text{ W}$$

Using Equation (11.6), we obtain

$$P_{\text{ave}} = \frac{2e^2}{R} = \frac{2(5^2)}{75} = 0.67 \text{ W}$$

With the sideband system of electronic communications we have shown how the power used for a transmitter and the bandwidth requirements can be substantially reduced. The amount of sideband technology used is dependent on the particular application. Also, the range of the overall system has to be a prime consideration when adopting a sideband system.

11.6 Summary

This chapter has introduced the topic of sideband systems. We have shown how various sideband methods can be used to conserve power in a communications system.

We covered the balanced modulator, which is the key to these systems and then illustrated two transmission methods—filter and phase shift.

Finally, we discussed the reception of sideband signals. Once again trigonometry was used to determine frequencies at various points of the receiver as well as the output.

Questions

11.1 Introduction

1. Why is SSB transmission more efficient than DSB?
2. Define **SSB**.

11.2 Theory

3. What does **suppressing** the carrier mean?
4. What does **reducing** the carrier mean?
5. Draw the spectral representation of DSB-FC, DSB-SC, and SSB-RC.
6. What is different about ISB systems?
7. Where can ISB be used?
8. Explain **vestigial sideband**.
9. What is SSB-SC? Draw the spectrum.
10. Why doesn't the S/N ratio of SSB improve over that of the DSB-FC for AM?
11. Compare the bandwidths of DSB-FC, FM, and SSB.
12. What does **vestigial** mean?
13. Why is some lower sideband left in the vestigial sideband system after filtering?

11.3 Balanced Modulator

14. Mathematically show that a balanced modulator is a product detector.
15. Name two types of product detectors (modulators).
16. What type of balanced modulator resembles a Wien bridge oscillator?
17. Why must the carrier be larger than the modulating signal in a balanced modulator?
18. In the FET modulator what class of operation is used?

11.4 Transmission Methods

19. Why are crystal and mechanical filters used in the **filter method** of SSB transmission?
20. Name, and explain, the four elements of a mechanical filter.
21. What is the major advantage of using the phase-shift method for SSB transmission?
22. Why is a rapid roll off needed for filters in a SSB system?
23. In the crystal-lattice filter why are some crystals operated in a series mode while others are operated in a parallel mode?
24. In the crystal-lattice filter to what frequency are the input and output filters tuned?
25. Explain how the crystal-lattice filter response is achieved.
26. How is a ceramic filter like a crystal filter?
27. What material is used for ceramic filters?
28. Name one advantage and one disadvantage of the ceramic filter over the crystal filter.
29. How does the phase-shift method of SSB transmission produce modulation?

11.5 Reception Methods

30. Explain the difference between **coherent** and **noncoherent** receivers.
31. What is a BFO?
32. What is the purpose of the carrier recovery–synthesizer circuit in a coherent receiver?

33. How does the recovery circuit eliminate frequency offset errors?
34. Why are two tones used for measuring SSB?
35. What units are used for rating SSB transmitters?

Problems

11.1 Introduction

1. An AM communications system presently uses 100 W of power. It is found that only 75 W is available. What percentage of modulation must be used when suppressing the carrier?
2. Determine the power savings when the carrier is suppressed in an AM signal modulated to 67%.
3. Determine the percentage of power savings when the carrier is suppressed in an AM signal modulated to 25%. Is this a practical system?

11.2 Theory

4. If we have an rf carrier of 2.5 MHz and modulate it with a signal ranging from 2 to 4 kHz, what is the required bandwidth if DSB-FC AM is used? What is the bandwidth if SSB-SC is used?
5. In Problem 4 what would be the frequencies of a bandpass filter for the DSB-FC case?
6. A carrier of 100 MHz is modulated with a signal that ranges from 1 to 2 kHz. Show the frequency spectrum for AM, DSB-SC, and SSB-SC.

11.3 Balanced Modulator

7. If the following signals are applied to a balanced modulator: F_c = 1.2 MHz, and F_m = 10 kHz, what is the output of the balanced modulator?
8. If a bandpass filter that will pass 1.195 MHz to 1.220 MHz is placed at the modulator output, how does the output spectrum of the filter appear?
9. Verify mathematically that if we apply a carrier $\sin(\omega_c)t$ and a modulating signal $\sin(\omega_m)t$ to a balanced modulator with a low-pass filter at its output, the final output will be only the lower sideband.

11.4 Transmission Methods

10. For the single-frequency filter method, the modulation input is 1.5 kHz and the carrier is 200 kHz. If the bandpass filter will pass 198 to 200 kHz, what is the single-frequency output?
11. In the phase-shift method, the input to balanced modulators A and B is 3 kHz for the modulating signal and 100 kHz for the carrier. What is the output?

11.5 Reception Methods

12. The circuit in Figure 11.15 has an input that has a carrier of 25 kHz with a modulating frequency of 1.2 kHz. What is the output of the detector?

13. For a noncoherent SSB receiver the rf is 60 to 60.010 MHz, the local oscillator is 30 MHz, and the i-f is 30 to 30.01 MHz. What must the BFO be in order to have the demodulated information spectrum be 0 to 5 kHz?
14. For a two-tone test signal of 500 Hz and 1.5 kHz with a carrier frequency of 85 kHz, find the output frequency spectrum.
15. For the conditions in Problem 14, if the values of the tone are 2.65 V and the load resistance is 64 Ω, determine PEP and the average output power.

12

Outline

Objectives

- To provide an introduction to television and a detailed discussion of the television signal
- To discuss in detail the topics of monochrome and color television and how the circuitry works
- To provide an introduction to the area of satellite communications
- To provide an introduction to the area of fiber optics and fiber-optic communications

Key Terms

television
satellite communications
vestigial sideband
flyback time
interleaved scanning
raster
back porch
vidicon tube
photoconductive
color wheel
look angles
downlink
link budget
acceptance angle
monochrome
fiber optics
rest frequency
retrace
resolution
front porch
diplexer
photomissive
peaking coil
nonsynchronous
uplink
noise density
index of refraction
numerical aperture

Television, Satellite Communications, and Fiber Optics

12.1 Introduction

The three terms in the title of this chapter have become household words in today's world of high-technology communications systems. The satellite hookups that tie points around the world together for viewing world news or sports, and the fiber-optic advertising for telephone companies are familiar ideas to most people.

Television was nothing more than the dream of a few individuals in the early 1900s. This dream has come to be considered an absolute necessity in every household today. Many people plan their evenings around what is on TV; it amuses their children, and influences what they buy and who they vote for. The satellite has enhanced this love of television by enabling us to view events from around the world and even from out in space. Many people have seen television pictures from the moon as they were broadcast from the Apollo spacecraft in the late 1960s and early 1970s. Also, the space shuttle astronauts have no privacy at all when they are orbiting the earth and conducting scientific experiments that will benefit us all in one form or another.

Fiber optics is a buzzword of the 1990s, and is becoming one of the most efficient and most widely used transmission media. Once a laboratory curiosity, it is now essential to most communications system manufacturers.

Because these three terms affect the communications industry so profoundly, we will investigate the properties of each and how each fits into the total picture of the communications world.

12.2 Television

12.2.1 Television Signal

To this point we have studied systems with relatively simple and straightforward types of modulation schemes: amplitude modulation (DSB-FC or DSB-SC), angle modulation (FM or PM), or a well-defined sideband system. In most cases we dealt with one type of modulating signal, usually audio, which transmitted information from one point to another, received it at the proper receiver, and demodulated it to reproduce the original audio signal. With each of these systems there was a carrier with sidebands (generally) on both sides of the signal. The total bandwidth was $2F_m$. The television signal, however, does not follow these conventions because the modulation placed on a carrier is now a combination of video and audio. The video signal is complex in that it must have synchronous and blanking pulses and control the color of the image. This color may be various shades between black and white (monochrome TV) or the combination of red, blue, and green in a color system. A television signal does not, therefore, fall into the typical communications category but requires a special type of modulation scheme. Recall from Chapter 11 that the vestigial sideband scheme is used for TV applications. Now we need to examine exactly how this is done.

Recall from Chapter 11 that the vestigial-sideband system consisted of a full carrier and two sidebands. In this system the upper sideband is a full sideband, but the lower sideband is only a partial band. This arrangement causes a reduction in bandwidth because two full sidebands are not required, a point that comes in handy when dealing with TV signals. To understand this, we consider the composition of a TV signal. Figure 12.1 shows a standard TV broadcast channel and how the signal is placed within each channel. To provide adequate resolution of a TV signal, the video signal must include modulating frequency components from dc to 4 MHz. Since 4 MHz is the maximum modulating frequency and the video is to be amplitude modulated onto a carrier, it would seem logical that the bandwidth of the TV system should be 8 MHz ($2F_m$). Figure 12.1, however, shows that the bandwidth is only 6 MHz—the bandwidth allocated by the FCC for each channel. (This is true for all channels shown in Table 12.1.)

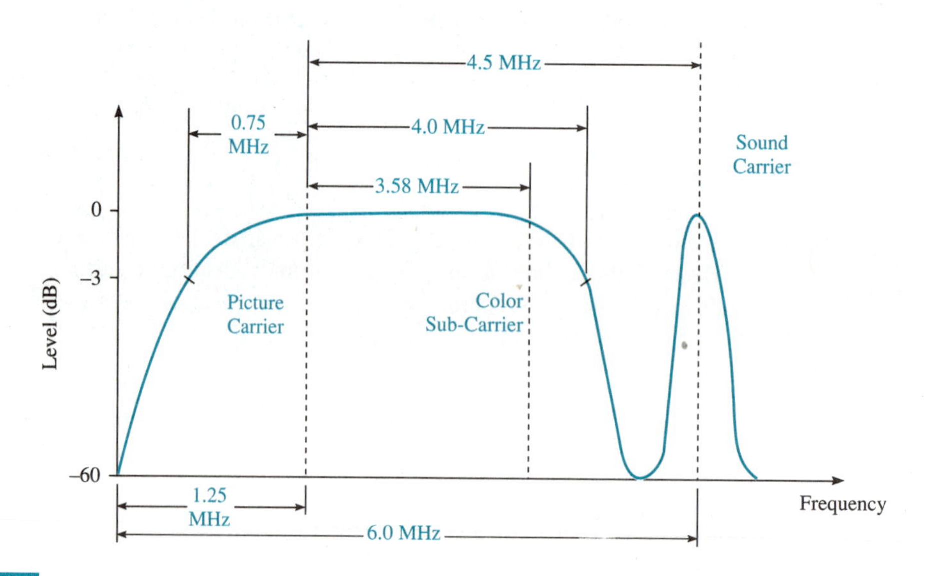

Figure 12.1 Standard TV channel.

Table 12.1 Television Channels

Channel Number	Band (MHz)	Picture Carrier (MHz)	Sound Carrier (MHz)
2	54–60	55.25	59.75
3	60–66	61.25	65.75
4	66–72	67.25	71.75
5	76–82	77.25	81.75
6	82–88	83.25	87.75
7	174–180	175.25	179.75
8	180–186	181.25	185.75
9	186–192	187.25	191.75
10	192–198	193.25	197.75
11	198–204	199.25	203.75
12	204–210	205.25	209.75
13	210–216	211.25	215.75
14	470–476	471.25	475.75
15	476–482	477.25	481.75
16	482–488	483.25	487.75
17	488–494	489.25	493.75
18	494–500	495.25	499.75
19	500–506	501.25	505.75
20	506–512	507.25	511.75
21	512–518	513.25	517.75
22	518–524	519.25	523.75
23	524–530	525.25	529.75
24	530–536	531.25	535.75
25	536–542	537.25	541.75
26	542–548	543.25	547.75
27	548–554	549.25	553.75
28	554–560	555.25	559.75
29	560–566	561.25	565.75
30	566–572	567.25	571.75
31	572–578	573.25	577.75
32	578–584	579.25	583.75
33	584–590	585.25	589.75
34	590–596	591.25	595.75
35	596–602	597.25	601.75
36	602–608	603.25	607.75
37	608–614	609.25	613.75

Channel Number	Band (MHz)	Picture Carrier (MHz)	Sound Carrier (MHz)
38	614–620	615.25	619.75
39	620–626	621.25	625.75
40	626–632	627.25	631.75
41	632–638	633.25	637.75
42	638–644	639.25	643.75
43	644–650	645.25	649.75
44	650–656	651.25	655.75
45	656–662	657.25	661.75
46	662–668	663.25	667.75
47	668–674	669.25	673.75
48	674–680	675.25	679.75
49	680–686	681.25	685.75
50	686–692	687.25	691.75
51	692–698	693.25	697.75
52	698–704	699.25	703.75
53	704–710	705.25	709.75
54	710–716	711.25	715.75
55	716–722	717.25	721.75
56	722–728	723.25	727.75
57	728–734	729.25	733.75
58	734–740	735.25	739.75
59	740–746	741.25	745.75
60	746–752	747.25	751.75
61	752–758	753.25	757.75
62	758–764	759.25	763.75
63	764–770	765.25	769.75
64	770–776	771.25	775.75
65	776–782	777.25	781.75
66	782–788	783.25	787.75
67	788–794	789.25	793.75
68	794–800	795.25	799.75
69	800–806	801.25	805.75
70	806–812	807.25	811.75
71	812–818	813.25	817.75
72	818–824	819.25	823.75

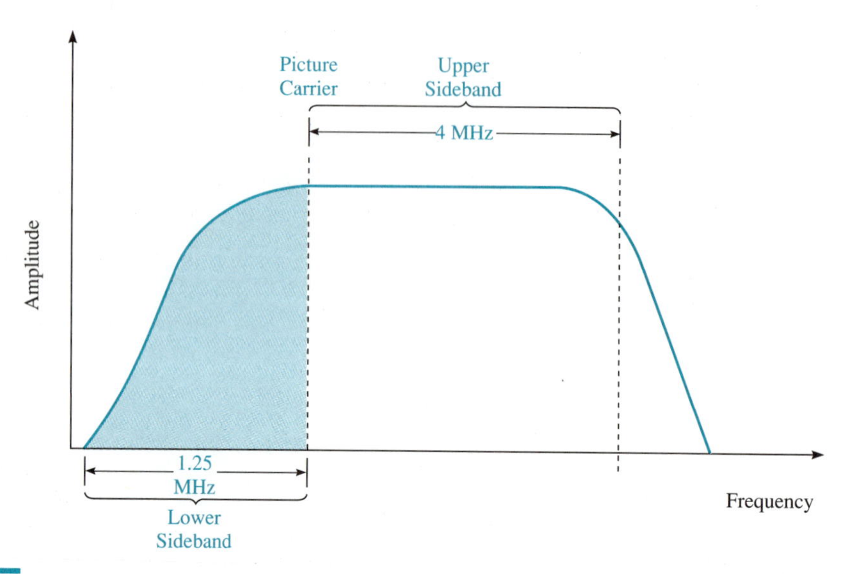

Figure 12.2 Vestigial sidebands for TV.

In Table 12.1, we will refer to the picture carrier frequency throughout the discussions on TV and any related technologies.

The 6-MHz bandwidth carries not only the video signal but also, as can be seen in Figure 12.1, the audio signal. The vestigial sideband comes into play in the reduction of the bandwidth from the conventional 8 MHz to the allocated 6 MHz. The lower sideband of the vestigial-sideband system is shown as 1.25 MHz (Figure 12.2) and the upper sideband is the full 4 MHz. This satisfies all conditions for vestigial-sideband operation. The upper sideband is a full band, and the lower one is a reduced band, less than one-third that of the upper sideband in this case.

Let us now look closer at what is contained in the channel shown in Figure 12.1. As previously stated, the carrier (picture carrier) is placed 1.25 MHz above the lower limit of the assigned-channel frequency band and is partially attenuated as can be seen by the roll-off of the response at the left of the figure. The upper sideband is left intact as a full 4-MHz band in which the video information is placed. The attenuated characteristics of the lower sideband can be explained by first pointing out that the amplitude for the lower sideband is equal to the carrier amplitude of 0.75 MHz below the carrier. At this point it drops to a level of -3 dB. At 1.25 MHz below the carrier, the level is then reduced to -60 dB below the carrier level. This creates the narrower, attenuated lower sideband that was discussed in Chapter 11 and is necessary for vestigial-sideband operation.

The upper sideband remains at the carrier level for the required 4 MHz, with a drop off to a -3 dB level at the 4 MHz point. The most stringent requirement is that the response must fall to the -60 dB level, 4.475 MHz above the carrier frequency. This means that there must be a 57 dB drop in the span of only 475 kHz, a steep skirt for a response curve. This requirement is necessary because the sound response has its carrier 4.5 MHz above the picture carrier and is only 0.5 MHz wide. This sharp drop off of the video sideband ensures that there is no interference with the sound response.

We stated previously that the video portion of a TV signal was amplitude-modulated. In contrast, the sound portion of the signal is frequency-modulated, a fact that further substantiates that the TV signal is not a conventional electronic communications signal. The carrier frequency of the sound signal, or the **rest frequency**, is 4.5 MHz above the picture carrier, as shown in Figure 12.1. The audio signal ranges from 50 Hz to 15 kHz and will modulate the carrier to 100% for a deviation of the full bandwidth shown, ± 25 kHz. Recall from Chapter 10 that the deviation ratio was the carrier deviation divided by the modulation range. That is, for this case, 25 kHz/15 kHz = 1.67 radians. Also recall that a narrow-band application was described as having a ratio of $\pi/2$, or 1.57 radians. The sound portion of a TV signal is, therefore, a narrow-band FM application.

Although the video portion of the TV signal appears to be a rather straightforward response, in actuality it is a very complex signal. This signal contains three parts: picture information, synchronization pulses, and blanking pulses. The **picture information**, sometimes called the **luminance signal**, is the portion of the TV signal that you probably think about first when the term *video signal* is used because it is the part of the signal that contains the video information. We have been calling this information intelligence throughout this text. It may be a black-and-white or a color picture. Details of how this is displayed and how the signal is constructed will be covered when we look specifically at monochrome (black and white) and at color TV. For now, it is sufficient to say that picture information is the portion of the signal that most people consider to be the important part of the TV signal—the picture.

The second part of the video signal comprises the **synchronization pulses**, which are also called the "instructors." They tell the TV receiver that it is time to start a trace, end a trace, and start the retrace so that the next scan line can be synchronized as needed. The scanning on a TV receiver is what makes the transmitted picture reproducible at the receiver. It begins in the upper left corner of the screen as shown in Figure 12.3 (point 1). The electron beam goes diagonally from 1 to 2. This is called the *active* portion of the scan because the images are being converted to electrical signals. The beam now returns to the left side of the screen. This turn of the beam, or **retrace**, or **flyback, time**, is very short. When the beam is in its retrace mode, it is turned off, or **blanked**, so that there is no video on the display at that time. When the beam reaches the bottom of the screen (point 6), it retraces back to the top (point 1) again, and the process repeats itself. This type of scanning is called **sequential scanning**.

Another type of scanning is the **interleaved** scanning. In sequential and interleaved scanning a total of 525 horizontal scan lines constitute one picture frame. One horizontal scan line is made up of the active and blanked portion of a single horizontal scan; that is, one horizontal scan line is the beam scanning from left to right and retracing back to the left side. In **interleaved scanning** the picture frames are divided into two fields, each with 262.5 horizontal lines. This concept is shown in Figure 12.4. Horizontal scanning produces a left-to-right movement of the electron beam,

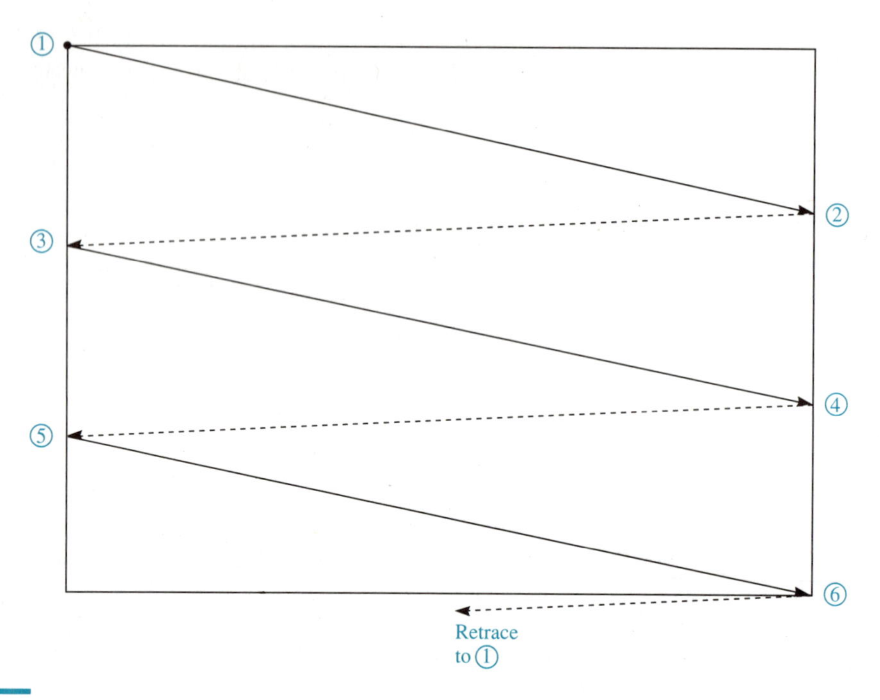

Figure 12.3 TV scanning.

and vertical scanning produces a downward movement. The vertical scanning rate is 30 Hz, which means that 30 frames are produced per second. Since there are two fields, each field is vertically scanned at a 60-Hz rate. The horizontal scanning frequency is $30 \times 525 = 15{,}750$ Hz, or 15.75 kHz. This means that the horizontal scanning *time* is 1/15,750, or 63.5 μs, and includes the time it takes for the trace to move across the screen and the retrace time.

You will see a variety of timing schemes presented in different texts. Some say that the trace takes 52 μs and the retrace 11.5 μs, others say 53.5 μs for the trace and 10 μs for the retrace, still others say simply that the total scan takes 63.5 μs. This last case will always appear because it is always true. Unfortunately, it does not give you the information you need to analyze a TV signal for its scanning properties. For our discussions we will use 52 μs for the trace and 10 μs for the retrace, numbers that are important in determining the system **resolution**, or the system's ability to resolve picture elements.

There are two types of resolution: vertical and horizontal. **Vertical resolution** is actually the number of horizontal lines that can be resolved (separated). Only about 70% (0.7) of the horizontal lines constitute the vertical resolution. Thus, the vertical resolution for a typical TV system is 0.7×525 lines $= 367.5$.

Similarly, the **horizontal resolution** depends on the number of vertical lines that can be resolved. As stated previously, the maximum modulating frequency is 4 MHz. It, therefore, stands to reason that the more vertical lines there are to resolve,

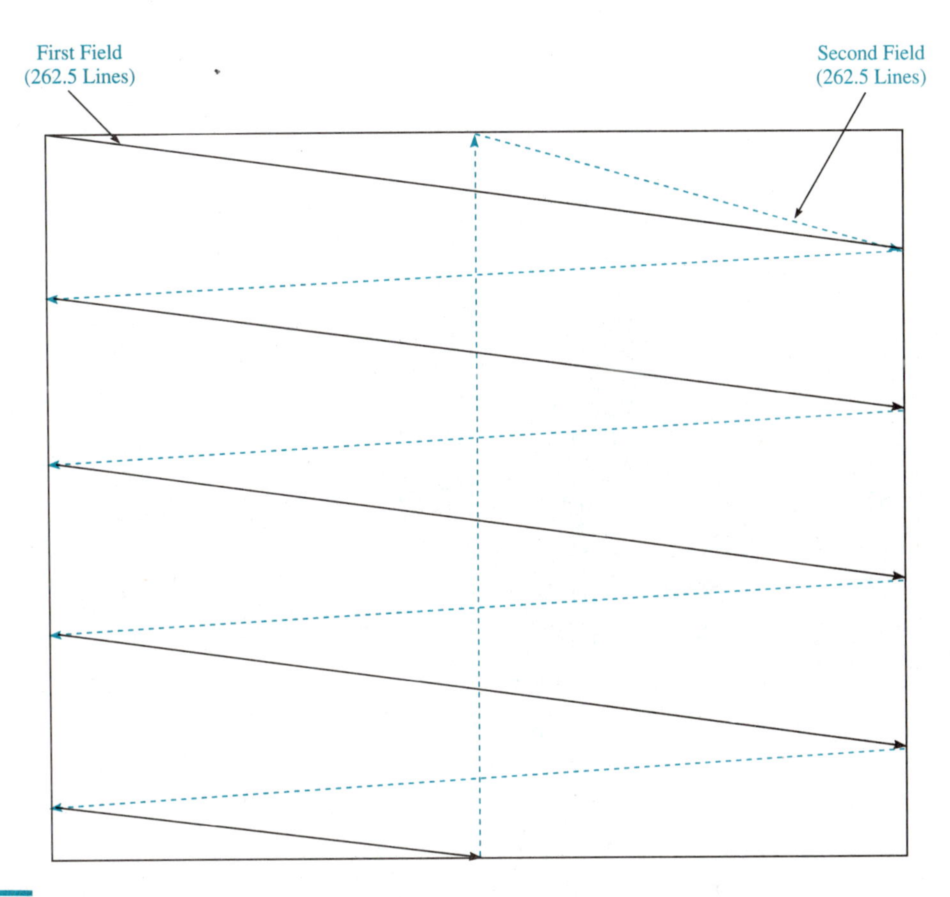

Figure 12.4 Interlaced scanning.

the higher will be the frequency of the video signal. As we said, the horizontal scanning rate is 15.75 kHz, and the total scan time is 63.5 μs. The blanking time is 11.5 μs, with 52 μs for horizontal trace time. Because two lines can be converted into one cycle of video signal,

$$\text{Number of vertical lines resolved} = \text{Highest frequency} \times \text{Scan time} \times 2 \quad (12.1)$$

Using Equation (12.1) and the numbers we have been discussing, we obtain, 4 MHz $\times$ 52 μs $\times$ 2 = 416 lines. If different scan times or different video frequencies are used, the resolution obviously will be different.

Example 12.1

A high-density TV system has a scanning rate of 35 kHz. What is the horizontal resolution of this system for a 4-MHz video signal?

Solution

$$\text{Scan time} = \frac{1}{\text{Frequency}} = \frac{1}{35 \times 10^3} = 28.6\ \mu\text{s}$$

From Equation (12.1) we obtain

$$\begin{aligned}\text{Resolution} &= \text{Frequency} \times \text{Scan time} \times 2\\ &= 4\ \text{MHz} \times 28.6\ \mu\text{s} \times 2\\ &= 228.8\ \text{lines}\end{aligned}$$

Example 12.2

For a system with a scan rate of 30 kHz and a video modulating-signal bandwidth of 5.2 MHz, what is the horizontal resolution?

Solution

$$\text{Scan time} = \frac{1}{\text{Frequency}} = \frac{1}{30 \times 10^3} = 33.3\ \mu\text{s}$$

$$\begin{aligned}\text{Resolution} &= \text{Frequency} \times \text{Scan time} \times 2\\ &= 5.2\ \text{MHz} \times 33.3\ \mu\text{s} \times 2\\ &= 346.32\ \text{lines}\end{aligned}$$

The previous explanation of scanning is intended to lead to an understanding of the nature and importance of the **synchronous (sync) pulses** on a TV signal. The arrangements and timing for both horizontal and vertical sync pulses are presented in Figure 12.5, so as to emphasize the timing between them. The pulses will not look like this if observed on an oscilloscope. We will present the total signal when all of the sections of the video signal have been covered. Figure 12.5 shows the numbers developed earlier in this section. The total scan time between horizontal sync pulses is 63.5 μs. This is as it should be because the total trace time is one sweep plus a retrace. After this time the system must again be pulsed to begin the next sequence of trace–retrace operations. Also, if there are 262.5 horizontal scan lines for each field (odd and even), the total time for each field will be 262.5 $\times$ 63.5 μs, or 16.7 ms. If these relationships are maintained, the original image at the transmitter will be reproduced at the receiver in exact-time synchronization.

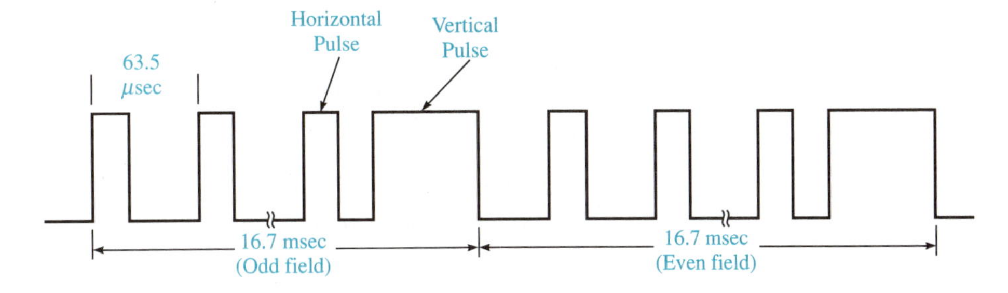

Figure 12.5 Horizontal and vertical sync pulses.

The final portion of the TV signal, the **blanking pulses**, are added to the video signal and the sync pulses to ensure that the receiver is blanked out (the beam of electrons that illuminate the CRT is cut off) during the vertical and horizontal retrace times. Consider the blanking pulses as video signals with amplitudes that do not produce illumination on the screen, in other words, *negative* signals. Blanking pulses are shown in Figure 12.6. It can be seen that the duration of these pulses is the retrace time, or 11.5 μs.

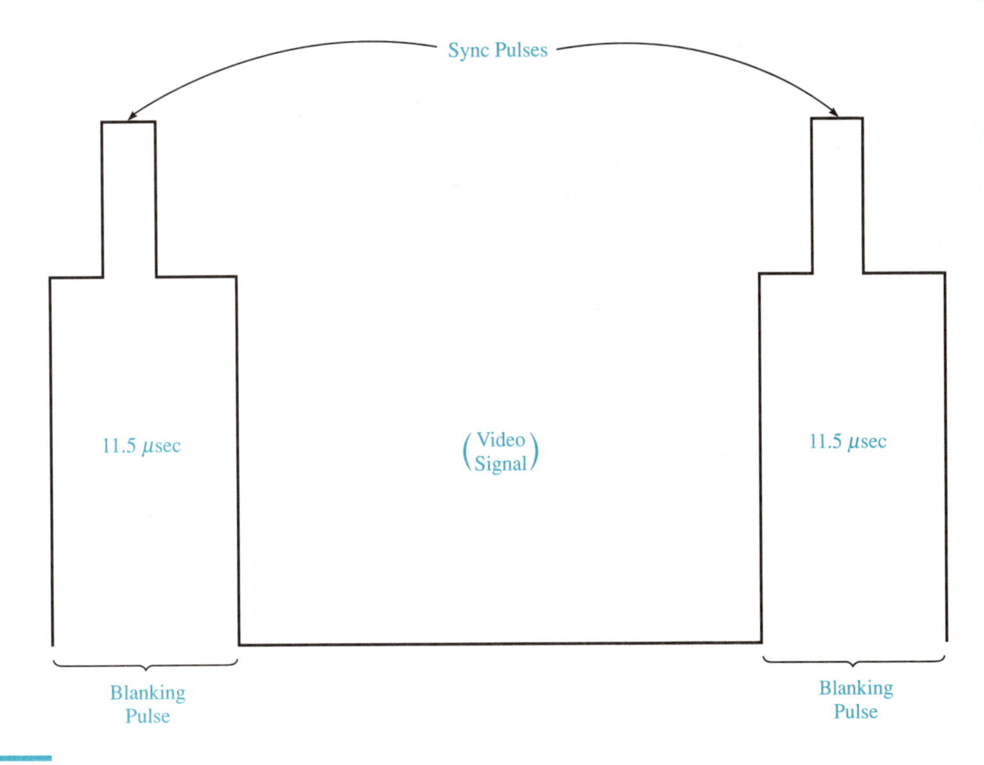

Figure 12.6 Blanking pulses.

When the screen is illuminated but there is no picture on it, it is called a **raster**. The horizontal lines on a display are the **raster lines**, and 525 raster lines make up the raster itself. You have all encountered the situation of setting your TV on a channel with no assigned station, or turning on the TV early in the morning before the programming day begins. No picture or sound is broadcast, but rather a white raster, which some people refer to as *snow*.

Having discussed all three sections of the TV signal, we now look at the overall signal and how it operates within a TV system. Figure 12.7 shows the complete three-part TV signal: the video signal (only one line of video is shown in the figure), the sync pulses (1.85 μs wide), and the blanking pulses (11.5 μs wide). The total horizontal scan time is shown as 63.5 μs.

There are two terms in Figure 12.7 that have not yet been discussed: the front porch and back porch of the blanking–sync pulse combination. The **front porch** is the time between the beginning of the blanking pulse and the leading edge of the sync pulse. This time is short—around 2 μs in Figure 12.7. The **back porch** is the time between the trailing edge of the sync pulse and the end of the blanking pulse. This time is longer than the front porch time—around 7 μs in the figure. There is then approximately 2 μs for the sync pulse, more than sufficient time to start the trace since the sync pulse need only be a trigger to start the cycle. The sync pulse need not have any width at all. In a more general form the front porch, back porch, and sync pulse width may be defined in relation to the total blanking pulse width designated as H. Using this method, the front porch is approximately $0.2H$, the back porch is $0.64H$,

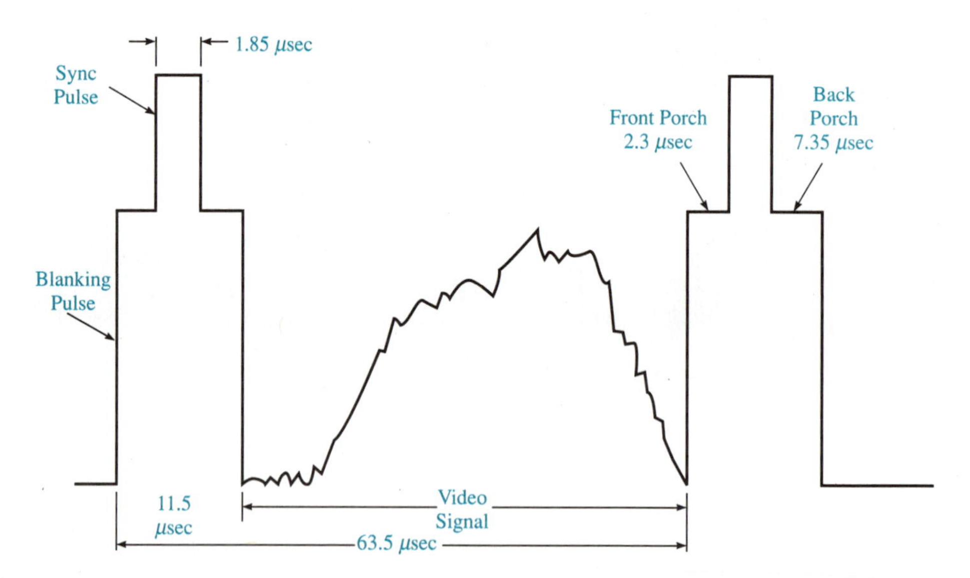

Figure 12.7 TV signal.

and the sync pulse $0.16H$. These relationships can be verified in Figure 12.7. Thus, if a blanking width is specified other than the 11.5 μs used here, you should be able to calculate all three values.

Example 12.3

The blanking width for a TV signal is reduced from 11.5 μs to 9.5 μs. How wide is the sync pulse for this system?

Solution

In this system, $H = 9.5\ \mu s$

$$\text{Sync width} = 0.16H$$
$$= 0.16(9.5)$$
$$= 1.52\ \mu s$$

Figure 12.8 is a variation of Figure 12.7, showing the TV signal with the rf envelope. The sync pulse, blanking pulse, and the video (or luminance information) signals are all indicated, but with the rf signal on each of them. Figure 12.8(a) is a negative transmission and part (b) is positive transmission. In the **negative transmission** the lower amplitude signals correspond to the light, or white, portions of the transmitted image, and the higher amplitude signals correspond to the dark portions of the transmitted image. Thus, the picture that appears on the TV screen gets darker as the amplitude of the transmitted signal increases. In contrast, in the **positive transmission** the lower amplitude signals are the dark images, and the higher amplitude signals are the lighter images. This means that the higher the amplitude of the transmitted image, the lighter the picture will be. The negative transmission is the FCC standard for the final picture carrier, meaning that you will see Figure 12.8(a) on the oscilloscope when looking for the final carrier out of a transmitter or into a TV receiver. This does not mean, however, that the positive transmission is not used. There are various areas throughout the transmitter where both positive and negative transmission signals can be found. Some areas are much easier to work with if there is a positive signal present. Remember, nevertheless, that the *final* carrier must be negative.

There are two reasons why the negative transmission signal is used for the final transmission carrier.

1. The *average* voltage in the wave is smaller for bright pictures. Since most scenes are generally bright rather than dark, the required power to transmit these scenes will be smaller, and thus the overall power requirements will be lower.
2. If the signal is overmodulated, it will not cause the sync pulses to be compressed. This is what happens with positive transmission and can cause the synchronization of the transmit and receive signals to be lost. This does not happen with a negative transmission signal.

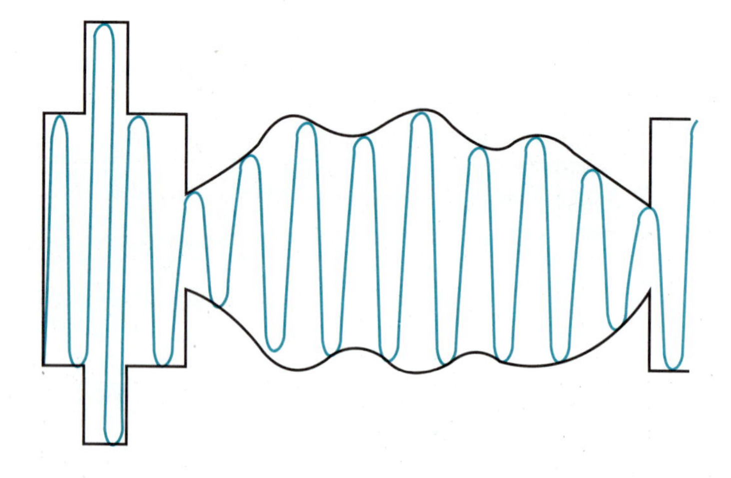

(a) Negative Transmission

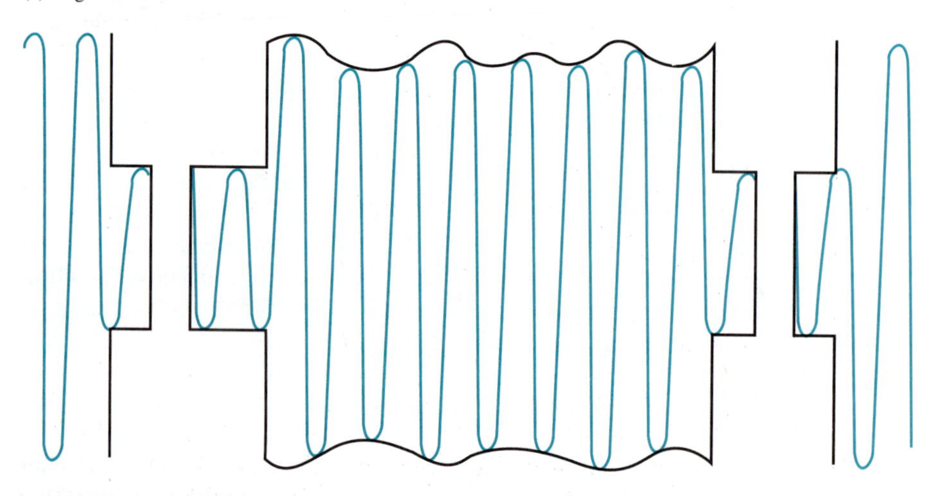

(b) Positive Transmission

Figure 12.8 TV signal with rf envelope.

Having described the basic look of the TV signal when it is transmitted and how it must be synchronized with the receiver, we now examine specific types of TV systems: monochrome and color.

12.2.2 Monochrome Television

Many of you may never have seen a program on a monochrome television set. This was the original form of television as broadcast in the late 1940s and 1950s. Such programs as ''Howdy Doody,'' ''Playhouse 90,'' and ''I Love Lucy'' were seen by your parents and grandparents in **monochrome**, or **black and white**. This type of system, however, is not merely an old-fashioned type of TV. It has many applications today; for example, the TV security monitors in banks, large parking lots, or department stores are black-and-white (monochrome) television systems. Monochrome is much more economical and does the job every bit as well as color TV. Thus, monochrome TV is a necessary area of study in communications.

Monochrome TV is based on the camera scanning an object and converting that object into a series of various shades from white (for no image) to black (for a completely dark image). Various shades of gray complete the image between the two extremes of black and white. Figure 12.9 shows the scanning of a simple object (the letter "E") for a TV system. For our initial discussions we will assume that the letter is completely black or some very dark color. This will result in all of the electrical signals being at the highest level. Bearing in mind that this is the negative transmission system, let us look at what happens as the camera scans the letter and gets it ready to transmit to a receiver.

Figure 12.9(a) shows the letter to be scanned. Figure 12.9(b) represents how the camera will see that same letter. You can see that the camera sets up a matrix that breaks the letter (image) into a series of vertical and horizontal blocks. In this case there are five spaces for the horizontal scan and seven spaces for the vertical. As you will remember, the camera scans from left to right. As the camera begins its scan in the upper left corner of the matrix it finds nothing but a blank display across

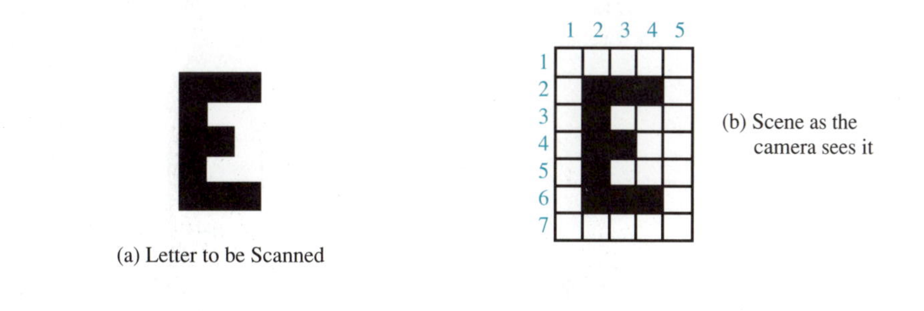

(a) Letter to be Scanned

(b) Scene as the camera sees it

(c) Scanned sequence

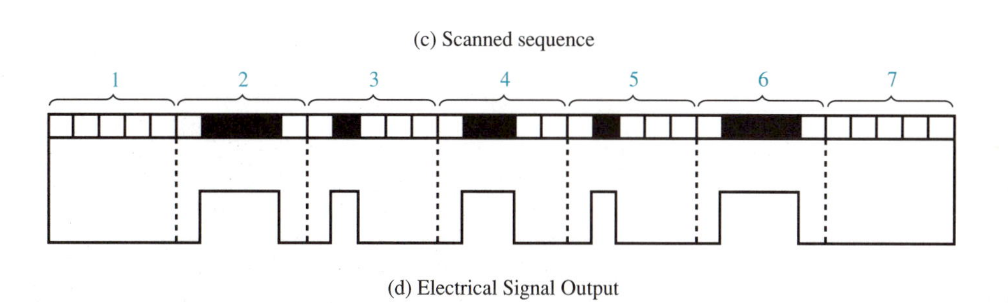

(d) Electrical Signal Output

Figure 12.9 Scanning of an object.

the entire matrix on the first line. The scan now retraces back to the left side and begins again. This time it encounters no image until it reaches the second block. The image is there for three spaces and then is gone. The camera retraces and continues through the entire scene and then begins again at the upper left corner. The resulting scanning sequence is shown in Figure 12.9(c). The first vertical scan has nothing in it, the second has three of the five spaces filled, and so on. This sequence is converted to an electrical output in Figure 12.9(d). There is a signal output every time a space is filled in. Where no dark area is scanned, there is no electrical signal output. The result is a reproduction of the original image at the output of the camera, which can be transmitted to an appropriate receiver.

Note that the amplitude of all the electrical signals in Figure 12.9(d) are the same. This can be attributed to the fact that we made the color of the "E" dark or black, creating an image of maximum amplitude because a negative transmission is being used. If we made the center of the "E" (vertical scan 4) a light shade of gray, for example, a smaller amplitude would be present for the two spaces shown in section 4 of Figure 12.9(d). The amplitude of these signals depends on what shade of gray the portion of the image is to be. Thus, the amplitude of the electrical signal determines the shade between black and white, and the position of the signal determines whether an image is present at that time.

Figure 12.10 shows a basic monochrome TV transmitter. In the first channel, the TV camera picks up the image to be transmitted. It then runs it through a video amplifier to ensure that the signal is at the proper level for transmission and later reception. As stated earlier, the video portion of a TV signal is an amplitude-modulated signal and therefore is sent to a conventional AM transmitter, such as those covered in Chapter 9.

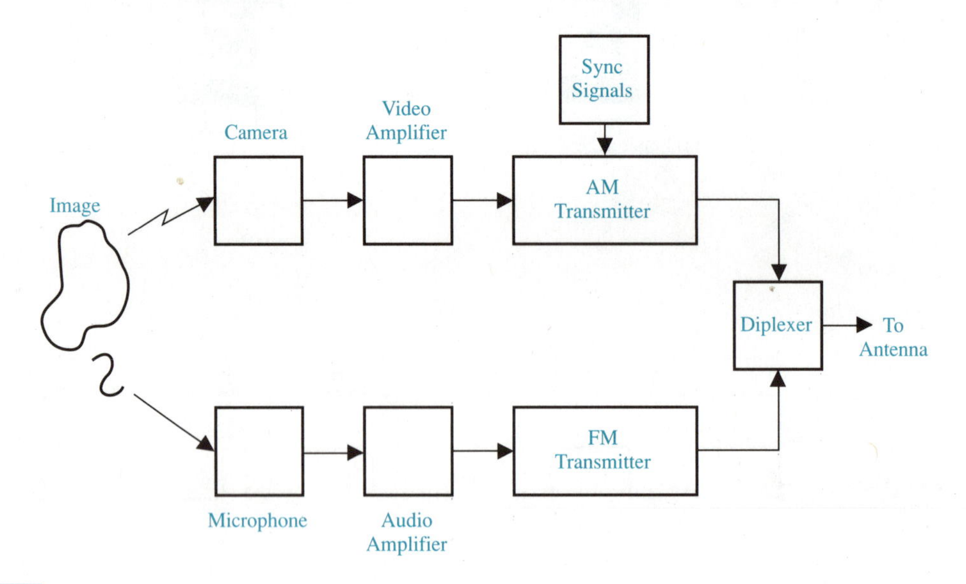

Figure 12.10 Basic monochrome TV transmitter.

The second channel in Figure 12.10, the audio channel, begins with a microphone in a studio sending the audio tones to an audio amplifier, like the circuits covered in Chapter 7. Recall also that the audio portion of a TV signal is frequency-modulated, and therefore, the amplified audio signal is sent to an FM transmitter, similar to those discussed in Chapter 10. The video and audio channels are combined at their outputs by means of a **diplexer**—a circuit that takes the AM video signal and the FM audio signal generated within the transmitter and combines them into the single composite signal discussed earlier in this chapter.

The diplexer has the unique property of combining two different signals, such as those we have presented, and sending them on to an antenna while completely isolating the signals from one another. You can imagine what type of TV program you would get if you simply tied the outputs of the video and audio channels together with a transmission line and sent them out on an antenna. To begin with, the impedance mismatch would probably be unbearable. In addition, the video would disrupt the audio, distorting the sound, while the audio would add an additional amount of amplitude modulation, scrambling the picture and creating the wrong shades of gray, and many things other than the desired picture. The diplexer, therefore, plays a vital part in the accurate reproduction of the original image at a TV station or on a remote location.

We have not yet discussed the sync signals block in Figure 12.10. This block is very important for the proper operation of the TV signal. Without it there would be no synchronization of the transmitted and received signal; that is, there would be no command to the receiver about when to begin each scan of the TV screen or when to blank the screen for a retrace. The viewer would receive a confused signal, containing various pieces of unrelated pictures as well as many additional retrace lines over the images. Thus, this group of circuits is very important to the overall operation of the system.

We need to investigate the video camera more closely in order to fully understand what happens when a TV picture is broadcast. Figure 12.11 is a diagram of a typical TV camera tube, commonly referred to as a **vidicon tube**. The image to be transmitted is projected on the surface of the tube by a lens whose focal point F is

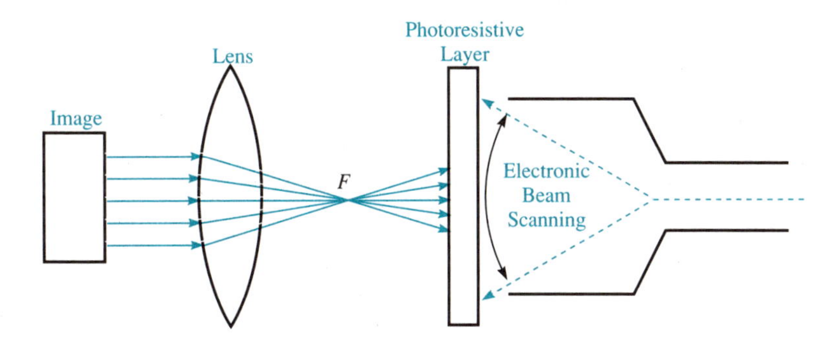

Figure 12.11 Black-and-white TV camera tube.

such that the image is projected within the surface area of the tube. The image is projected to a photoresistive layer at the face of the tube. These photoresistive materials may be either photomissive or photoconductive materials. **Photomissive** materials emit photoelectrons proportional to the intensity of the light striking the surface of the material, meaning, the brighter the picture, the more electrons. This is the *positive* transmission mentioned earlier.

The resistance of a **photoconductive** material is inversely proportional to the intensity of the light striking it. When excited by an electron beam, a picture element outputs a signal proportional to the intensity of the light striking it. If the elements are individually scanned in sequence, as illustrated in Figure 12.9, the amplitude of the output signal will vary according to the intensity of the beam being scanned.

This sensitive layer on the face of the tube consists of multiple light-active areas: the transparent conductive film, a semiconductor photoresistive layer deposited on the conductive film, and the photoconductive, or **mosaic layer**, deposited onto the semiconductor layer. The first, or transparent, layer allows the image to be impressed on the face of the tube. The second layer, the semiconductor photoresistive layer, exhibits a high resistance when it is dark; the resistance being reduced when struck by photons of light. The photoconductive layer has many separate areas that make up a mosaic arrangement similar to that in Figure 12.9(b). The layer within the tube, however, contains millions of areas as opposed to the 35 shown in Figure 12.9. This will give a much more fine-grained picture.

The electron beam scans across the mosaic areas in order to charge up each of the many tiny capacitors that are formed by the photoresistive layer, which acts as a dielectric; the conductive film which acts as one electrode; and photoconductive mosaic areas, which act as the other plate. Light on the mosaic portion of the material discharges these capacitors through a load resistor that is external to the tube shown in the figure. The scanning electron beam recharges the capacitors and produces a video-signal voltage drop across this load resistor, which is proportional to the light intensity at the individual areas being scanned. With this arrangement, the video output is a signal whose instantaneous output is the result of scanning just one of the capacitors at a time. Thus, a high degree of resolution is obtained. The interaction of the electron beam and the photons created by the image to be transmitted interact to form an electronic signal that can be used to reproduce the image at a remote TV set.

Although we have presented audio and rf amplifiers, we have said nothing specifically about video amplifiers. Recall from our previous discussions that the video portion of the signal shown in Figure 12.1 was characterized as going from dc to 4 MHz. This 4 MHz is not a high frequency but does require special precautions when designing and fabricating the circuits to be used at this frequency. We can, therefore, classify it as a low-rf amplifier circuit and take all of the previous statements about rf amplifiers into consideration when working with video circuits.

One area that will look somewhat different when working with video amplifiers is the collector circuit. There is a coil in this circuit, called the **peaking coil**, shown in Figure 12.12(a). This coil lies across the collector resistor, and as the frequency of operation increases, the parallel impedance applied by the coil aids in restoring the gain of the system. Remember, that as the frequencies of operation increase, the

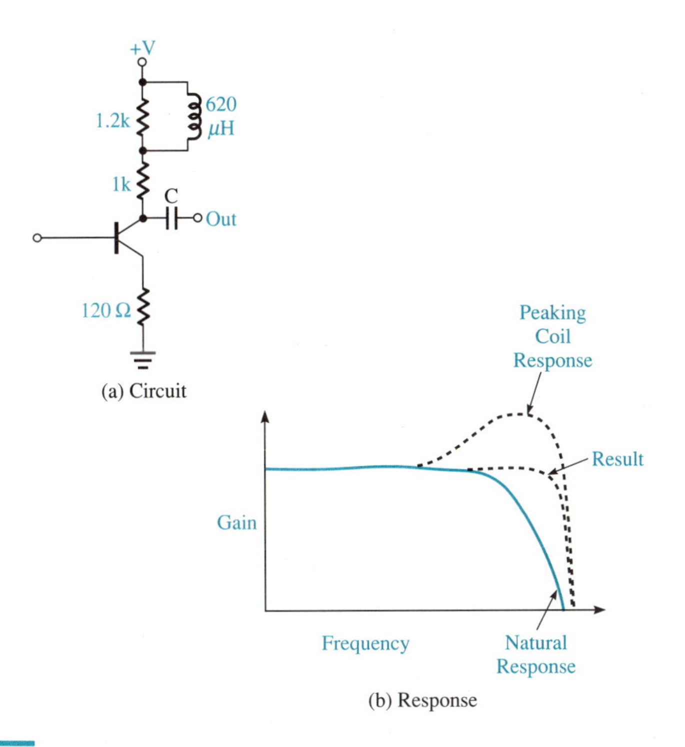

Figure 12.12 Video peaking-coil circuit.

interelement capacitances of the transistors as well as other stray capacitances will cause the gain of the system to roll off drastically at the high end of the response, as shown in Figure 12.12(b). The coil by itself will produce a ''peaking'' of the high-frequency response because of its high degree of control as opposed to that of the collector resistor at high frequencies. When this peaked response is combined with that of the natural roll-off response of the circuit without a peaking coil, the resulting response is flat out to, and somewhat beyond, the required 4-MHz bandwidth for proper video operation.

Having described the TV transmitter, we now look at a more familiar device, the TV receiver. The monochrome, or black-and-white, receiver is required to reproduce the image and sound from the transmitter exactly as it appears and sounds, exactly at the time it is occurring. Although this sounds like a large order, the information already presented tells us otherwise.

To understand the reception of a TV signal, recall that the signal consists of sync pulses, blanking pulses, and video. The picture is amplitude-modulated, the sound is frequency-modulated, and the frequency bands have a picture carrier and a sound carrier with specific bandwidths for each (as illustrated in Figure 12.1). Thus, we are receiving more than a singular AM or FM signal; it is a complex signal that requires special processing to achieve the picture on your TV set.

Figure 12.13 is a block diagram of a black-and-white TV receiver. We can break the diagram up into five distinct sections according to the individual task performed by each: rf section (tuner), i-f section, sound section, video section, and the horizontal and vertical deflection section. These five sections are more than adequate to receive and process the signal arriving at the antenna of the receiver. The rf section is tuned to the proper band to receive the required channel (as shown in Table 12.1); the i-f section processes this rf signal into a more usable signal at a lower frequency; the sound system processes the audio according to the sound carrier frequency (as shown in Table 12.1); the video section processes the picture, which has its own carrier frequency and is an AM signal similar to those discussed in Chapter 9: and, finally, the horizontal and vertical deflection uses the sync and blanking pulses to horizontally and vertically scan the picture and reproduce it as it was transmitted.

Having explained the basic functions of the five sections of a TV receiver, let us now look at each individual section and discuss the operation and features of each.

The rf section in Figure 12.13 shows an rf amplifier and a mixer–local oscillator combination. This is familiar as a typical superheterodyne receiver input. The rf

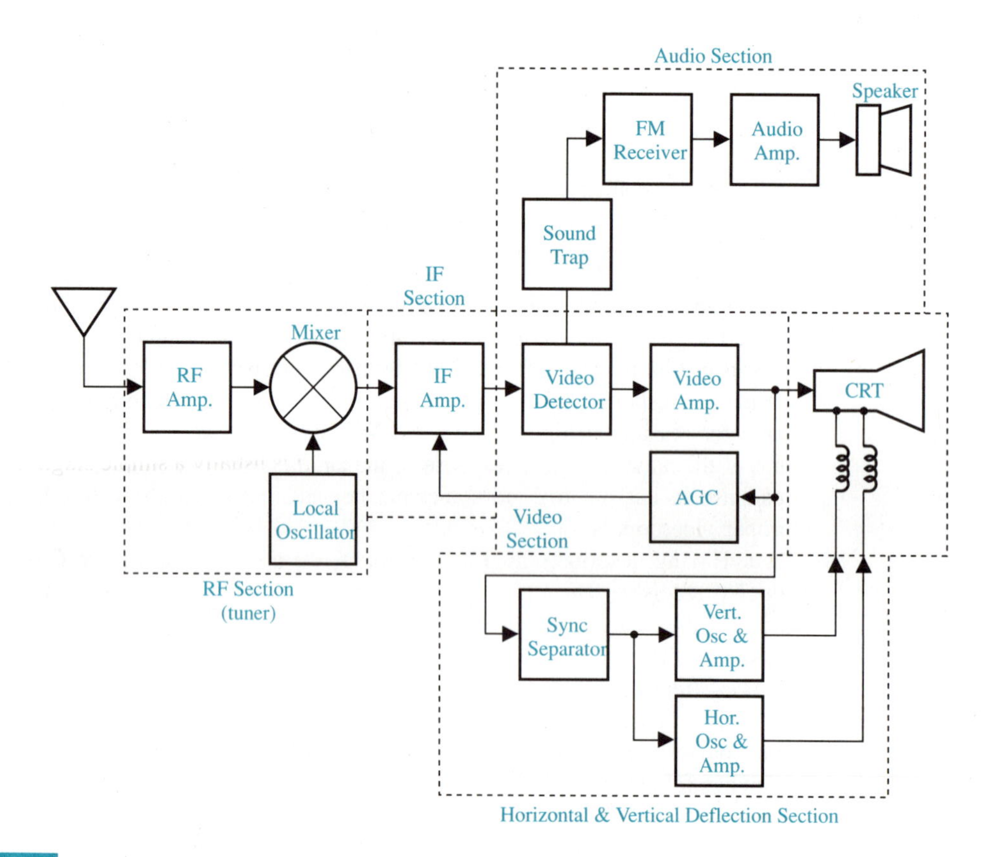

Figure 12.13 TV receiver block diagram.

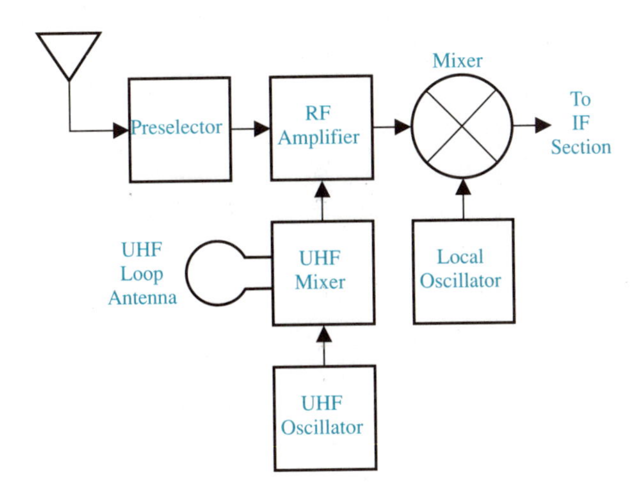

Figure 12.14 Revised rf section.

section provides channel selection, image frequency rejection, and converts the rf input to an appropriate i-f signal. Although the rf section is the basic input to the receiver, there is actually more to it than the figure shows. Most TV sets have VHF (Channels 2–13) and UHF (Channels 14–72) channels (these frequencies are listed in Table 12.1). The VHF range is from 54–216 MHz (with the 88 to 108-MHz band removed for commercial FM), and the UHF range is from 470–806 MHz. Therefore, most TV's usually have a switch that allows movement between VHF and UHF, as well as having separate antenna terminals on the back of the set. Including these exceptions produces the rf section that appears in Figure 12.14. There now are two separate inputs. One is the antenna used for VHF (Channels 2–13), which goes through a preselector (a series of filters, as described in earlier chapters on AM and FM) to an rf amplifier, and is mixed with a local oscillator signal to form an i-f. The second path is from the loop antenna, so familiar on TV sets, which detects the UHF signal. The signal is sent through a UHF mixer, which is usually a simple single diode mixer, and is mixed with the UHF oscillator output. The resulting output of the mixer is an rf signal compatible with the rf amplifier. When UHF signals are received, we can actually consider this rf amplifier to be another i-f amplifier, although higher in frequency than those in the following circuits.

As we have stated, one of the functions of the rf section is the selection of channels. To many people that is the *only* function, and undoubtedly the most important. Early TV sets had a turret switch for changing channels. Individual resonant circuits would be switched in to make the front end of the receiver a narrow-band receiver to receive only that station. Later only the VHF channels were switched in this manner and the UHF channels had a continuously varying control in which only one parameter of the L–C tank was varied over the UHF range. Today, digital systems have replaced these mechanical tuners. One representative electronic tuner, shown in Figure 12.15, uses varactor diodes with the appropriate dc voltage applied to them to

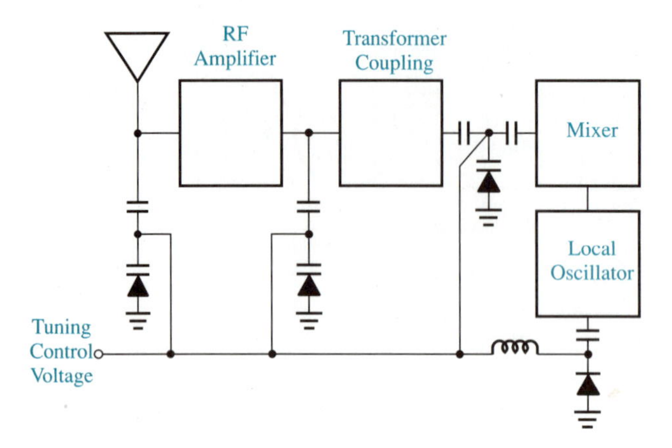

Figure 12.15 Electronic tuner.

tune to a specific channel. In this circuit a dc voltage (usually 1–27.5 V) controls the frequency-determining components for the rf amplifier, mixer, and local oscillator.

A phase-locked loop synthesizer to change the frequencies is a second method of channel selection. Figure 12.16 is a block diagram of such a system. This system works very well with remote controls because the microprocessor circuitry is compatible with the remote circuitry. Recall from Chapter 8 that this is an example of indirect synthesis. This system works well with remote control circuitry, but it is also used for front-panel touch controls for channel selection, volume control, and on–off switching of the TV.

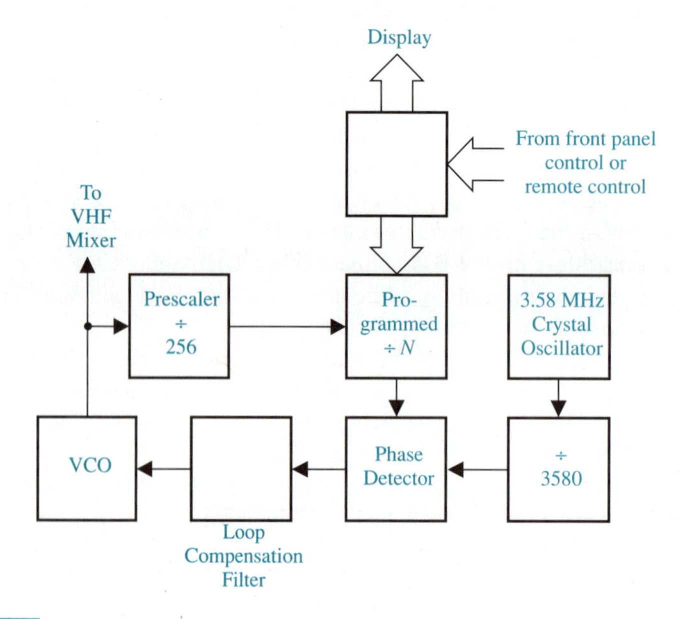

Figure 12.16 PLL synthesizer for channel selection.

The i-f section is usually a multistage system with AGC applied to each stage. These stages are required to amplify both the video and audio signals in the TV system. For commercial systems the frequencies used are 45.75 MHz for the picture and 41.25 MHz for the sound. This results in a 4.5-MHz separation, which you recall, is the exact separation between the picture carrier and the sound carrier (as presented in Table 12.1 and shown in Figure 12.1). There are two points of note in the response curve of the i-f shown in Figure 12.17. First, we notice the frequencies: the sound carrier is at 41.25 MHz, and the picture carrier is at 45.75 MHz. There is also a 0.75-MHz frequency difference shown below the picture carrier (45 MHz). This difference compensates for the vestigial sideband, which is 3 dB down (0.75 MHz below) the picture carrier (Figure 12.1). This will mean that there is twice as much voltage in the signal for the first 0.75 MHz than for all higher frequencies. To compensate for this, the picture carrier is set halfway down the i-f skirt (−3 dB) at the high-frequency end of the response curve, as shown. The second point to note is the levels of the individual frequencies: on the high end (−3 dB at 45.75 MHz), and on the lower gain level for the sound carrier (−10 dB). We've already explained the first level. The second level occurs because a TV system needs more gain in the video portion of the signal than in the audio. For this reason the gain of the higher picture carrier is more than that in the sound portion.

Instead of the 4-MHz band we have been referring to throughout this text, a 3-MHz i-f passband is sometimes used for color applications where the interference from the color signal is possible. **Wavetraps** (narrow-band band-stop filters) are used to narrow the i-f passband.

The video section, as seen in Figure 12.13, consists of video amplifiers, a video detector, AGC circuitry, and the CRT that displays the video. A typical block diagram for the video section is shown in Figure 12.18. The output of the i-f section is sent to a video detector, which is usually a single-diode peak detector similar to the type described in Chapter 9. The output of the detector is sent first to a video amplifier and then to the sound section. The sound output is centered at 4.5 MHz.

The output of the first video amplifier also has two paths to take. One goes to an additional video amplifier, and the other sends the composite video signal to the

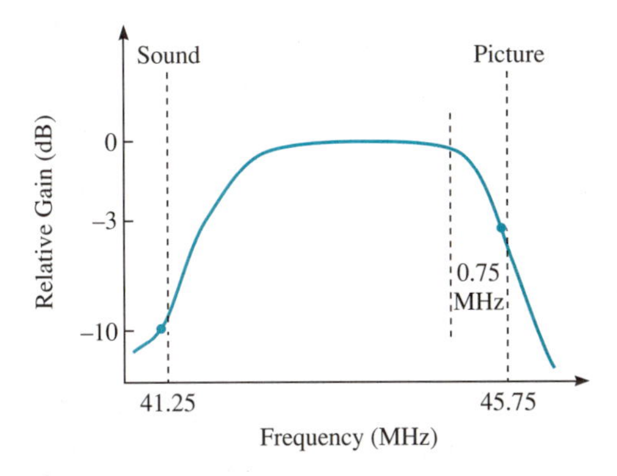

Figure 12.17 I-f response curve.

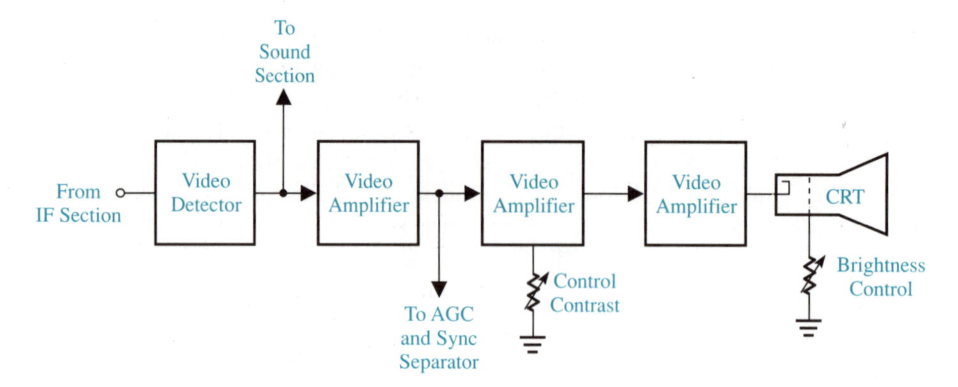

Figure 12.18 Video section block diagram.

AGC circuitry and to the sync separator, where the vertical and horizontal deflection signals are developed. The second stage of video amplification is where the *contrast* control for the TV is located, whereas the *brightness* control is a potentiometer on the control grid of the CRT.

The audio section, with its 4.5-MHz input from the video detector, is basically an FM receiver. Figure 12.19 shows an expanded block diagram of the section presented in Figure 12.13. The 4.5-MHz trap ensures that a 4.5-MHz signal with a narrow bandwidth will be present at the input to the audio section. The limiter, as described in Chapter 10, is a standard circuit in many FM receivers because the amplitude of the signal must be held constant. This is the circuit that ensures a constant amplitude is maintained. An additional stage of amplification is inserted prior to the FM detector. This detector may be any form of FM discriminator covered in Chapter 10, but Foster-Seely or ratio detectors are popular choices in many TV applications. The final stage is a simple audio amplifier, like the circuits covered in Chapter 7, whose function is to raise the audio to a level sufficient to drive the speaker of the TV.

The final section is the horizontal and vertical deflection section, which consists of the sync separator, the vertical oscillator and amplifier, and the horizontal oscillator and amplifier. Its input comes from the video section, and its outputs are applied to the vertical and horizontal deflection coils of the CRT. This section takes the complex TV signal and makes it presentable on the screen.

The sync separator does exactly what the name implies, it separates the vertical and horizontal sync pulses. Figure 12.20 shows a sync separator circuit. A clipper

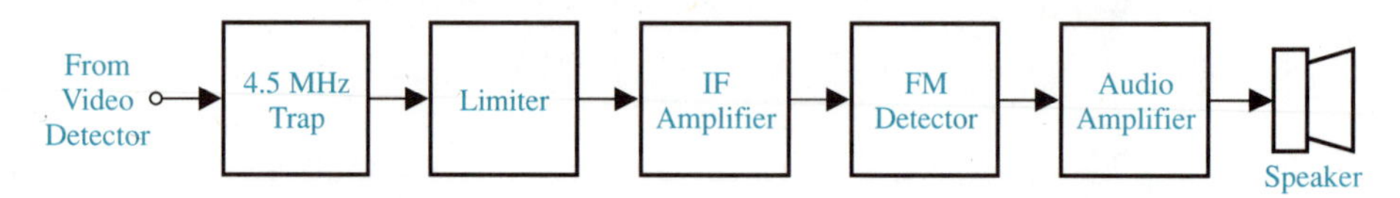

Figure 12.19 Audio section block diagram.

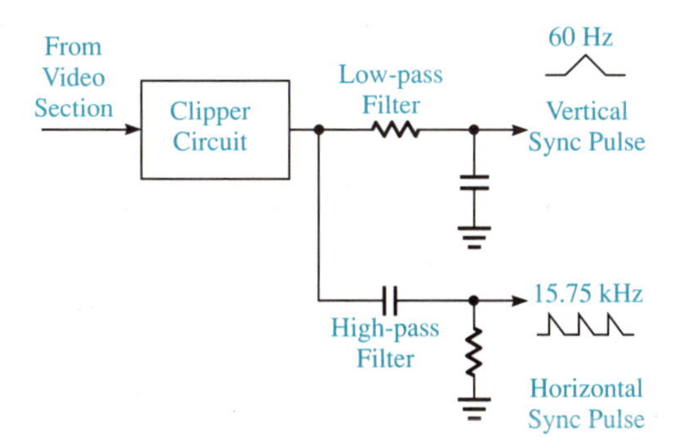

Figure 12.20 Sync separator.

circuit removes the blanking level and the video from the incoming signal and sends the resulting signals on to two filters. A low-pass filter allows the 60-Hz vertical sync pulses to pass, and a high-pass filter passes the 15.75-kHz horizontal sync pulses.

The first output of the sync separator, the vertical sync pulse, is fed to a vertical deflection oscillator that produces a 60-Hz linear, sawtooth-shaped deflection voltage, resulting in the vertical scan present on the screen of the TV. This circuit is often a blocking oscillator which is a triggered oscillator that produces a sawtooth output synchronized with the incoming sync pulse rate. This circuit is ideal for TV applications because they require a very linear sawtooth waveform at the 60-Hz vertical sync rate. The output of the vertical oscillator is then amplified to produce the high-voltage sawtooth required to drive the vertical deflection coils on the CRT.

The horizontal output (15.75-kHz pulse) of the sync separator is sent to a horizontal oscillator. When dealing with the horizontal circuits, efforts must be made to improve noise immunity, and thus eliminate false sync pulses. For this reason AFC is used for the horizontal deflection oscillator. The oscillator is held in sync by comparing the horizontal sync pulses to a signal fed back from the horizontal output. This comparison is done with phase-detector diodes. Any difference in the two signals is detected as a phase difference and is applied as a dc level to the input of the horizontal oscillator (usually the base of a transistor). This dc voltage then corrects any differences.

Similar to the vertical output, the output of the horizontal oscillator is sent to an amplifier to raise the level of the signals sufficiently to operate the next circuit—the **flyback transformer**. This high-voltage component allows horizontal scanning of the CRT and can be heard as a high-pitched signal within the TV chassis. Technicians will often listen for this sound to verify that the horizontal circuit is working properly. The flyback circuitry operates as follows:

1. The sawtooth level builds up on the input of the horizontal amplifier, causing a current to build up in the transformer primary and in the damper diode.
2. The sawtooth reaches its peak and suddenly changes during the retrace cycle, allowing the current to drop to zero.

3. The magnetic field around the horizontal yoke coils collapses and induces an electromotive force (emf), called a flyback, across the transformer secondary.
4. A pulse of current is now induced in the transformer secondary, and a flyback is induced in the primary.
5. The high voltage (in the kilovolt range) is rectified and applied to the CRT as an anode voltage.

A damper diode is a protective device. During the flyback period, a damped oscillation is produced by the collapsing magnetic field, which may interfere with the following sawtooth as it begins to rise again. The **damper diode** is placed at the output of the horizontal amplifier and serves as a short circuit during the flyback time, thus serving to damp out any unwanted oscillations.

From our examination of the apects of five monochrome TV receiver sections, we see that TV systems are complex pieces of equipment that require much in the way of design effort and manufacturing expertise.

12.2.3 Color Television

Many of the circuits described in the previous section for monochrome TV are used for color TV, as well. By knowing the basics of monochrome, you should be able to grasp the ideas behind color TV simply by adding a few more fundamentals that apply to color TV alone.

Recall that specific levels of black and white and different shades of gray go into creating a black-and-white picture. These combinations produced an acceptable monochrome display, which was used for many years on millions of TV sets. The production of color, however, is somewhat different. All of the colors produced on a TV screen are created by combining three primary colors: red, green, and blue, as illustrated in Figure 12.21, a Venn diagram. The Venn diagram is a diagram, named for English logician John Venn, in which circles or ellipses are used to give a graphic representation of basic logic relations. In this case the logic relations are primary colors. Notice in the diagram that the center is designated as W for white. This means that if all three colors are emitted, white will be the result. This is not to say that equal amounts of red, green, and blue will result in white. Actually a combination of

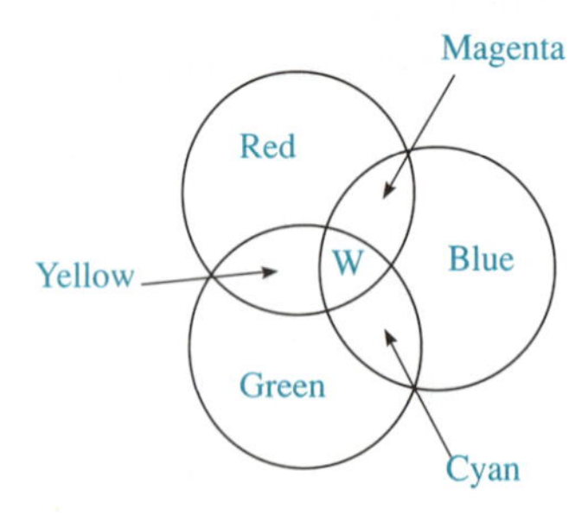

Figure 12.21 Venn diagram.

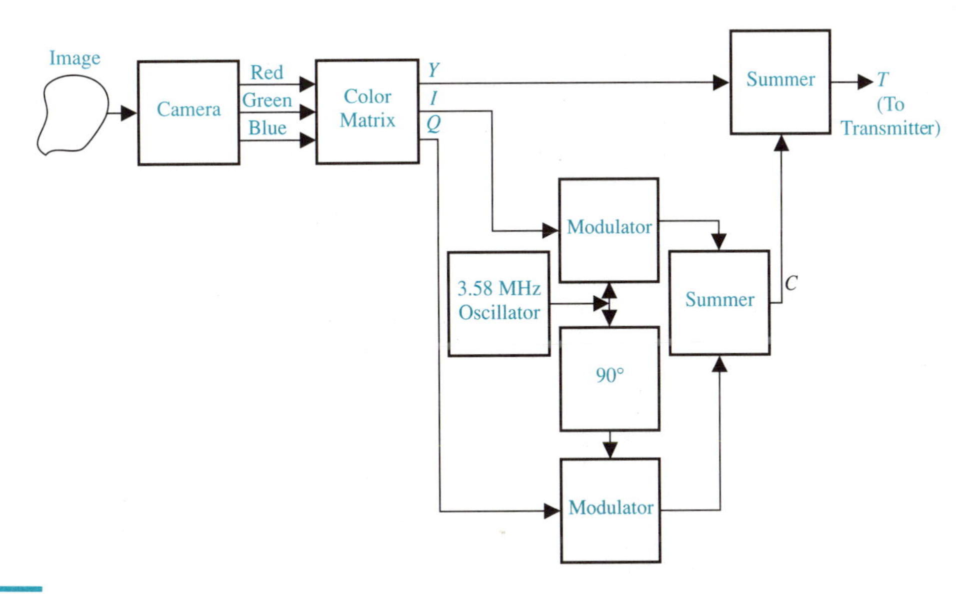

Figure 12.22 Color TV camera and transmitter.

11% blue, 59% green, and 30% red will result in white. This is another important point to remember: It is not necessarily which colors are emitted but rather the percentage of each color which results in a particular color being present on the CRT.

You may wonder why red, green, and blue where chosen to be the colors for the color TV systems. The reason goes back to the sensitivity of the human eye. Red is the electromagnetic radiation of longest wavelength to which the human eye is sensitive. Similarly, blue is a short wavelength (the only color that is shorter is violet). And, finally, green is approximately between red and blue, so that all the colors of the rainbow can be produced by using these three colors.†

The color TV camera differs from the monochrome camera in that it is actually three cameras in one case. When an image is scanned, separate tubes are used for each of the primary colors. The colors are then combined in an encoder to produce the composite color signal.

To understand the idea of color encoding, refer to Figure 12.22, a basic block diagram for a color TV camera and transmitter. The image to be projected is picked up by the camera and produces a certain combination of red (R), green (G), and blue (B). This combination is sent to a color matrix, which produces the brightness (**luminance**), or Y, video signal, and the I and Q **chrominance**, or color, video signals. The luminance signal is the same as the monochrome signal discussed in detail in Section 12.2.2. The I (in-phase) and Q (quadrature) signals are each sent to a modulator where they are multiplied with a 3.58-MHz oscillator signal. This signal is called the **color-burst signal**, designated in Figure 12.1 as the color subcarrier. It was

† In the art world, the primary colors are red, yellow and blue, which refer to pigments, rather than light.

shown as 3.58 MHz above the picture carrier in the figure. The I and Q signals are combined in a summer to form the composite signal designated as C in Figure 12.22. This combined signal is a **quadrature-amplitude-modulated signal** (QAM), a form of digital modulation where the digital information is contained in *both* the amplitude and phase of the transmitted carrier. This is considerably different from any type of modulation scheme we have covered in this text. Recall that whenever we described a type of modulation previously, we varied only one parameter: the amplitude, frequency, or phase. With QAM, both the amplitude and phase are varied. This type of modulation is digital, and is covered extensively in digital texts.

The QAM signal C is finally combined with the original Y signal to form the total composite video signal T.

Although you may not have realized it, the luminance signal is the white signal that we arrived at by adding together the appropriate amounts of red, green, and blue. Recall that the formula was 11% blue, 59% green, and 30% red. If we change this to an equation we have

$$Y = 0.11B + 0.59G + 0.30R \tag{12.2}$$

If we used only the Y signal to produce a pattern on a TV receiver, we would see seven bars on the screen. The bar on the extreme left side would be white and the bar on the extreme right would be black. The five bars between would be varying shades of gray going from a light shade on the left to a dark shade on the right. The bandwidth for a monochrome picture is 4 MHz; however, this bandwidth is usually reduced to eliminate any interference with the 3.58-MHz color signal. For this reason, the Y signal is usually run with a 3.2-MHz bandwidth.

The I signal is transmitted with a bandwidth of 1.5 MHz and the Q signal with a 0.5-MHz bandwidth, producing the frequency spectrum shown in Figure 12.23.

The chrominance signal, or C, is the most important signal in the color TV system. This is the one, as you recall, that is a combination of the I and Q color signals and is the actual color combination of the image to be transmitted. The I signal is fabricated by combining 60% of the red signal, 28% of the inverted green signal,

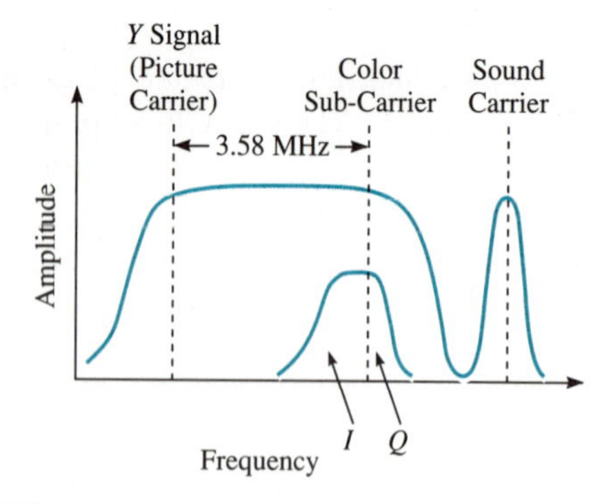

Figure 12.23 Color TV frequency spectrum.

and 32% of the inverted blue signal. If we express as we did for the Y signal in Equation (12.2), we have

$$I = 0.60R - 0.28G - 0.32B \qquad (12.3)$$

Notice the difference in this signal as compared with the Y signal in Equation (12.2). Rather than the direct addition of colors in the I signal computation, there are subtractions in the equation that are representative of inverted colors.

The Q signal is fabricated by combining 21% of the red signal, 52% of the inverted green signal, and 31% of the blue signal. Once again, we can express this mathematically as

$$Q = 0.21R - 0.52G + 0.31B \qquad (12.4)$$

Notice in this expression that there is a combination of additions and subtractions, indicating that some regular colors are being added to other inverted colors. The result is the Q signal, which is now ready to be added to the I signal to form the complete composite color signal.

The I and Q signals have been said to be in quadrature (90° apart), and as such they are combined by taking their phaser sum. That is,

$$C = \sqrt{I^2 + Q^2} \qquad (12.5)$$

Equation (12.5) shows us that the amplitude and phase of the complete composite color signal that is developed in the transmitter from the camera and the transmitter circuitry is dependent on the amplitude of the I and Q signals. These amplitudes, in turn, are proportional to the percentage of red, green, and blue, thereby allowing us to produce all the colors of the rainbow from three basic wavelengths of light.

The **color wheel** used for standard broadcast television is shown in Figure 12.24. The R–Y and B–Y signals are used for demodulating the individual colors in a TV receiver. The **hue**, or **color tone**, of the signal is determined by the phase of the C signal, with the depth being proportional to the magnitude. The outside edge of the color wheel in the figure corresponds to a relative value of 1.

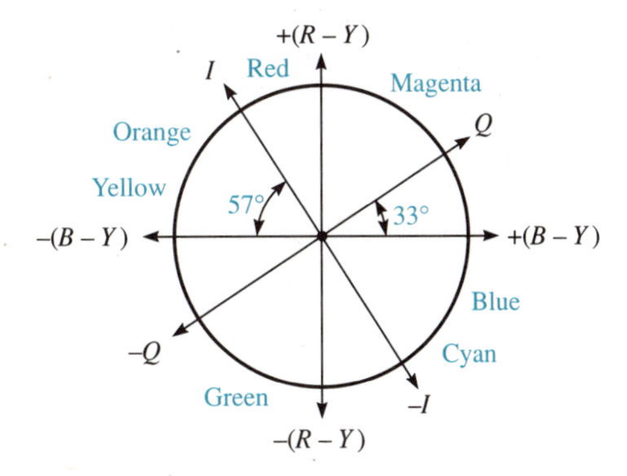

Figure 12.24 Color wheel.

When we used the term **color burst** before, we associated a frequency of 3.58 MHz with it. Because the phase of this 3.58-MHz signal is the reference for color demodulation, this subcarrier must be transmitted with the composite video signal so that the receiver can reconstruct the original signal to reproduce and correct color at the proper time in the transmission. Eight to ten cycles of the subcarrier are inserted on the back porch of each horizontal blanking pulse, as shown in Figure 12.25. The term *color burst* comes from this burst of eight to ten cycles that is removed in a receiver to help synchronize and reproduce the colors. Recall that we allowed approximately 7 μs for the back porch time of the blanking pulse. Given that the time for one cycle of the 3.58-MHz color carrier is 0.27 μs, we discover that ten cycles easily fit within the 7-μs time frame.

The receiver used for color TV is basically the same as that used for monochrome TV, except for the picture tube and the color-decoding circuits. The picture tube is very different from the conventional CRT. In the typical CRT the inside surface of the screen is coated with a special mixture of phosphors that emit white light after high-energy electrons from the beam have raised the electrons in the phosphor atoms to a high-potential state. A photon of light is released during this transition from excited to normal state. The varying levels of the signal determine the black, white, or gray regions of the picture. The color CRT, however, has three separate phosphors placed on the screen, one each for red, green, and blue. These phosphors emit the appropriate color when struck by electrons. The phosphors are kept separate but close

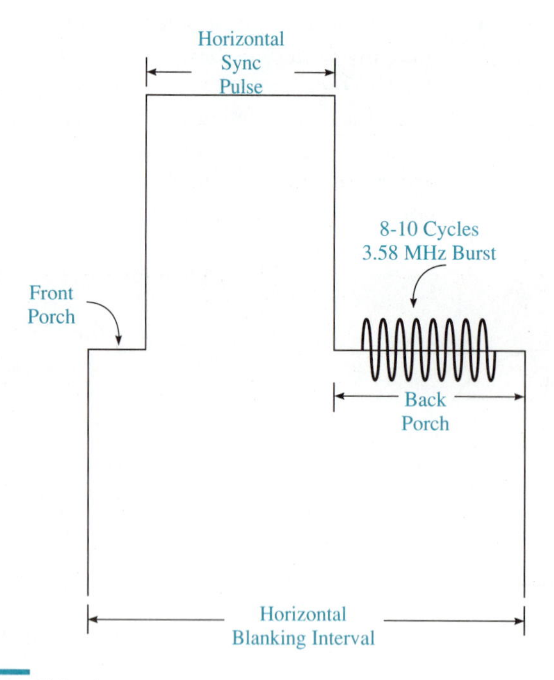

Figure 12.25 Color burst.

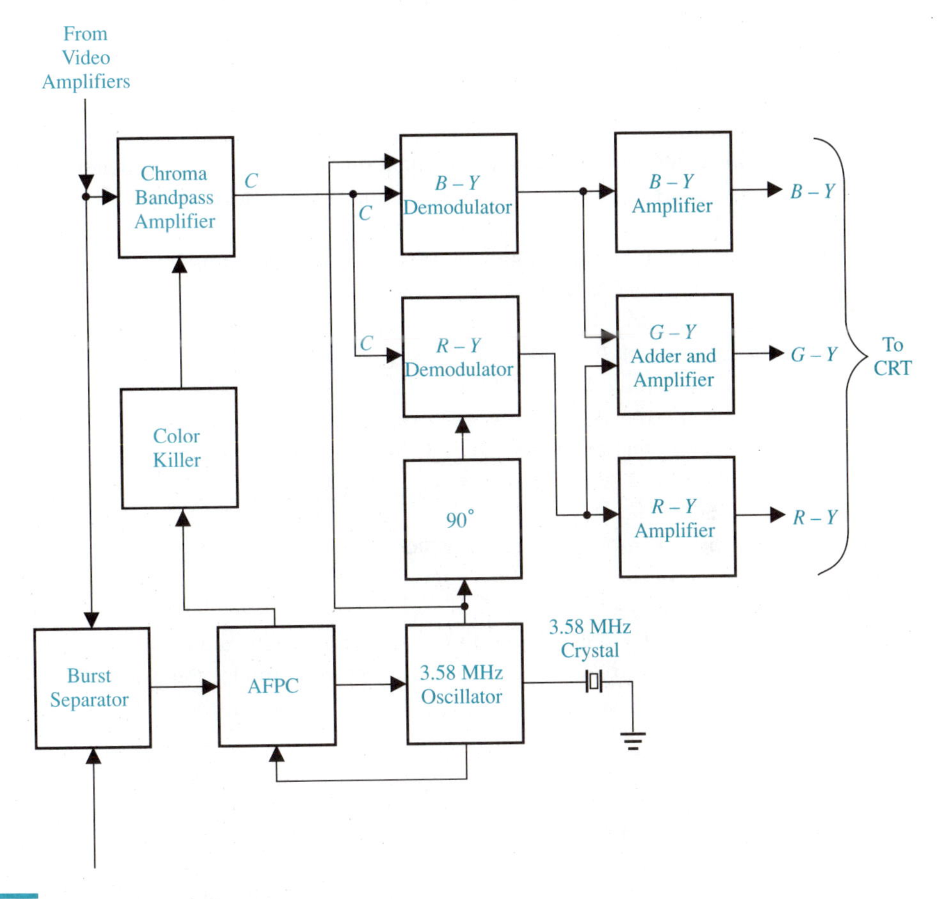

Figure 12.26 Color-demodulator circuit.

together in alternating stripes or as small dots in a tight matrix. This allows the electron beam that has been modulated with the color signal to activate the appropriate stripe or dot and combine them for the proper colors on the screen face.

A typical color-decoding, or demodulation, scheme (Figure 12.26) shows that the composite video is supplied from the video circuits and applied to a chroma bandpass amplifier—basically a tuned amplifier centered at 3.58 MHz with a bandwidth of 0.5 MHz. With this type of circuit present, only the C signal can be amplified and sent on to the demodulator circuits. The color burst described above is removed from the blanking pulse by means of the burst separator by switching on the separator only during the horizontal flyback time. The 3.58-MHz subcarrier is reproduced using the color AFC circuitry, which consists of the 3.58-MHz oscillator shown in Figure 12.26. This oscillator is controlled by a 3.58-MHz crystal and the AFPC (automatic frequency and phase control) circuit. This circuit combination ensures that the color reference stays on frequency, which is of vital importance because the color combination depends on everything being synchronized when it gets to the receiver and to the screen of the CRT.

The circuit called the **color killer** turns off the chroma amplifier during any monochrome operation. This circuit is used to ensure that either the right color or no color is received. This is an excellent diagnostic for video problems with a TV. If the color is not right from the video section, the color killer will cut out all color and the TV displays a black-and-white picture.

The full demodulation process occurs by combining the C signal with the phase-coherent, 3.58-MHz subcarrier. The B–Y and R–Y signals produce the red (R) and blue (B) video signals by combining them with the Y signal (which is the luminance signal consisting of 30% red, 59% green, and 11% blue) in the relationship of B–Y + Y = B, and R–Y + Y = R. The green signal is produced by also using the B–Y and R–Y signals in the proper proportions to result in a green signal. If we relate these combinations to the I and Q signals, we find that

$$\text{Green} = -I - Q + Y$$

$$\text{Blue} = -I + Q + Y$$

$$\text{Red} = +I + Q + Y$$

Now that the appropriate colors have been regenerated, they are sent to the face of the CRT to activate the proper phosphors and produce the exact color picture that was picked up by the TV camera in the studio or out on location.

12.3 Satellite Communications

The world of **satellite communications** was literally launched in 1957 when the Soviet Union put *Sputnik I* in space. Although it only operated and transmitted information for 21 days, it shocked other nations into getting their own satellites into orbit. From that point on the satellite boom was under way. Today there are volumes filled with the satellites launched every year by different countries.

It is difficult to imagine our world without the conveniences provided by satellites and satellite communications. Besides providing a large number of channels for television, the communications satellite provides accurate and efficient communications for both civilian and military applications. There is also the application the original *Sputnik I* satellite was designed for, *telemetry*. Whatever the motivation for launching the first satellite, it has greatly improved many aspects of life throughout the world.

The following sections are intended merely to introduce satellite communications; they are not a comprehensive coverage of the topic. It is important that you be aware of this type of communications in order to round out the discussions of analog communications. Many texts have been written specifically on this topic and can be consulted for more in-depth study.

12.3.1 Satellites and Link Models

Communications satellites come in a variety of sizes and shapes, their physical dimensions dependent on their functions and the size of the propulsion system available to put them into orbit. Their function also will determine which of two basic types of orbit they will finally rest in: orbital (or nonsynchronous) and geostationary (or geosynchronous). The **orbital (nonsynchronous)** orbit revolves around the earth. The satellite rotates around the earth in a low-level elliptical or circular path with its angular velocity either greater or less than that of the earth. If the velocity is greater than the earth's, it is called **prograde**. If the velocity is less than that of the earth, it is called **retrograde**, which results in the satellite never being stationary relative to any point on the earth. This can be rather inconvenient for some applications, but if the operator only needs to use the satellite for a short time each day, it may be the best solution. This type of orbit, however, is seldom used today.

Geostationary, or **geosynchronous**, satellites are used almost exclusively today for satellite communications. With this type of orbit the velocity of the satellite exactly matches that of the earth, allowing the satellite to remain in the same location above the earth 24 hours a day, thereby providing for constant use. Although this type of orbit sounds ideal, there are advantages and disadvantages with the geosynchronous orbiting system.

Advantages

1. Tracking equipment is not needed because the satellite remains basically in one spot all the time.
2. Several different satellites are not necessary because the same one is used at all times. Thus, there are no interruptions in transmission.
3. The high-altitude orbit possible with geosynchronous satellites allows very wide coverage on earth as compared with the nonsynchronous satellites, which achieve only low-altitude orbits.

Disadvantages

1. Because geosynchronous satellites are at such a high altitude (22,500 miles), propagation delays occur that are not encountered with lower level orbits. Delays in the order of 300 to 500 ms may be encountered for round-trip transmission.
2. Higher power ground transmitters and higher sensitivity receivers are needed in geosynchronous satellites.
3. Spacecraft requirements are more sophisticated for launching geosynchronous satellites, and propulsion systems are needed to keep them in their assigned orbit.

Once the geosynchronous satellite is in orbit, we need to have a way of finding it. To determine the location of a satellite we need to know two parameters: the elevation and the azimuth. The term used to denote these is the **look angles**. The **elevation angle** θ_E is formed between the plane of the wave radiated from the earth station and the satellite, that is, the angle between the satellite and the horizon of the

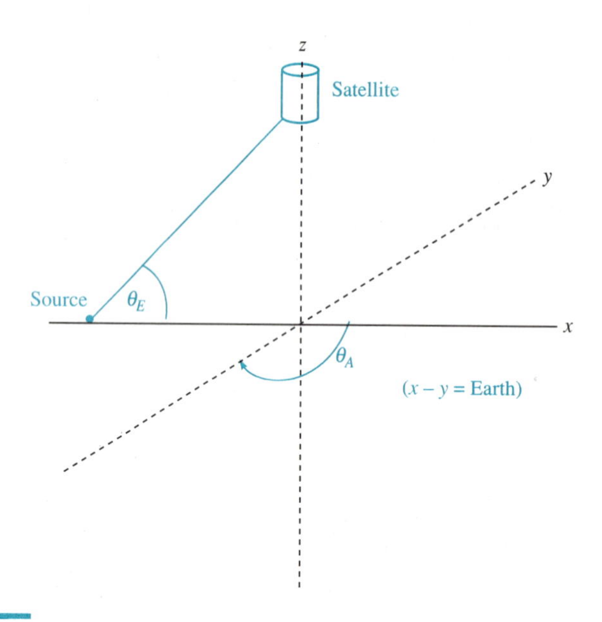

Figure 12.27 Look angles.

earth (as shown in Fig. 12.27). It is obvious that it is important to know the elevation angle if you are trying to locate a satellite. If this angle is too small, the satellite may be out of range or may be starting to appear over the earth's horizon, making communications impossible. Thus, the minimum tolerable elevation angle is 5°; any angle below this will result in a rapidly increasing attenuation of the signals.

Knowing only the elevation angle of a satellite is like telling a friend you will meet them in the mall. The mall has over 100 stores and you like to browse through all of them. Your friend may eventually find you, but may be a bit upset at having to spend so much time doing so. If only the elevation angle of a satellite is given, it is going to take some time to find it. In the case of your friend it is much easier if you suggest a specific store as a meeting place. Similarly, it is much easier and faster to locate a satellite if you are also given the satellite's azimuth. The **azimuth** is simply the horizontal angle (in degrees of arc) of an antenna measured in a clockwise direction from true north, shown in Figure 12.27 as θ_A. Figure 12.27 demonstrates that the satellite can be found very accurately if both angles are known.

Recall from Chapter 4 that every antenna has a radiation pattern. Since there are antennas on satellites, there also are patterns, that are formed which depend on many factors. These patterns for the satellite are called **footprints**, which fall into three categories: spot, zonal, and earth (as shown in Fig. 12.28). The **spot pattern** covers a specific geographic location, metropolitan New York City, for example. **Zonal patterns** cover less than one-third of the earth's surface, and the **earth pattern** covers approximately one-third of the earth's surface. The earth pattern is achieved when the beam width of the antenna is approximately 17°.

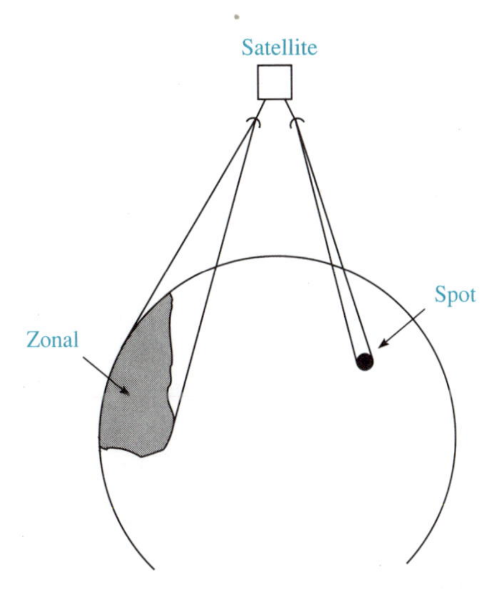

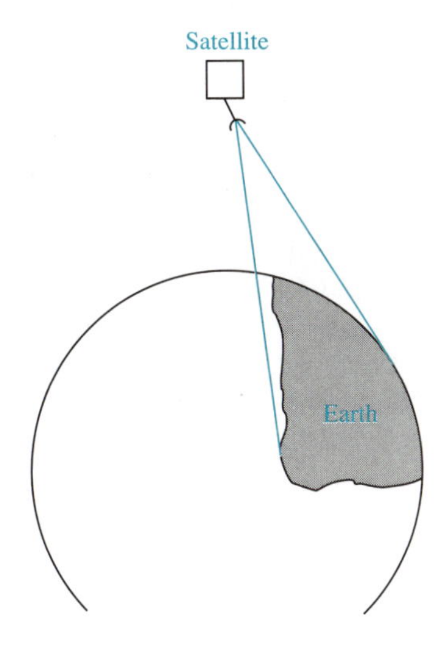

Figure 12.28 Radiation patterns-footprints.

Having presented some of the terms used for satellite communications, we now look at a satellite system and describe the links that make up a communications system.

A basic system for satellite communications consists of three sections: an uplink, a satellite transponder, and a downlink (as shown in Figure 12.29). The **uplink** is the ground-station transmitter, through which the intelligence is placed on an rf carrier and sent to a high-power amplifier to be transmitted to a satellite. Modulation schemes

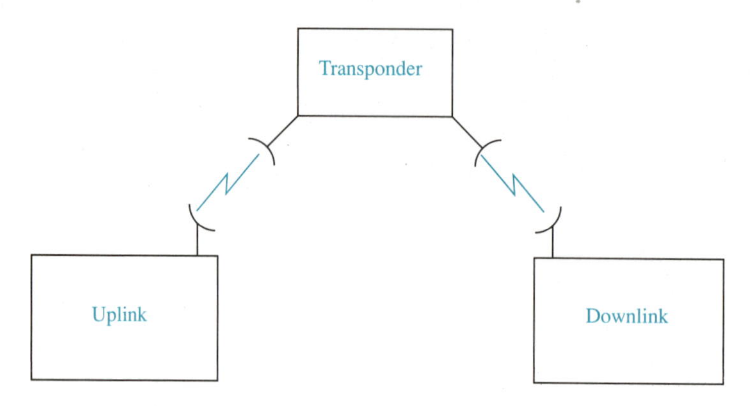

Figure 12.29 Satellite communications system.

that may be used are FM, QAM (introduced in Section 12.2), or PSK (phase-shift keying, a digital modulation). The high-power amplifier in this link is necessary because the modulated signal must be transmitted a great distance to the geosynchronous satellite, which is usually greater than 22,000 miles above the earth. Recall from Chapter 2, that many layers of the atmosphere must be penetrated to get to the satellite, thus requiring a great deal of power.

The **satellite transponder** can be considered as a conversion unit, whose function is to receive the signal sent from the ground-station transmitter, convert it to a different frequency, and retransmit this new frequency. Some typical conversions for satellites presently in operation are shown in Table 12.2.

The **downlink** is the earth-station receiver, whose task is to recover the very small signal the satellite transponder sends back to earth and reproduce the original intelligence that was placed on the rf carrier at the earth-station transmitter (uplink). The primary requirement for this link is that it have very low noise characteristics at the input to the receiver. A signal coming from 22,000 miles above the earth is very small, and a considerable amount of noise may accompany it, making it difficult to distinguish between the signal and the noise. The receiver must, therefore, have excellent noise properties.

Although not mentioned in the beginning of this section, **cross-links**, which are transmission links *between* satellites, are most important when discussing long-dis-

Table 12.2

Uplink (GHz)	Downlink (GHz)
5.9–6.4	3.7–4.2
7.9–8.4	7.25–7.75
14–14.5	11.7–12.2
27–30	17–20
30–31	20–21
50–51	40–41

tance satellite communications. When most people think of satellite communications, they imagine a signal going to a satellite and being either reflected off it or retransmitted from it to another station on the ground. Often, however, the signal is sent into space, received by a satellite, retransmitted to one or more satellites in space, and then sent back to earth. These cross-links are vital to many types of transmission system because they greatly increase the range over which a signal may be transmitted.

12.3.2 System Parameters

As with any system, satellite communications systems have specific parameters that must be understood and calculated before the system can be realized. Some of the parameters we will look at are transmit power, bit energy, EIRP, noise properties, and carrier-to-noise ratio, all of which lead to the important part of any satellite link, the link equations for the uplink and downlink. These equations are part of the link budget that governs the operation of satellite communications systems.

The first parameter listed above is **transmit power**, referred to in the uplink portion of our link models in Section 12.3.1. There we emphasized how important the transmitted power was to the operation of the uplink system. Associated with the transmitted power is the term **back-off loss** (designated as L_{bo}), which refers to the amount that a final power amplifier is backed off in power to keep it in a linear range of operation. By keeping the amplifier in this linear range, intermodulation distortion is greatly reduced. Recall our discussions of linearity and distortion in previous chapters. Thus, there usually is a term (expressed in decibels) which is inserted into calculations for the back-off loss of a system.

A parameter that is meaningful when using digital forms of modulation is energy per bit, designated as E_b. It is directly related to the power of the carrier, as

$$E_b = P_t T_b \qquad (12.6)$$

Where: E_b = Energy per bit (in joules per bit)
P_t = Total carrier power (in watts)
T_b = Time for a single bit (in seconds)

If we go one step further and realize that $T_b = 1/F_b$, Equation (12.6) can be written as

$$E_b = \frac{P_t}{F_b} \qquad (12.7)$$

Example 12.4

A communications system has a transmission rate of 60 Mbit/s (megabit per second). For a transmitted power of 2500 W, determine the energy per bit.

Solution

We can approach this problem in one of two ways: either use Equation (12.6) and calculate T_b, or use F_b and use Equation (12.7).

1. Using Equation (12.6), we need to find the time for a single bit T_b.

$$T_b = \frac{1}{F_b}$$

$$= \frac{1}{60 \times 10^6}$$

$$= 0.016 \times 10^{-6} \text{ s}$$

Substituting into Equation (12.6), we have

$$E_b = P_t T_b$$

$$= 2500 \text{ W (or joules)} \times 0.016 \times 10^{-6}$$

$$= 40\ \mu\text{W, or } 40\ \mu\text{J}$$

2. Using Equation (12.7), we have

$$E_b = \frac{P_t}{F_b}$$

$$= \frac{2500}{60 \times 10^6}$$

$$= 40\ \mu\text{J}$$

Example 12.5

In the problem presented in Example 12.4, the energy per bit must remain at 40 μJ/bit. If the power is reduced to 1500 W (J), what is the new transmission rate?

Solution

Using a modified form of Equation (12.7), we have

$$F_b = \frac{P_t}{E_b}$$

$$= \frac{1500}{40 \times 10^{-6}}$$

$$= 37 \text{ Mbit/s}$$

Covered in detail in Chapter 4, effective isotropic-radiated power (EIRP) was defined as the product of the total radiated power and the transmit antenna gain. If we change this to decibels and take the losses into consideration, we find that the EIRP is now

$$\text{EIRP} = P_t - L_{\text{bo}} - L_{\text{bf}} + A_t \quad (12.8)$$

Where: P_t = Power output of the transmitter [in decibels above 1 W (dBW)]
L_{bo} = Back-off losses (in decibels)
L_{bf} = Branching and feeder losses (in decibels). These losses occur between the power amplifier and the antenna.
A_t = Transmit antenna gain (in decibels)

The noise characteristics of a satellite communications system are of great importance. One parameter used to analyze those characteristics is the **noise density**, the total noise power normalized to a 1-Hz bandwidth. This parameter is used to find other noise properties within a system. Expressed mathematically, noise density is

$$N_0 = \frac{N}{B} \quad (12.9)$$

since

$$N = KT_eB \quad (12.10)$$
$$N_0 = kT_e$$

Where: N_0 = Noise density (in watts per hertz)
N = Total noise power (in watts)
B = Bandwidth (in hertz)
K = Boltzmann's constant (1.3806×10^{-23} J/K)
T_e = Equivalent noise temperature (K). This quantity was defined in Chapter 6.

If we express the noise density in decibels above 1 W,

$$N_0 = 10 \log N - 10 \log B \quad (12.11)$$

or in decibels above 1 W per hertz

$$N_0 = 10 \log \text{K} + 10 \log T_e \quad (12.12)$$

Two terms very important in the characterization of satellite systems are **carrier-to-noise density ratio** (C/N_0), and energy of bit-to-noise ratio (E_b/N_0). The first is the average wide-band carrier power-to-density ratio; the second term is a convenient method of comparing digital systems that use different transmission rates, modulation schemes, or encoding techniques. They are expressed as

$$\frac{C}{N_0} = \frac{C}{KT_e} \quad (12.13)$$

and

$$\frac{E_b}{N_0} = \frac{C/F_b}{N/B} = \frac{CB}{NF_b} \quad (12.14)$$

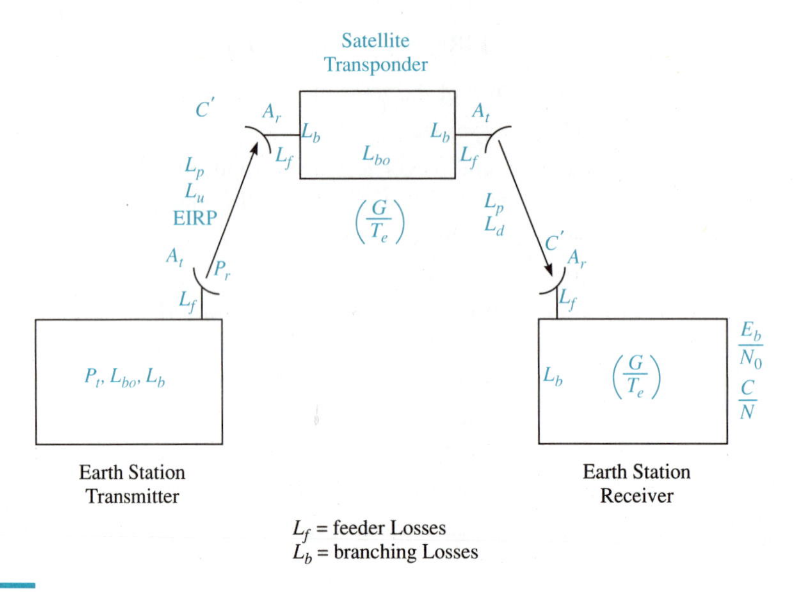

Figure 12.30 Satellite system.

Looking closer at Equation (12.14), we can break it up into the familiar parameters of carrier-to-noise ratio (C/N) and bandwidth per transmission rate (B/F_b). This relationship allows us to find the E_b/N_0 ratio in decibels by simply subtracting the carrier-to-noise ratio from the bandwidth per transmission rate, both expressed in decibels.

Bearing in mind the critical terms presented so far, we can now get to the heart of all satellite communications systems, the link equations for both the uplink and downlink. Figure 12.30 shows the earth-station transmitter, the satellite transponder, and the earth-station receiver and their critical parameters. It is necessary to calculate and account for all of these parameters before the system design is completed. The listing of these parameters is called the **link budget** for a particular system.

The uplink equation for the system shown is

$$\frac{C}{N_0} = \frac{A_t P_r (L_p L_u) A_r}{KT_e}$$

or (12.15)

$$\frac{C}{N_0} = \frac{A_t P_r (L_p L_u)}{K} \times \frac{G}{T_e}$$

Where: L_p = Path losses

L_u = Additional uplink losses due to the atmosphere

If we express this equation in log form, we have

$$\frac{C}{N_0} = \underbrace{10 \log A_t P_r}_{\text{EIRP of earth station}} - \underbrace{20 \log\left(\frac{4\pi D}{\lambda}\right)}_{\text{Free space path loss}} + \underbrace{10 \log\left(\frac{G}{T_e}\right)}_{\text{Satellite } \frac{G}{T_e}} - \underbrace{10 \log L_u}_{\text{Additional atmospheric losses}} - \underbrace{10 \log K}_{\text{Boltzmann's constant}} \quad (12.16)$$

When Equation (12.16) is used with all the appropriate numbers, you will be able to completely characterize everything between the earth-station transmitter and the satellite transponder. Next we repeat this process for the downlink, or the path from the satellite transponder to the earth-station receiver.

The downlink equation is similar to the uplink equation, which is not surprising if you look at the parameters used between the satellite and the earth-station receiver. The equation is

$$\frac{C}{N_0} = \frac{A_t P_r (L_p L_d) A_r}{K T_e}$$

or

$$\frac{C}{N_0} = \frac{A_t A_r (L_p L_d)}{K} \times \frac{G}{T_e} \quad (12.17)$$

Where: L_p = Path loss
L_d = Additional downlink losses due to the atmosphere

Once again, in log form the equation becomes more understandable as

$$\frac{C}{N_0} = \underbrace{10 \log A_t P_r}_{\text{EIRP of satellite}} - \underbrace{20 \log\left(\frac{4\pi D}{\lambda}\right)}_{\text{Free space path loss}} + \underbrace{10 \log\left(\frac{G}{T_e}\right)}_{\text{Earth station } \frac{G}{T_e}} - \underbrace{10 \log L_d}_{\text{Additional atmospheric losses}} - \underbrace{10 \log K}_{\text{Boltzmann's constant}} \quad (12.18)$$

The combination of Equations (12.16) and (12.18) takes into account all the parameters in Figure 12.30. You should now be able to make up a link budget that will enable you to determine all the gains, losses, and net gains between assemblies. In other words, you will be able to characterize the complete satellite communications system.

12.4 Fiber Optics

12.4.1 Introduction

Fiber optics—probably the number one buzzword for the 1990s—has found its way into communications, where it is used as a parallel network for microwave systems, in telephone applications, and a multitude of other applications where large amounts of information are transmitted and received. In some cases fiber optics represents an improvement over previous media, but in others it is just an expensive experiment. Whatever the case, fiber optics and its applications will be a part of the communications industry for many years.

12.4.2 Fiber-Optic Characteristics

In many of the previous chapters the common parameter that we addressed in a variety of systems was the bandwidth of each system. Bandwidth is a primary limiter of the quantity of information a system can process and transmit to another location. If, for example, we are dealing with a system that needs a 10% bandwidth, we may have to operate the system at 150 MHz and its bandwidth would be 15 MHz. If, on the other hand, we moved up into the microwave range and still required a 10% bandwidth for a 5-GHz system, we now have 500 MHz to work with.

In a fiber-optics system light is used to transmit information from one point to another. The frequencies used in this case are 100,000 to 1,000,000 GHz (10^{14} to 10^{15} Hz). If we use the center of this range and operate at 500,000 GHz, we have a 10% bandwidth of 50,000 GHz, a large bandwidth. Commonly used systems do not actually handle this much information, nor do they use 10% bandwidths. The point is that a significantly large improvement can be realized by using a fiber-optic communications system if the conditions and financing are right.

There are advantages and disadvantages to using fiber-optic systems just as with any other system.

Advantages

1. Large bandwidths result in a much higher information capacity for the system.
2. Fiber-optic systems are immune to crosstalk between cables because they do not carry electricity and thus do not set up the magnetic field that causes crosstalk.
3. Fiber-optic cable operates over a larger temperature range than a metallic cable.
4. There is no problem with noise in a fiber-optic system because there is no electrical conductivity.
5. Fiber-optic cables are very difficult to tap into, thereby preventing eavesdropping on conversations.
6. Fiber-optic cables are smaller in size and lighter in weight than are metallic conductors.

Disadvantages

1. The cost of the fiber-optic system is significantly high, a disadvantage that may disappear over the years as systems improve and become more widely used.
2. Construction techniques for fiber-optic systems are more difficult, and many of the personnel presently working on such systems are not trained in the construction of cables and connectors.

Now that you know some advantages and disadvantages of fiber-optic systems, it is time to see how such a system works. Basically, in a fiber-optic system light is transmitted down a glass or plastic fiber, with a modulation applied to it. Light travels down this fiber at 186,000 miles/s (3×10^{10} cm/s), the same speed that electromagnetic energy travels within a cable before it encounters any type of dielectric that causes its velocity to decrease. One misconception many people have about fiber-optic systems is thinking that a light source is placed at one end of an optical fiber and basically sent down the line. An analogy is a flashlight placed at one end of a garden hose. It does not take too much imagination to realize that the flashlight beam would not make it very far down the hose before it could not be seen anymore. This would be the case in the fiber-optic system if a light source were simply placed at the end of a cable parallel to the cable's axis. In fact, there is a critical angle at which the light must enter the cable for transmission to occur down the entire length of the system. This means that the fiber-optic systems rely on some of the most basic principles taught in high school physics: reflection and refraction. We looked at these terms when we discussed electromagnetic energy and its transmission in Chapter 4. Remember, that in reflection a ray of energy (light or electromagnetic energy) comes from one medium to a denser medium. The second medium causes some, or all, of the energy to be sent back toward the original medium. If total reflection occurs, no energy is transmitted. If partial reflection occurs, some percentage of the energy is reflected back, and the remainder is transmitted into the second medium.

Refraction is best explained by looking at a prism, as shown in Figure 12.31. A source of white light is applied to the prism. If the prism had the same density as

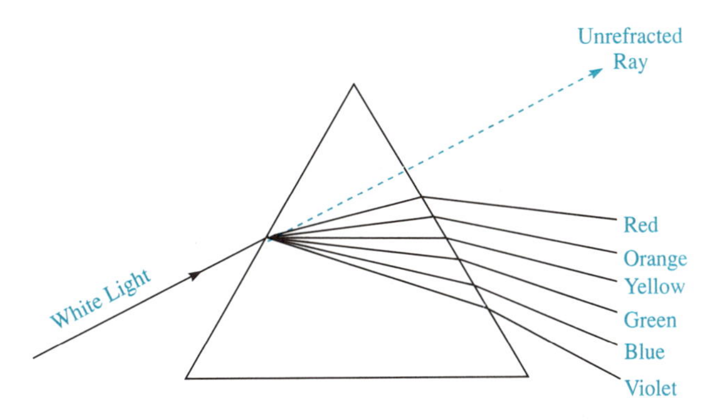

Figure 12.31 Refraction of light.

air, the unrefracted ray shown would result. But because the prism exhibits a different obstruction to each one of the colors, it bends each color by a different degree. This is the idea of refraction. The amount of bending that occurs at a particular frequency because of a particular medium is called the **index of refraction** (or refractive index). This value is determined by the velocity of the light v in a particular medium compared with the velocity of light in free space c. That is,

$$n = \frac{c}{v} \tag{12.19}$$

Where: c = Velocity of light in free space
v = Velocity of light in the medium

Some typical values of refractive indexes are

- vacuum = 1.0
- air = 1.003
- fused quartz = 1.46
- glass = 1.5–1.9
- silicon = 3.4
- gallium arsenide = 3.6

Other values for specific materials are available from individual manufacturers. The exact value of the material being used should be found because it is a critical parameter that should be as accurate as possible.

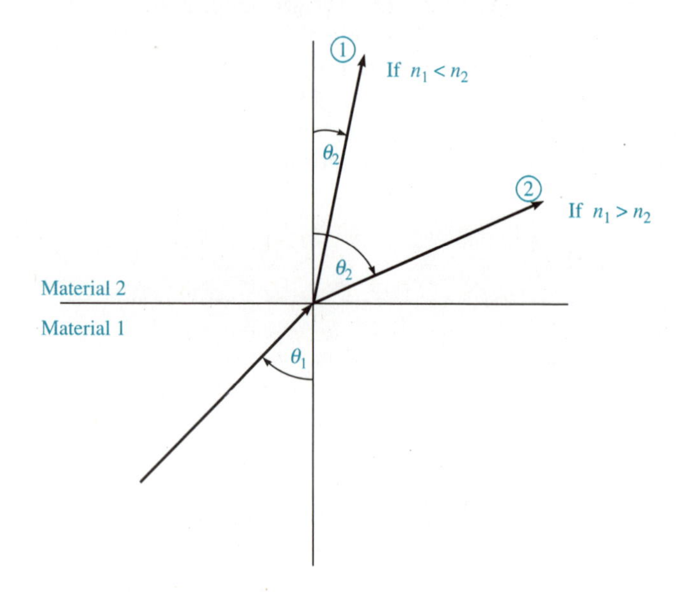

Figure 12.32 Snell's law.

Another area concerned with refraction, covered previously, was Snell's law. Recall that

$$n_1 \sin \theta_1 = n_2 \sin \theta_2 \qquad (12.20)$$

Where: n_1 = Index of refraction of material 1
n_2 = Index of refraction of material 2
θ_1 = Angle of incidence
θ_2 = Angle of refraction

The relationships shown in Equation (12.20) are represented graphically in Figure 12.32. Notice that two refracted rays in the figure are labeled ① and ②. In the case of ray #1, the index of refraction of the original medium is less than the index of refraction of the second medium. The refracted ray is therefore bent toward the normal as shown. For ray #1, the index of refraction of the second medium is greater than that of the first, and the ray is bent away from the normal. Keep this concept in mind when choosing material for optical fibers.

Example 12.6

A light ray is incident on a material from air into a silicon material. The angle of incidence is found to be 45°. What is the angle of refraction, and does the ray bend toward or away from the normal?

Solution

The refractive indexes are $n_1 = 1.0$, and $r_2 = 3.4$. If we apply Snell's law, we have

$$\begin{aligned} \sin \theta_2 &= \left(\frac{n_1}{n_2}\right) \sin \theta_1 \\ &= \left(\frac{1}{3.4}\right)(0.707) \\ &= (0.294)(0.707) \\ &= 0.206 \\ &= 11.88° \end{aligned}$$

This angle indicates that the ray is bent *toward* the normal, which confirms that when the refractive index of the first medium is less than that of the second medium, the ray is bent toward the normal.

Before we delve into the actual transmission of light in a fiber-optic cable, we need to discuss what type of cable to use. The optical fibers currently available are either glass, plastic, or a combination of the two. Three common types of fibers are:

1. Plastic core and cladding
2. Glass core with plastic cladding (plastic-clad silica, PCS)
3. Glass core and glass cladding (silica-clad silica, SCS)

Plastic fibers are generally more flexible, more rugged, easier to install, less expensive, and weigh less than glass fibers. They exhibit, however, higher attenuation properties than the glass fibers.

Cables used for optical transmission basically consist of a core, a cladding, a protective tube, buffers, strength members, and a variety of types of protective jackets. Every cable does not contain all of these elements, but the ones that need all the strengthening and protection elements will contain all of them. Basic types of cable configurations are: the loose tube, constrained fiber, multiple strand (probably the most widely used), telephone cable (which used fiber ribbons), and the plastic-clad silica cable. Once again, the type of cable you use will depend upon the application.

There are two types of modes, or paths, in an optical fiber: the single mode and the multimode. In **single-mode operation** the analogy of the flashlight and the garden hose comes into play. The path is down the center of the fiber and can transmit energy only for a relatively short distance. In **multimode operation**, the most common mode used for fiber-optic systems, the light is "bounced" back and forth through the cable.

As shown in Figure 12.33, the energy propagates down the cable at specific angles. The angles the energy will make with the cladding in the cable in Figure 12.33 depend on the material being used. This can be a single-mode step, as shown in Figure 12.34(a); a multimode step, as in Figure 12.34(b); or a multimode graded index, as in Figure 12.34(c). The single-mode step has a small core so that only a single path (mode) can propagate down the cable. This is a very simple cable, whose outside cladding is air. The multimode step shown in part (b) has a narrow cladding compared with the core, and thus creates a much different step function for the energy. This concept is used in many fiber-optic systems. The final type, the multimode graded index, is widely used and a more expensive type of construction for optical fibers. In this cable the density of the material is a continuously varying parameter from the outside edge to the cable to its maximum point at the center of the core. This type of cable is excellent for optical use because there are no abrupt changes in the refractive index, only a smooth, continuous change.

To understand the concept of a light ray propagating down an optical fiber and how the energy gets transferred from a source and down the line, we need to define two terms: acceptance angle (acceptance cone) and numerical aperture.

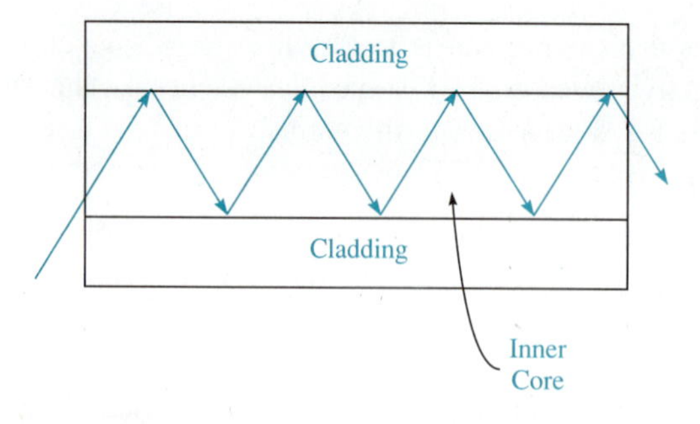

Figure 12.33 Multimode propagation.

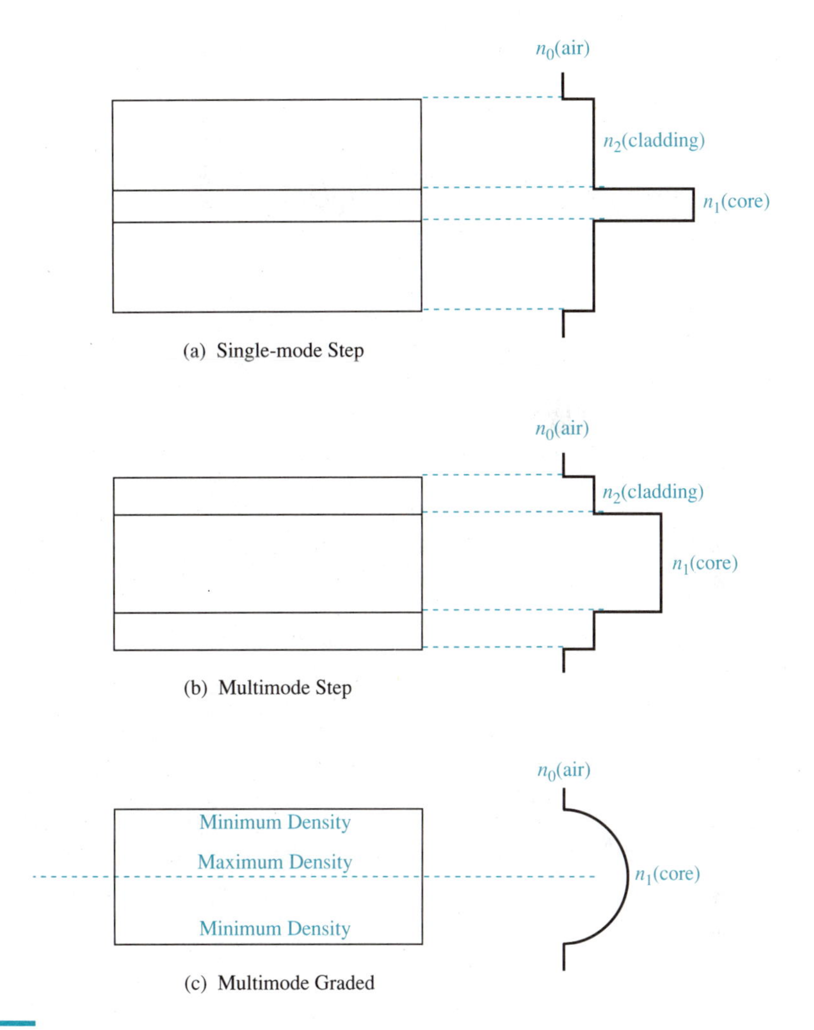

Figure 12.34 Core index profiles.

The acceptance angle for an optical fiber is found by going back to Snell's law and taking certain factors into consideration. Some of these factors are that the incident angle of the ray is equal to 90° minus the critical angle (θ_c), the relation between sine and cosine functions and the Pythagorean theorem, and that in most cases the ray is entering the fiber from air. These factors result in the following relationship:

$$\theta_{\text{in(max)}} = \sin^{-1} \sqrt{n_1^2 - n_2^2} \qquad (12.21)$$

This quantity, $\theta_{\text{in(max)}}$, is called the **acceptance angle**, or acceptance cone half-angle. Figure 12.35 shows how this parameter is used at the entrance to the fiber.

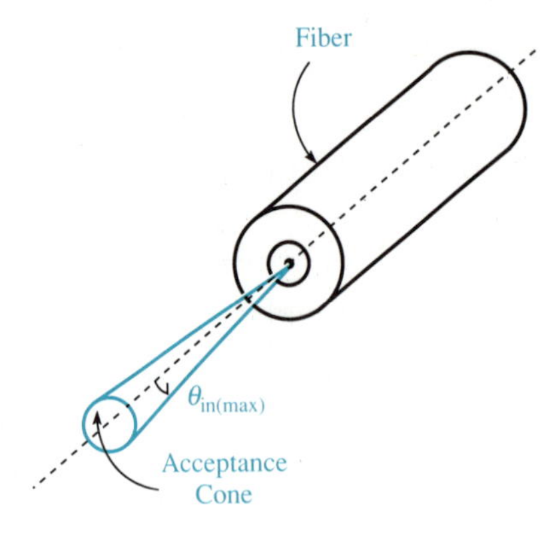

Figure 12.35 Acceptance cone.

The **numerical aperture** (NA) is called a figure of merit used to measure the light-gathering ability of the fiber. It is expressed as

$$NA = \sin \theta_{in} = \sqrt{n_1^2 - n_2^2} \tag{12.22}$$

For a graded index, $NA = \sin \theta_c$.

Example 12.7

For a multimode step-index fiber with $n_1 = 1.72$, and a cladding with $n_2 = 1.6$, find the critical angle, acceptance angle, and numerical aperture.

Solution

For the critical angle

$$\theta_c = \sin^{-1}\left(\frac{n_2}{n_1}\right) = \sin^{-1}\left(\frac{1.6}{1.72}\right) = 68.47°$$

The acceptance angle is

$$\theta_{in} = \sin^{-1}\sqrt{n_1^2 - n_2^2} = 39.1°$$

The numerical aperture is

$$NA = \sqrt{n_1^2 - n_2^2} = 0.398$$

Losses that can occur in optical fibers are absorption losses (similar to power dissipation); material, or Rayleigh scattering, losses (light rays escaping from a fiber because of defects); chromatic, or wavelength, dispersion (a distortion of the ray); radiation losses (caused by small bends in the fiber); modal dispersion (caused by the difference in the propagation times of light rays that take different paths down the fiber); and coupling losses (losses due to the connectors needed for the fibers). All of these losses, like any other losses, must be considered in the overall scheme of the system used.

Common sources used for fiber-optics communication are the **LED (light-emitting diode)** and the **ILD (injection laser diode)**. The LED is used for most applications.

To detect the ray at the end of a fiber a PIN diode, or APD (avalanche photodiode), may be used. These two diodes share the bulk of the fiber-optic applications.

12.5 Summary

In this chapter we introduced three important aspects of the communications industry. First we introduced television, discussing the TV signal in detail to enable the student to fully understand the complexity of such a signal compared to what had been presented to that point in the text. Next, the theory of monochrome (black-and-white) and color television was presented.

The last two sections introduced satellite communications and fiber optics, touching on the basics of each.

Questions

12.2 Television

1. How is a TV modulating scheme different from conventional AM and FM?
2. Describe a **vestigial sideband**.
3. Why are the sidebands different in a vestigial sideband system?
4. What is the bandwidth of a TV system?
5. Why is there a frequency gap between Channels 6 and 7?
6. Where is the frequency break for UHF and VHF TV?
7. Explain why the lower vestigial sideband is attenuated.
8. Why does the upper vestigial sideband need to fall off to -60 dB so fast?
9. What type of modulation is used for the video portion of a TV signal?
10. What type of modulation is used for the audio portion of a TV signal?
11. Define **rest frequency**.
12. Is the audio portion of a TV signal wide-band or narrow-band?
13. Why are sync and blanking pulses important in a TV signal?
14. Describe a TV scanning system.
15. What is **interleaved scanning**?

16. Define **flyback time**.
17. Define **sequential scanning**.
18. What is **horizontal** scanning time?
19. Define **resolution**.
20. What factors determine the number of vertical lines resolved in a TV system?
21. What is the purpose of the blanking pulse?
22. Define **front porch**.
23. Define **back porch**.
24. Define **raster**.
25. How many raster lines make up a raster?
26. What is a **negative transmission**?
27. What is a **positive transmission**?
28. Why is negative transmission used for the final transmission carrier?
29. What are two applications of **monochrome** TV?
30. Explain the scanning operation of a monochrome TV system.
31. Why is the amplitude of the electrical signals in Figure 12.9 the same?
32. Define **diplexer**.
33. What is the function of the diplexer in a TV system?
34. What is a **vidicon** tube?
35. Distinguish between **photomission** and **photoconductive** material.
36. Why is a peaking coil used for video circuits?
37. What are the five basic sections of a monochrome TV?
38. What are the different methods used to select a TV channel?
39. How is the varactor diode used in a TV system?
40. Draw a block diagram and explain how UHF and VHF can use the same rf section.
41. Draw the i-f response for a TV system.
42. Define wavetrap.
43. Why is the video section so important in a TV receiver?
44. Name two types of FM detectors used in TV.
45. Describe a sync separator.
46. What is the function of the flyback transformer?
47. Describe the function of the damper diode.
48. What three basic colors are used for color TV?
49. How does the color TV camera differ from the monochrome camera?
50. Define the I, Q, C, and T signals in color TV.
51. Describe a color wheel.
52. What is a color burst and where does it appear?
53. How is the color TV CRT different from a monochrome CRT?
54. What is a color killer?

12.3 Satellite Communications

55. What is a **nonsynchronous orbit**?
56. Define **geostationary orbit**.
57. Define **prograde**.
58. Define **retrograde**.
59. What is one advantage of a geostationary orbit?
60. What is one disadvantage of a geostationary orbit?
61. What two paramcters constitute the **look angles**?
62. What is considered to be a minimum elevation angle?

63. Name three radiation patterns for satellites.
64. Draw a block diagram for a basic satellite communications system.
65. What is a **cross-link**?
66. Define **bit energy**.
67. What is a **link budget**?
68. Define **back-off loss**.
69. Define **noise density**.

12.4 Fiber Optics

70. List two advantages of using fiber optics.
71. Define **reflection**.
72. Define **refraction**.
73. How is the index of refraction related to velocity?
74. What does **Snell's law** say?
75. Name two types of optical fibers.
76. Define **cladding**.
77. Define **core**.
78. What is an acceptance cone in fiber optics?
79. What is the numerical aperture in fiber optics?
80. What is a common source for fiber optics?
81. What is an APD?

Problems

12.2 Television

1. If a TV system has 400 scanning lines, what is the vertical resolution?
2. A TV system has a problem with its display. The horizontal scanning time is measured at 75 μs. What has happened to the TV?
3. A TV has a scanning rate of 28 kHz and the video signal is 4 MHz. What is the horizontal resolution?
4. We need to have a horizontal resolution of at least 350 lines. Our TV has a signal bandwidth of 5 MHz. What must the scan rate be?
5. A TV system has a scan rate of 18 kHz and is known to have a resolution of 400 lines. What is the video bandwidth?
6. We have a scan rate for a TV system of 15 kHz, what is the scan time?
7. A TV system has a scan rate of 20 kHz and 262.5 horizontal scan lines. What is the total time for the signal?
8. A blanking pulse is 11.5 μs wide. If the sync pulse must be a minimum of 1 us wide, what are the approximate widths of the front and back porch of the pulse?
9. The blanking width of a TV signal is 10 μs. How wide should the sync pulse be for this signal?

12.3 Satellite Communications

10. A system has a carrier power of 500 W and a transmission rate of 40 Mbit/s. What is the energy per bit?

11. The requirements for a satellite communications system say that we must maintain an energy per bit of no more than 75 μJ/bit or less than 30 μJ/bit. If the power transmitted is constant at 1500 W, what is the range of the transmission rate?
12. A system has a transmitted power of 60 dBW, branching and feeder losses of 3 dB, and back-off losses of 2 dB. If an EIRP of 79 dB is needed, what must be the antenna gain?

12.4 Fiber Optics

13. If $n_1 = 1.3$, $n_2 = 1.6$, and the incident angle is 37°, what is the angle of refraction?
14. In Problem 13, does the ray bend toward or away from the normal?
15. A system with two materials has a ray measured as being refracted to an angle of 41°. If $n_1 = 1.3$ and $n_2 = 1.75$, what is the angle of incidence?
16. For a multimode step index with $n_1 = 1.3$ and $n_2 = 1.6$, find the critical angle and numerical aperture.
17. For the system in Problem 16, the index of refraction n_1 is changed to 1.7. Find the new critical angle and numerical aperture.

Appendix A
SPICE Program Instructions

The SPICE program (Simulated Program with Integrated Circuit Emphasis) is one of the most powerful programs available for analyzing circuits used in communications systems. It is not, however, the easiest program you will use; you can say that it is not user friendly. This does not mean that you cannot use it easily and obtain good results. It does mean that you must study the program and its requirements and use it wisely. This section will present the program, its capabilities, and some very useful hints for its use. At the end of the section we will present some examples that can be very helpful to the beginner and also to those who have used the program before.

The SPICE program is defined as a general purpose circuit simulation program for nonlinear dc, nonlinear transient, and linear ac analysis. The circuits may consist of the following components:

- □ Resistors
- □ Capacitors
- □ Inductors
- □ Mutual inductors
- □ Independent voltage and current sources
- □ Dependent sources
- □ Transmission lines
- □ Diodes
- □ Bipolar transistors
- □ JFET's
- □ MOSFET's

The program has built-in models for semiconductor devices. The student should consult the manuals to see which devices are contained on his system. With this feature the student needs only to specify the model parameter values and the correct device is entered into the program. Additional models of semiconductor devices are shown in Appendix A-1.

DC Analysis

The first step in understanding the SPICE program is to study the types of analysis that are possible. The first type is the *dc analysis*. This analysis determines the dc operating point of the circuit with all of the inductors shorted and the capacitors open. This operation is automatically performed prior to a transient analysis to determine the transient analysis initial conditions. The dc analysis is performed prior to an ac small-signal analysis to determine the linearized, small-signal models for non-linear devices. The dc analysis can also be used to generate dc transfer curves. With this option, a specified independent voltage or current source is stepped over a specified range and the dc output variables are stored for each source value. Additionally, the program can determine the dc small-signal sensitivities of specified output variables with respect to circuit parameters. The options available with the dc analysis are shown below. Note that each of these options has a period before it. This is very important. The program will *not* recognize the command without this period. Be sure that you place it on the proper line when you make up your files.

DC Analysis Options

.DC: A nonlinear analysis which determines the dc operating point of the circuit. As previously noted, all capacitors are open and inductors shorted during this operation. The sources may be stepped over a range.

.TF: Performs a small-signal transfer function analysis of a circuit. The input resistance, output resistance, and the transfer function are printed. The transfer function may be voltage gain, current gain, transresistance, or transconductance.

.OP: This command results in the program calculating and printing the dc operating point of the circuit. This function is automatically accomplished when the ac analysis is performed. It allows the program to find the ac small-signal model parameters. This is also performed when a transient analysis is requested to determine the initial operating conditions.

.SENS: This command tells the program to find the dc small-signal sensitivities of one or more output variables with respect to every parameter in the circuit. This can be a very time consuming operation if a large circuit is to be analyzed. You should, therefore, limit this operation to only the smaller circuits if time is an important element to you.

The format for using the dc analysis is:

.DC SRC START STOP INCR ⟨SRC2 START2 STOP2 INCR2⟩

Where: SRC is the name of the independent voltage or current source being varied

START is the starting value

STOP is the final value

INCR is the increment size in volts or amps

As an option a second source may be specified (SCR2). The parameters associated with this source (START2, STOP2, INCR2) are placed in their appropriate positions on the format line. If a second source is used, the first source will be swept over its entire range, as specified, for each value of the second source.

Examples of the dc analysis are shown below.

.DC VBASE −3 5 0.65

This example tells the operator that the independent voltage source, VBASE, is incremented from −3 volts to +5 volts in 0.65 volt steps.

.DC IBASE 0 85U 5U VCE 0 10 5

This example tells the operator that the independent current source IBASE, will step from 0 to 85 microamps in 5 microamp steps while the independent voltage source, VCE, is fixed at 0 volts. Then VCE jumps to 5 volts and IBASE will step again from 0 to 85 microamps. This process is repeated until VCE reaches 10 volts.

AC Analysis

The next type of analysis to be considered is the *ac small-signal analysis*. This analysis computes the ac variables as a function of frequency. As previously stated, it computes the dc operating point of the circuit and determines linearized, small-signal models of all the nonlinear devices in the circuit.

The desired output of an ac small-signal analysis is usually a transfer function. This may be a voltage gain, current gain, transresistance, or transconductance. If the circuit has only one ac input, it is usually convenient to set that input equal to unity and zero phase. This makes the input/output transfer functions much more meaningful.

The different types of noise can be taken into consideration using the ac analysis. The generation of white noise by resistors and semiconductor devices can be simulated using the ac analysis. (This type of noise is covered in Chapter 5 of this text.) Equivalent noise source values are determined automatically from the small-signal operating point and the contribution of each noise source is added together. Flicker noise (which is covered in Chapter 6 also) sources can be simulated by including values of the parameters KF (flicker noise coefficient) and AF (flicker noise exponent) for the appropriate semiconductor (BJT's and FET's).

Distortion characteristics can also be simulated. When the analysis is done, it is assumed that one or two signal frequencies are imposed at the input.

For the ac analysis the frequency range, noise, and distortion parameters are specified as follows:

.AC: Linear small-signal analysis. SPICE determines the dc operating point and calculates values for small-signal models of all nonlinear devices (semiconductors, inductors, capacitors, dependent sources). Then the circuit is analyzed at each specified frequency and printed or plotted. Examples of the format to use are shown below.

.AC LIN NP FSTART FSTOP

.AC DEC ND FSTART FSTOP

.AC OCT NO FSTART FSTOP

Note that there are three types of ac analysis listed above: LIN, DEC, and OCT, or **linear, decade,** and **octave,** respectively. The three types indicate the frequencies at which the analyses are done and they are defined as follows:

LIN NP: caused the analysis to occur at a number of frequencies equal to NP (number of points), linearly spaced between FSTART and FSTOP (The lowest and the highest frequency). If a single frequency analysis is desired, simply make NP = 1 and let FSTART = FSTOP.

DEC ND: This command divides the frequency range into decades with ND the number of points per decade. A decade is a ratio of frequencies where the highest frequency is ten (10) times that of the lowest frequency. Thus, a range of 1.0 MHz to 10 MHz or 10 MHz to 100 MHz would both be decades. Frequencies will be logarithmically spaced. This type of spacing is most useful for wide ranges of frequencies.

OCT NO: This command divides the frequency range into octaves with NO the number of points per octave. Whereas the decade was a ratio of 10:1 the octave is a frequency ratio of 2:1. That is, frequency ranges of 1.0 MHz to 2.0 MHz, 5 MHz to 10 MHz, and 2 GHz to 4 GHz are all octaves. Again, frequencies will be logarithmically spaced as with decades. And, like the decade, the octave command is most useful for wide frequency ranges.

Examples of the previously discussed ac analysis are given below for each of the frequency range cases: LIN, DEC, and OCT.

Example 1:

.AC LIN 101 6000 7000

This example says that independent ac sources in the circuit will start at 6 kHz and rise in frequency in 10 Hz increments until 7 kHz is reached.

Example 2:

.AC LIN 1 5.1 MEG 5.1 MEG

This example, as referred to above, says that all independent ac sources are set to a single frequency, 5.1 MHz.

Example 3:

.AC DEC 5 1 10K

This example says that there are 4 decades of frequency to be covered [log (10k/1)] = 4, and we have 5 points per decade to be used.

Example 4:

.AC OCT 7 10 5120

This example says that the number of octaves is equal to [log(5120/10)]/log 2 = 9. Thus, we have 7 points per octave for 9 octaves spaced logarithmically between 10 Hz and 5.12 kHz.

.NOISE: When used with the .AC control line the program will find the equivalent output and equivalent input noise at specified output and input points in the circuit.

The .NOISE control line format is as follows:

.NOISE OUTPUTV INPUTSRC NUMSUM

(This control line format must be used with the .AC control line.)

OUTPUTV: A voltage which will be considered the output summing point for noise.

INPUTSRC: The element name of an independent voltage or current source which will be the noise input reference.

NUMSUM: The interval at which a summary printout of the contributions of all noise generators (semiconductors and resistors) is printed. If NUMSUM is omitted or set equal to zero, no summary printout is done. If numsum is an integer, such as 4, then a summary printout will occur at every fourth ac analysis frequency.

Example 1:

.NOISE V(1,5) VIN2 7

In this example the noise analysis will be done with the ac analysis. The summary printout will list at every seventh frequency, the output noise will be measured as the difference between nodes 1 and 5, and the equivalent input noise will be referenced to voltage source VIN2.

Example 2:

.NOISE V(15) IBIAS

In this example, the noise analysis is done with the output noise measured at node 15 and the equivalent input noise referenced to current source IBIAS. You will notice that no summary points are called out. Even though there are no points called out, it is possible to have the output noise and equivalent input noise printed or plotted at each frequency of the ac analysis.

.DIST: This control line can be present with the .AC control line. It will determine several kinds of distortion of the circuit in small-signal analysis.

Control line format for the distortion analysis is:

.DISTO RLOAD INTER ⟨SKW2 ⟨REFPWR ⟨SPW2⟩⟩⟩

(This control line must be used with a .AC control line.)
Definitions of the control line are as follows:

RLOAD: The element name of the output resistor into which all the distortion power will be calculated.

INTER: The interval at which a summary printout of the contributions of all nonlinear devices to the total distortion is printed. If this term is omitted, no summary printout is done.

SKW2: The ratio of frequency $f2$ to frequency $f1$ (to be defined below). If SKW2 is omitted, a value of 0.9 is used. That is, $f2 = 0.9\ f1$.

REFPWR: The reference power level used in computing the distortion products. If this number is omitted, a value of 1.0 Mw (0 dBM) is used.

SPW2: The amplitude of $f2$. This defaults to a value of 1.0.

The distortion analysis is performed assuming one or two signal frequencies are at the input. The two frequencies are $f1$, which is the ac analysis frequency, and $f2$, which is, taken from above, equal to SKW2 $\times$ $f1$. The output that results from these frequencies and the .DISTO control line will contain the following parameters:

HD2: The magnitude of the second harmonic of $f1$, assuming that $f2$ is not present.

HD3: The magnitude of the third harmonic of $f1$, assuming that $f2$ is not present.

SIM2: The magnitude of the frequency component $f1 + f2$.

DIM2: The magnitude of the frequency component $f1 - f2$.

DIM3: The magnitude of the frequency component $2(f1) - f2$.

The following examples illustrate the .DIST control used with the .AC control line:

Example 1:

.DISTO ROUTPUT 4 0.65

This example shows that the distortion analysis will occur every fourth frequency of the ac analysis and the frequency, $f2$, will be 0.65 $f1$. Also, the reference power level will be 1.0 Mw (0 dBM).

Example 2:

.DISTO RDRAIN 1

This example shows that the distortion analysis will occur at each frequency of the ac analysis, the second frequency, $f2$, will be 0.9 $f1$, and the reference power level will be 1 Mw (0 dBM).

Transient Analysis

This function computes the transient output variables as a function of time over a specified interval. The initial conditions to be used are automatically determined by a dc analysis as described earlier in this text. Any sources that are not time dependent (power supplies for example) are set to their dc value during this analysis. For large signal sinusoidal simulations, a Fourier analysis of the output waveform can be specified. The frequency domain Fourier coefficients will be obtained. The transient time intervals and Fourier analysis are specified on .TRAN and .FOUR lines.

.TRAN: Transient analysis, determines the output variables (voltage and current) as a function of time over a specified interval.

.FOUR: Performs a Fourier analysis of an output variable when done with the transient analysis. Computes the amplitudes and phases of the first nine frequency components (harmonics) of a specified fundamental frequency and the dc component.

The control line format for .TRAN is as follows:

.TRAN TSTEP TSTOP ⟨TSTART ⟨TMAX⟩⟩ ⟨ UIC⟩

Definitions of the control line are as follows:

TSTEP: Printing or plotting increment. This is the suggested computing increment.

TSTOP: Time for the last transient analysis.

TSTART: Time of the initial transient analysis; if it is omitted it is assumed to be zero.

TMAX: The maximum step size that the program will use.

UIC: (Use Initial Conditions) This tells the program that it should not solve for the quiescent operating point; it is to use the values placed in each line that is designated as IC = ... values.

Example 1:

.TRAN 1U 80U

This example says that a transient analysis will take place every 1 microsecond from time zero to 80 microseconds. This means that 81 points will be printed or plotted.

Example 2:

.TRAN 5M 500M 100M 2M UIC

This example says that every 5 msec from time zero up to 500 msec a transient analysis will occur. Results, however, will not be stored until 100 msec. The computing time will be 2 msec. Initial conditions will be used for the first transient analysis.

Analysis at Different Temperatures

All input data is assumed to have been measured at +27°C (300K). Simulations in the program assume +27°C. The circuit can be simulated at other temperatures by using the .TEMP line. This feature of the program is used many times for *worst case analysis*.

Control line format is as follows:

.TEMP T1 ⟨T2 ⟨T3 . . . .⟩⟩

Definitions of the control line format is as follows:

T1,T2: Temperatures to be used for simulation in degrees centigrade. Analysis is performed at each of the temperatures given.

Example 1:

TEMP 0 34 75 150

This example says that all analysis specified in the circuit input file will be performed at 0, 34, 75, and 150 degrees centigrade.

Input Format

The program uses a free format which sets up a file to run your particular circuit. The fields (sections) are separated by one or more BLANKS, a COMMA, an EQUAL SIGN (=), or a LEFT or RIGHT PARENTHESIS. Any extra spaces are ignored. A line may be continued by entering a plus (+) in column 1 of the following line, the program continues reading with column 2.

The NAME line must begin with a letter (A to Z) and cannot contain any delimiters (term that sets boundaries). Only the first eight (8) characters of the name are used. A number line may be an integer (10, −15), a floating point (3.1456), either an integer or floating point number followed by an integer exponent (1E-6, 2.55 E5), or an integer or floating point number followed by one of the following scale factors:

T = 1E12

G = 1E9

MEG = 1E6

k = 1E3

MIL = 25.4E-6

M = 1E-3

U = 1E-6

N = 1E-9

P = 1E-12

F = 1E-15

Letters immediately following a number that are not scale factors are ignored. Letters immediately following a scale factor are ignored. For example, the following all represent the same number: 10, 10V, 10 volts, and 10 Hz. Also, these terms all represent the same number: 1000, 1000.0, 1000 Hz, 1E3, 1.0E3, 1 kHz, and 1 k.

The SPICE program uses a Nodal representation to describe the circuits to be analyzed. This means that the student should draw out the circuit and define where the nodes are to be. An example of this is shown in Figure A.1. Notice that all of the ground connections are labeled with a "0" designation. This designation is the COMMON NODE.

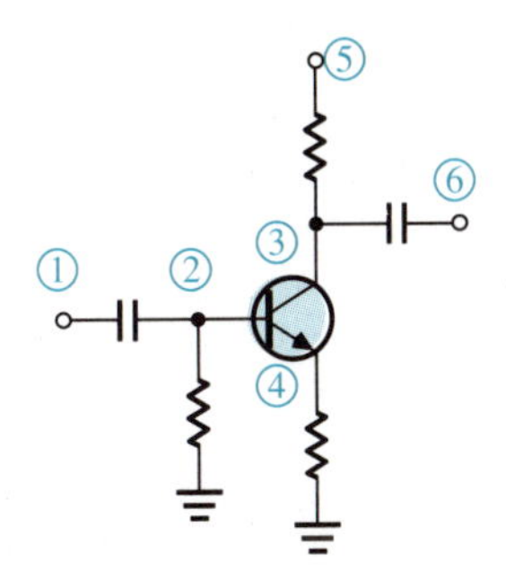

Figure A.1

Circuit Configuration

The circuits to be analyzed are described by a series of lines in a file format. The first line of the file is the TITLE line. The last line of the file is the .END line. Most of the literature on the SPICE program states that the order in between the TITLE line and the .END line may be arbitrary except for particular lines which must follow one another. Although this statement is made numerous times, it is advantageous to keep everything in order as the circuit flows from left to right. That is, (referring to Fig. A.1) list the input capacitor first, the shunt resistor next, the transistor next, and so on. This will be a great help in troubleshooting files and looking for errors. An arbitrary order would cause confusion and would increase the troubleshooting time. So, to make life much easier for yourself, list all of the circuit elements in order as they appear in the circuit going from left to right.

Each circuit element is specified by a line that must contain the following:

1. The element name.
2. The circuit nodes to which the element is connected.
3. The value of the element.

The first letter of the element name specifies the element type. The capacitor, for example, must begin with the letter C. The identification of an element may contain from one (1) to eight (8) characters. For example, the designations C, C1, COUT, and C3AZC2RY are all valid to use to describe a particular capacitor.

Any data fields are enclosed as follows: ⟨ . . . . ⟩, and are optional.

Branch voltages and currents follow the associated reference convention with the current flowing in the direction of the voltage drop.

All of the nodes must be *nonnegative* numbers and, once again, much of the literature states that the nodes need not be numbered sequentially. We suggest that you number all of your nodes sequentially, for the same reason that you place the lines in order—to find any problems that may be in your file. It is much easier to find mistakes if the schematic and the files are in sequential order. And again, the ground nodes are always zero.

One other line should be mentioned before we look at specific element lines. The COMMENT line can be an important line in a file because it can be placed anywhere in the file. It can be used to break up sections of a large circuit or it may be used to write reminder notes of changes that have been made or conditions that have been simulated in a particular file or line of that file. Therefore, these lines can be very important to a successful file and analysis. The comment line looks like this:

* ⟨ Any comment you desire ⟩

The asterisk (*) is the important part of this line. It tells the program that this is a comment and that it need not operate on the line that follows the asterisk. Some typical comment lines might be:

* R2 = 10k The gain should be about 100

* May the force be with my circuit

As stated previously, these lines may be placed anywhere in the file. Remember, they can be very valuable.

Element Lines

The following is a list of element lines that can be used in the SPICE program. As stated before, the element can have from one (1) to eight (8) characters in its description. Many times this will be portrayed as CXXXXXXX, indicating the seven additional characters that are allowed. We will present each of the elements and show them as C-------, for example, to indicate the seven additionally allowed characters.

Resistor

General form: R------- N1 N2 VALUE ⟨TC = TC1, ⟨,TC2⟩⟩

N1 N2	Element nodes
VALUE	Resistance in ohms
TC1,TC2	Temperature Coefficient (optional)

Example 1:

R1 1 2 100

This example says that we have a resistor connected between nodes 1 and 2 that is 100 ohms.

Example 2:

RC1 12 17 1k TC = 0.001,0.015

This example says that we have a 1000 ohm resistor connected between nodes 12 and 17 and the temperature coefficients are 0.001 and 0.015.

Capacitors and Inductors

General form: C------- N+ N− VALUE ⟨IC = INCOND⟩
L------- N+ N− VALUE ⟨IC = INCOND⟩

N+, N−	Positive and negative nodes
VALUE	Capacitance in farads, Inductance in henries

INCOND For the capacitance it is the initial value of voltage in volts; for the inductance it is the initial value of inductor current in amperes that flows from N+ through the inductor to N−. (Recall that the initial conditions only apply if the UIC option is specified on the .TRAN line.)

Example 1:

CBYP 13 0 1UF

This example says that we have a 1 microfarad bypass capacitor connected between node 13 and ground (node 0).

Example 2:

COSC 17 23 10U IC=3V

This example says that we have a 10 microfarad capacitor in an oscillator connected between nodes 17 and 23 with an initial value of 3 volts on it.

Example 3:

LOSC 42 69 1UH

This example says that we have a 1 microhenry inductor used in an oscillator that is connected between nodes 42 and 69 (N+ = 42, N− = 69).

Example 4:

LSHUNT 23 51 10U IC=15.7MA

This example says that we have a shunt inductor which is 10 microhenries, connected between nodes 23 and 51 (N+ = 23, N− = 51), and has a current of 15.7 milliamps flowing as an initial condition.

Coupled (Mutual) Inductors:

General form: K------- L------- L------- VALUE

L,L Names of the two coupled inductors

VALUE Coupling coefficient, K

Example 1:

K43 L1 L2 0.87

This example says that we have coupling between inductors L1 and L2 with a coupling coefficient of 0.87.

Transmission Lines (Lossless):

General form: T------- N1 N2 N3 N4 Zo = VALUE ⟨TD = VALUE⟩ ⟨F = FREQ⟩ ⟨NL = NRM LEN⟩ ⟨IC = V1,I1,V2,I2⟩

N1,N2 Nodes at port 1

N3,N4 Nodes at port 2

Zo Characteristic Impedance

Line length may be expressed in either of two ways:

1. Transmission Delay (TD) may be specified directly (TD = 10Ns).
2. Frequency is given with NL (normalized length of the transmission line with respect to the wavelength in the line at the frequency F). If F is given and NL is not, the value of .25 (a quarter wavelength) is assumed for analysis purposes.

Example 1:

TSTUB 1 2 3 4 Zo = 50 TD = 60N

This example says that we have a 50 ohm transmission line that is used as a stub connected at nodes 1 and 2 at the input, nodes 3 and 4 at the output, and a time delay of 60 nanoseconds.

Example 2:

TSTUB 4 5 6 7 Zo = 70 F = 25MEG NL = 1.4

This example says that we have a 70 ohm transmission line that is used as a stub connected at nodes 4 and 5 at the input, nodes 6 and 7 at the output, operates at 25 MHz and has a length that is 1.4 wavelengths at 25 MHz.

One point should be noted about SPICE and transmission lines. When transmission lines are used, it is common for them to be either open circuited or short circuited. This is where the problem arises with SPICE. The program is not amenable to either an open or a short circuit in the files, so it is necessary to make a few changes. If there is a short circuited transmission line in a circuit, connect a very small resistor between the transmission line and ground (approximately 0.001 ohms). This will satisfy the program and provide usable results. Similarly, for an open circuit, connect a very large resistor between the transmission line and ground (approximately 100 Megohms). This, again, will allow the program to analyze the circuit and give the proper results.

Linear Dependent Sources

There are four (4) equations that characterize the sources:

$$i = g * v, \ v = e * v, \ i = f * i, \ v = h * i$$

Where: g = Transconductance
e = Voltage gain
f = Current gain
h = Transresistance

Linear Voltage Controlled Current Source

General form: G------- N+ N− NC+ NC− VALUE

N+N−	Positive and Negative Nodes (Current flows from N+ through the source to N−)
MC+,NC−	Positive and Negative Controlling Nodes
VALUE	Transconductance in MHOS

Example:

G1 2 0 5 0.1

Linear Voltage Controlled Voltage Source

General form: E------- N + N− NC+ NC− VALUE

N+,N−	Positive and Negative Nodes
NC+,NC−	Positive and Negative Controlling Nodes
VALUE	Voltage Gain

Example:

E1 2 3 14 1 2.0

Linear Current Controlled Current Source

General form: F------- N+ N− VNAM VALUE

N+,N−	Positive and Negative Nodes
VNAM	Name of the voltage source through which the controlling current flows
VALUE	Current Gain

Example:

F1 13 5 VSENS 5

Linear Current Controlled Voltage Source

General form: H------- N+ N− VNAM VALUE

N+,N− Positive and Negative Nodes

VNAM Name of voltage source through which the controlling current flows

VALUE Transresistance in ohms

Example:

H4 3 24 VS 0.5k

Independent Sources

General form: V------- N+ N− ⟨⟨DC⟩ DC/TRAN VALUE⟩
⟨AC⟨AC MAG ⟨AC PHASE⟩⟩⟩
I------- N+ N− ⟨⟨DC⟩ DC/TRAN VALUE⟩
⟨AC⟨⟨AC MAG ⟨AC PHASE⟩⟩⟩

N+,N− Positive and Negative Nodes

DC/TRAN DC and TRANSIENT analysis value of the source. If the source value is zero for both DC and TRAN, the value may be omitted. (If the source value is Time-invariant (Power Supply), the value may optionally be preceded by the letters DC.)

AC MAG AC Magnitude

AC PHASE AC Phase

(The AC elements of the line set the source for AC analysis. If no AC MAG is specified, unity is used. If no AC PHASE is specified, zero is used. If AC is not needed, then AC is omitted.)

Example 1:

VCC 12 0 DC 5.5

This example says that we have a V_{CC} for our circuit that is a positive 5.5 volts dc from node 12 to ground.

Example 2:

VIN 7 4 0.001 AC 1 SIN(0 1.5 50k)

This example says that we have an input voltage source that is a positive 1 volt at node 7, a transient analysis value of 0.001, has an ac component that is sinusoidal, 0 volts dc offset, has a peak amplitude of 1.5 volts, and varies at a 50 kHz rate.

Independent Source Functions

There are five (5) different types of independent sources that can be used with the SPICE program:

- PULSE
- EXPONENTIAL
- SINUSOIDAL
- PIECE-WISE LINEAR
- SINGLE FREQUENCY FM

Let us now look briefly at each of these sources.

PULSE

General form: PULSE (V1 V2 TD TR TF PW PER)

V1	Initial Value (volts or amperes)
V2	Pulsed Value (volts or amperes)
TD	Delay Time
TR	Rise Time
TF	Fall Time
PW	Pulse Width
PER	Period

Example:

VIN 3 0 PULSE (−1 2NS 50NS 100NS)

This example shows a pulse that goes from −1 volt to +1 volt, is delayed by 2 nanoseconds, has a rise and fall time of 2 nanoseconds, has a pulse width of 5 nanoseconds and a period of 100 nanoseconds. Figure A.2 shows this pulse example.

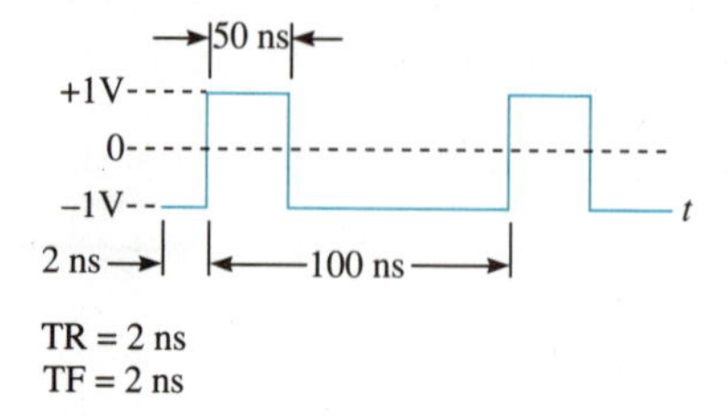

Figure A.2

Sinusoid:

General form: SIN (VO VA FREQ TD THETA)

VO	Offset (volts or amperes)
VA	Amplitude (volts or amperes)
FREQ	Frequency (Hz)
TD	Time Delay (second)
THETA	Damping Factor (1/seconds)

Example:

VIN 5 0 SIN (0 1 150MEG 1.5 NS 1E10)

This example shows a sinusoidal input to the circuit from node 5 to ground, has a 0 volt offset, an amplitude of 1 volt, a frequency of 150 MHz, a time delay of 1.5 nanoseconds, and a damping factor of one billion which says that it is virtually undamped. By damping we mean that there is a decaying sinusoidal signal whose amplitude decreases with time. This example has very little of that.

Exponential:

General form: EXP (V1 V2 TD1 TAU1 TD2 TAU2)

V1	Initial Value (volts or amperes)
V2	Pulsed Value (volts or amperes)
TD1	Rise Time Delay Time (seconds)
TAU1	Rise Time Constant (seconds)
TD2	Fall Delay Time (seconds)
TAU2	Fall Time Constant (seconds)

Example:

VIN 4 0 EXP (−4 −1 2NS 30NS 60NS 40NS)

This example presents an exponential signal from node 4 to ground with an initial value of −4 volts, a pulsed value of −1 volt, rise delay time of 2 nanoseconds, fall delay time of 60 nanoseconds, a rise time constant of 30 nanoseconds and a fall time constant of 40 nanoseconds.

Piece-Wise Linear:

General form: PWL (T1 V1 ⟨T2 V2 T3 V3 .....⟩)

Each pair of values (Ti, Vi) specifies the values of the source (volts or amperes) at time Ti.

Example:

VCLOCK 4 2 PWL (0 −7 10NS −7 11NS −3)

This example shows a piece-wise linear source which is from nodes 4 to 2, has a value of −7 volts at 0 time, has a value of −7 volts after 10 nanoseconds, and has a value of −3 volts after 11 nanoseconds.

Single Frequency FM:

General form: SFFM (VO VA FC MDI FS)

VO	Offset (volts or amperes)
VA	Amplitude (volts or amperes)
FC	Carrier Frequency (Hz)
MDI	Modulation Index
FS	Signal Frequency (Hz)

Example:

V3 8 0 SFFM (0 3M 35K 4 1.5K)

This example shows a single frequency FM source that has a 0 volt offset, an amplitude of 3 millivolts, a carrier frequency of 35 kHz, a modulation index of 4, and a signal (or modulating) frequency of 1.5 kHz.

Semiconductor Devices

Many parameters are required to describe the semiconductor devices used for circuit simulation and analysis. This section will present only the basic parameters needed. The detailed parameters are presented in Appendix A-2 of this text.

It is often necessary to provide a separate .MODEL line to fully describe the circuits. This method does away with the need to specify all of the model parameters on each device line.

Each device element line contains the device name, the nodes to which it is connected, and the device model name. The initial conditions may also be specified, if necessary. Two forms of initial conditions may be specified for a device:

1. Included to improve the DC convergence for circuits that contain more than one stable state,
2. Specified for use with the transient analysis. (These are true initial conditions as opposed to the convergence in form #1.)

(The .IC and .TRAN lines specify these conditions.)

Junction Diode

General form: D------- N+ N− MNAME ⟨AREA⟩ ⟨OFF⟩
⟨IC = VD⟩

N+,N−	Positive and Negative Nodes
MNAME	Model name
AREA	Area Factor (defaults to 1.0)
OFF	Indicates an optional starting condition in the device for DC analysis
IC = VD	Diode Voltage, used with .TRAN line when a transient analysis is desired starting from other than the quiescent operating point

Example 1:

DBRIDGE 3 7 DIODE 1

This example says that we have a diode used in a bridge circuit going from nodes 3 to 7, is termed "Diode" and has an area factor of 1.0.

Example 2:

DCLAMP 4 6 DMOD 3.0 IC = 0.2

This example says that we have a clamping diode from nodes 4 to 6, its name is DMOD, the area factor is three times that of the DMOD diode, and the desired starting point is 0.2 volts.

Bipolar Junction Transistor

General form: Q------- NC NB NE ⟨NS⟩ MNAME ⟨AREA⟩
⟨OFF⟩ ⟨IC = VBE,VCE⟩

NC,NB,NE	Collector, Base, and Emitter Nodes. (Be sure that these nodes are specified in this order.)
NS	Optional substrate node
MNAME	Model Name
AREA	Area Factor (defaults to 1.0)
OFF	Initial condition on the device for DC analysis
VBE,VCE	Base-to-Emitter Voltage, Collector-to-Emitter voltage

Example:

Q21 10 12 13 QMOD IC = 0.6 5.0

This example says that we have a Bipolar Junction transistor (#21) with its collector at node 10, base at node 12, and emitter at node 13. It is called QMOD and has initial conditions of VBE = 0.6 volts and VCE = 5 volts.

JFET

General format: J------- ND NG NS MNAME ⟨AREA⟩ ⟨OFF⟩
⟨IC = VDS,VGS⟩

ND,NG,NS Drain, Gate, and Source Nodes. (They must be presented in this order.)

MNAME Model name

AREA Area factor (Defaults to 1.0)

OFF Initial condition on the device for DC analysis

VDS,VGS Drain-to-Source Voltage, Gate-to-Source Voltage

Example:

J1 5 2 4 JM1 OFF

This example says that we have a JFET with the drain at node 5, the gate at node 2, the source at node 4, its model name is JM1, and the initial condition for DC analysis is Off.

MOSFET

General form: M------- ND NG NS NB MNAME ⟨L = VAL⟩
⟨W = VAL⟩ ⟨AD = VAL⟩ ⟨AS = VAL⟩ ⟨PD = VAL⟩
⟨PS = VAL⟩ ⟨NRD = VAL⟩ ⟨NRS = VAL⟩ ⟨OFF⟩
⟨IC = VDS,VGS,VBS⟩

ND,NG,NS,NB Drain, gate, source, and bulk (substrate) nodes

MNAME Model name

L,W Channel length and width in meters, respectively

AD,AS Areas of the drain and source diffusion in square meters, respectively

PD,PS Perimeters of the drain and source junctions, in meters, respectively

NRD,NRS Equivalent number of squares of the drain and source diffusion, respectively, that result in parasitic series drain and source resistances

OFF Indicates initial conditions

IC Specifies other initial conditions

VDS,VGS,VBS	Drain-to-Source Voltage, Gate-to-Source Voltage, Bulk-to-Source Voltage

Example 1:

M22 24 31 0 12 TYPE 1

This example says that we have a MOSFET called "Type 1" with the drain at node 24, the gate at node 31, the source at node 0 (ground), and the substrate (bulk) at node 12.

Example 2:

M13 3 5 12 10 MODM L=5U W=2U

This example says that we have a MOSFET termed MODM with the drain at node 3, the gate at node 5, the source at node 12, the substrate at node 10, with a channel length of 5 micrometers, and a channel width of 2 micrometers.

.MODEL LINE:

This line specifies a set of model parameters that will be used by one or more devices. MNAME is the model name as follows:

NPN	NPN Bipolar
PNP	PNP Bipolar
D	Diode
NJF	N-Channel JFET
PJF	P-Channel JFET
NMOS	N-Channel MOSFET
PMON	P-Channel MOSFET

(The SPICE manual will list parameters to be put in the model line.)

SUBCIRCUIT LINE:

This line can be used when there is a circuit within a circuit or a series of circuits that are repeated over and over again. This option allows you to put the circuit in a file once and then refer to the .SUBCKT name as many times as desired later on. The general form for .SUBCKT is:

.SUBCKT SUBNAME N1 ⟨N2 N3. . . .⟩

SUBNAME	Name of the subcircuit
N1,N2,etc	External node connections

The .ENDS line is used to end the subcircuit. Note that this is not the same command that is used to end the file. That command is .END, not .ENDS as is with this case. The general form is:

.ENDS ⟨SUBNAME⟩

PRINT LINE:

The designation is .PRINT and it defines the contents of a tabular listing of one (1) to eight (8) output variables. The general form is:

.PRINT PRTYPE OV1 ⟨OV2 OV3. . . .⟩

PRTYPE Type of analysis (DC,AC,TRAN,NOISE,DISTO) for which the outputs are desired

OV Output variables which are listed in the SPICE manual

PLOT LINE:

The designation is .PLOT and it defines the content of one plot from one (1) to eight (8) output variables. The general form is:

.PLOT PLTYPE OV1⟨(PLO1,PHI1)⟩ ⟨OV2 ⟨(PLO2,PHI2)⟩

PLTYPE Type of analysis (DC,AC,TRAN,NOISE,DISTO)

OV Output variables, same as for print

PLO,PHI Plot limits; Plot Low and Plot High

This general discussion of the SPICE program can be a tremendous help to the student when simulating and analyzing communication circuits. It is only a general discussion and it should be understood that more detailed instructions are needed, a text designed specifically to discuss SPICE should be consulted. An excellent text is listed in the references of this text.

In order to pull all of the information presented together, it would be useful to see how the student can put the statements together to form a file that can be used to simulate and analyze circuits. That is what will now be done. We will present two (2) representative examples of circuits and construct files for each of them.

Example 1: DC CIRCUIT

The circuit shown in Figure A.3 is a dc circuit that we wish to place in a SPICE file and check to see what the voltage across R_2 will be if $V_1 = 24$ volts. The values for all of the components of the circuit are as follows:

V_1	24 volts
L_1	10 microhenries
R_1	1000 ohms
C_1	10 microfarads
R_2	50 ohms

The first task to be performed when making up a SPICE file is to label the nodes so that the file can be written properly. The nodes to be used will

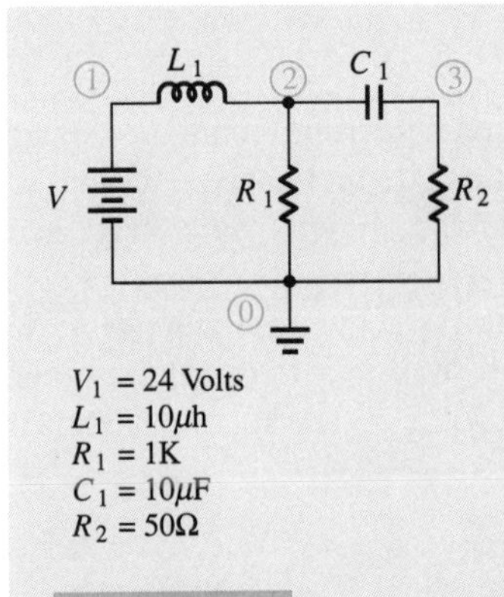

Figure A.3

be as shown in Figure A.3. Node 1 is at the intersection of V_1 and L_1; Node 2 at the intersection of L_1, C_1, and R_1; Node 3 is at the intersection of C_1 and R_2; and Node 0 is the common ground connections. Notice that the nodes are all numbered sequentially. Recall that we stated that this was not necessary, but is good practice because it is easier to keep track of a circuit if this is done. We will follow this same reasoning when we make up the file.

Now we are ready to make the file for the dc circuit. We will need to refer to previous sections and remember how the component (elements) is to be written, so that the program can recognize the elements and be able to operate on them. The file to be used for the dc circuit is shown below. We will have the program both print and plot the output voltage across R1.

```
DC CIRCUIT          (Title Line)
VIN 1 0 24          (DC supply)
L1 1 2 10U
R1 2 0 1K
C1 2 3 1U
R2 3 0 50
.PRINT DC V(2)
.PLOT DC V(2)
.END
```

You can see that all of the requirements for the SPICE file are met with this file. Remember, the words "title line" and "DC supply" are not put into this file. They are only put in here to show you what these lines represent. Also, remember that the PRINT, PLOT, and END commands *must have a period before them*.

Example 2:

AC CIRCUIT

In this example we will take the circuit shown in Figure A.4, place it into a SPICE file so that the program can simulate and analyze it, and have the program print and plot the magnitude and phase of the voltage across the output tank circuit (C_3 and L_1).

Again, we must set up the nodes as shown in Figure A.4. We have again placed them in order, and we will also write the file in order. We will sweep the input ac from 500 Hz to 1500 Hz. The file for the ac circuit is shown below.

```
AC CIRCUIT
VGEN 1 0 AC 1
R1 1 2 100
C1 2 0 1U
R2 2 3 100
.MODEL CLMP (RS=5 VJ=0.9 IBV=1.0E-3)
D1 3 4 CLMP
V1 4 0 5
R3 3 5 100
C2 5 0 1U
R4 5 6 100
C3 6 0 10P
L1 6 0 100U
.AC LIN 21 500 1500
```

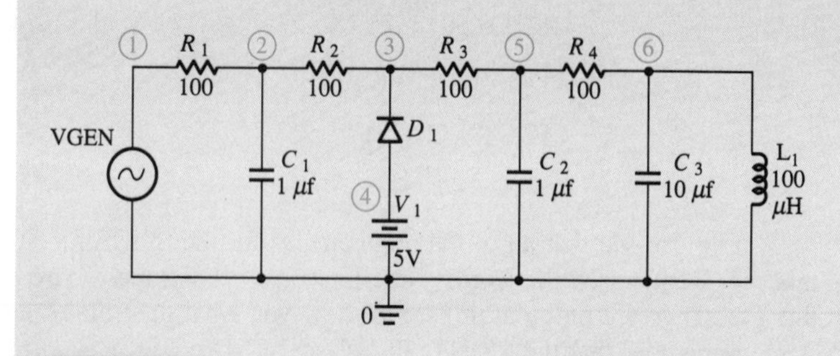

Figure A.4

```
.PRINT AC VM(6) VP(6)
.PLOT AC VM(6) VP(6)
.END
```

The file above needs a few comments before we leave it. First, the VGEN line has a 1 at the end of it. This indicates that the amplitude of the ac signal is 1 volt. Next, the .MODEL line is the one that describes the diode with its parameters. Then the D1 line is used to place this diode into the circuit. Next, the .AC line shows 21 points to be taken over a frequency range of 500 Hz to 1500 Hz, that is, there will be data taken every 50 Hz. Finally, the .PRINT and .PLOT lines show VM(6) and VP(6) for Voltage Magnitude at node 6 and Voltage Phase at node 6.

SPICE Questions

1. Name four (4) components that may be in your circuit to be simulated by SPICE.
2. Describe what the .OP command does in the SPICE program.
3. In ac analysis in SPICE, what does DEC ND mean?
4. In the following SPICE command, what is the step size of the input signal? Why?

   .AC LIN 101 6000 7000

5. What do each of the following scale factors stand for in SPICE: G, MIL, U, P?
6. Explain, in detail, what this SPICE line is describing.

   LXMR 16 18 15U IC=10.2M

7. Write the SPICE file line for a 10 volt input power supply connected between node 31 and ground.
8. Explain what this SPICE file line is describing.

   V2 7 0 SFFM (0 5.2M 16MEG 3 2.1K)

9. In the SPICE .PLOT line, what does PLTYPE represent?

Appendix B SPICE Semiconductor Models

```
* Library of model parameters

* Copyright 1987 by MicroSim Corporation
*   Neither this library nor any part may be copied without the express
*   written consent of MicroSim Corporation

* Reprint permission granted to Saunders College Publishing for the title
* "Analog Communications for Technology"

* Zener Diodes -----------------------------------------------------------

.model  D1N746  D(Is=5u Rs=14 Bv=2.81 Ibv=5u)
.model  D1N747  D(Is=5u Rs=12 Bv=3.15 Ibv=5u)
.model  D1N748  D(Is=5u Rs=12 Bv=3.45 Ibv=5u)
.model  D1N749  D(Is=1u Rs=11 Bv=3.82 Ibv=1u)
.model  D1N750  D(Is=1u Rs=10 Bv=4.24 Ibv=1u)
.model  D1N751  D(Is=0.5u Rs=9 Bv=4.65 Ibv=0.5u)
.model  D1N752  D(Is=0.5u Rs=6 Bv=5.20 Ibv=0.5u)
.model  D1N753  D(Is=0.05u Rs=4 Bv=5.79 Ibv=0.05u)
.model  D1N754  D(Is=0.05u Rs=3 Bv=6.41 Ibv=0.05u)
.model  D1N755  D(Is=0.05u Rs=3 Bv=7.11 Ibv=0.05u)
.model  D1N756  D(Is=0.05u Rs=4 Bv=7.79 Ibv=0.05u)
.model  D1N757  D(Is=0.05u Rs=5 Bv=8.67 Ibv=0.05u)
.model  D1N758  D(Is=0.05u Rs=9 Bv=9.49 Ibv=0.05u)
.model  D1N759  D(Is=0.05u Rs=15 Bv=11.37 Ibv=0.05u)

* 1N916 family -----------------------------------------------------------

.model  D1N914  D(Is=0.1p Rs=16 CJO=2p Tt=12n Bv=100 Ibv=0.1p)
.model  D1N4148 D(Is=0.1p Rs=16 CJO=2p Tt=12n Bv=100 Ibv=0.1p)
.model  D1N4531 D(Is=0.1p Rs=16 CJO=2p Tt=12n Bv=100 Ibv=0.1p)
.model  D1N916  D(Is=0.1p Rs=8 CJO=1p Tt=12n Bv=100 Ibv=0.1p)
.model  D1N4149 D(Is=0.1p Rs=8 CJO=1p Tt=12n Bv=100 Ibv=0.1p)
.model  D1N4446 D(Is=0.1p Rs=4 CJO=2p Tt=12n Bv=100 Ibv=0.1p)
.model  D1N914A D(Is=0.1p Rs=4 CJO=2p Tt=12n Bv=100 Ibv=0.1p)
```

```
.model  D1N4447 D(Is=0.1p Rs=4 CJO=1p Tt=12n Bv=100 Ibv=0.1p)
.model  D1N916A D(Is=0.1p Rs=4 CJO=1p Tt=12n Bv=100 Ibv=0.1p)
.model  D1N4448 D(Is=0.1p Rs=2 CJO=2p Tt=12n Bv=100 Ibv=0.1p)
.model  D1N914B D(Is=0.1p Rs=2 CJO=2p Tt=12n Bv=100 Ibv=0.1p)
.model  D1N4449 D(Is=0.1p Rs=4 CJO=1p Tt=12n Bv=100 Ibv=0.1p)
.model  D1N916B D(Is=0.1p Rs=4 CJO=1p Tt=12n Bv=100 Ibv=0.1p)

* 1N4150 ----------------------------------------------------------------------

.model  D1N4150 D(Is=10E-15 Rs=1.0 CJO=1.3p Tt=12n Bv=70 Ibv=0.1p)

* 1N4154 family ---------------------------------------------------------------

.model  D1N4151 D(Is=0.1p Rs=3 CJO=1p Tt=3n Bv=100 Ibv=0.1p)
.model  D1N3604 D(Is=0.1p Rs=3 CJO=1p Tt=3n Bv=100 Ibv=0.1p)
.model  D1N4152 D(Is=0.1p Rs=6 CJO=1p Tt=3n Bv=70 Ibv=0.1p)
.model  D1N3605 D(Is=0.1p Rs=6 CJO=1p Tt=3n Bv=70 Ibv=0.1p)
.model  D1N4533 D(Is=0.1p Rs=6 CJO=1p Tt=3n Bv=70 Ibv=0.1p)
.model  D1N4153 D(Is=0.1p Rs=6 CJO=1p Tt=3n Bv=100 Ibv=0.1p)
.model  D1N3606 D(Is=0.1p Rs=6 CJO=1p Tt=3n Bv=100 Ibv=0.1p)
.model  D1N4534 D(Is=0.1p Rs=6 CJO=1p Tt=3n Bv=100 Ibv=0.1p)
.model  D1N4154 D(Is=0.1p Rs=4 CJO=2p Tt=3n Bv=60 Ibv=0.1p)
.model  D1N4009 D(Is=0.1p Rs=4 CJO=2p Tt=3n Bv=60 Ibv=0.1p)
.model  D1N4536 D(Is=0.1p Rs=4 CJO=2p Tt=3n Bv=60 Ibv=0.1p)

* 1N4444 family ---------------------------------------------------------------

.model  D1N4305 D(Is=0.1p Rs=4 CJO=1p Tt=8n Bv=100 Ibv=0.1p)
.model  D1N3063 D(Is=0.1p Rs=4 CJO=1p Tt=8n Bv=100 Ibv=0.1p)
.model  D1N4532 D(Is=0.1p Rs=4 CJO=1p Tt=8n Bv=100 Ibv=0.1p)
.model  D1N4444 D(Is=0.1p Rs=2 CJO=1p Tt=12n Bv=100 Ibv=0.1p)
.model  D1N4454 D(Is=0.1p Rs=4 CJO=1p Tt=8n Bv=100 Ibv=0.1p)
.model  D1N3064 D(Is=0.1p Rs=4 CJO=1p Tt=8n Bv=100 Ibv=0.1p)

* 2N2222 family ---------------------------------------------------------------

.model Q2N2217  NPN(Is=3.108f Xti=3 Eg=1.11 Vaf=303.2 Bf=45.18 Ne=1.541
+               Ise=918.1f Ikf=1.296 Xtb=1.5 Br=16.28 Nc=2 Isc=0 Ikr=0 Rc=1
+               Cjc=14.57p Vjc=.75 Mjc=.3333 Fc=.5 Cje=26.08p Vje=.75
+               Mje=.3333 Tr=42.81n Tf=449.4p Itf=.1 Vtf=10 Xtf=2)
.model Q2N2218  NPN(Is=3.108f Xti=3 Eg=1.11 Vaf=303.2 Bf=90.37 Ne=1.541
+               Ise=459f Ikf=1.296 Xtb=1.5 Br=8.001 Nc=2 Isc=0 Ikr=0 Rc=1
+               Cjc=14.57p Vjc=.75 Mjc=.3333 Fc=.5 Cje=26.08p Vje=.75
+               Mje=.3333 Tr=49.92n Tf=449.4p Itf=.1 Vtf=10 Xtf=2)
.model Q2N2218A NPN(Is=3.108f Xti=3 Eg=1.11 Vaf=303.2 Bf=90.37 Ne=1.541
+               Ise=459f Ikf=1.296 Xtb=1.5 Br=8.001 Nc=2 Isc=0 Ikr=0 Rc=1
+               Cjc=14.57p Vjc=.75 Mjc=.3333 Fc=.5 Cje=26.08p Vje=.75
+               Mje=.3333 Tr=49.92n Tf=449.4p Itf=.1 Vtf=10 Xtf=2)
.model Q2N2219  NPN(Is=3.108f Xti=3 Eg=1.11 Vaf=131.5 Bf=217.5 Ne=1.541
+               Ise=190.7f Ikf=1.296 Xtb=1.5 Br=6.18 Nc=2 Isc=0 Ikr=0 Rc=1
+               Cjc=14.57p Vjc=.75 Mjc=.3333 Fc=.5 Cje=26.08p Vje=.75
+               Mje=.3333 Tr=51.35n Tf=451p Itf=.1 Vtf=10 Xtf=2)
.model Q2N2219A NPN(Is=3.108f Xti=3 Eg=1.11 Vaf=131.5 Bf=217.5 Ne=1.541
+               Ise=190.7f Ikf=1.296 Xtb=1.5 Br=6.18 Nc=2 Isc=0 Ikr=0 Rc=1
+               Cjc=14.57p Vjc=.75 Mjc=.3333 Fc=.5 Cje=26.08p Vje=.75
+               Mje=.3333 Tr=51.35n Tf=365.3p Itf=.1 Vtf=10 Xtf=2)
.model Q2N2220  NPN(Is=3.108f Xti=3 Eg=1.11 Vaf=303.2 Bf=45.18 Ne=1.541
+               Ise=918.1f Ikf=1.296 Xtb=1.5 Br=16.28 Nc=2 Isc=0 Ikr=0 Rc=1
+               Cjc=14.57p Vjc=.75 Mjc=.3333 Fc=.5 Cje=26.08p Vje=.75
+               Mje=.3333 Tr=42.81n Tf=449.4p Itf=.1 Vtf=10 Xtf=2)
.model Q2N2221  NPN(Is=3.108f Xti=3 Eg=1.11 Vaf=303.2 Bf=90.37 Ne=1.541
+               Ise=459f Ikf=1.296 Xtb=1.5 Br=8.001 Nc=2 Isc=0 Ikr=0 Rc=1
```

```
+               Cjc=14.57p Vjc=.75 Mjc=.3333 Fc=.5 Cje=26.08p Vje=.75
+               Mje=.3333 Tr=49.92n Tf=449.4p Itf=.1 Vtf=10 Xtf=2)
.model Q2N2221A NPN(Is=3.108f Xti=3 Eg=1.11 Vaf=303.2 Bf=90.37 Ne=1.541
+               Ise=459f Ikf=1.296 Xtb=1.5 Br=8.001 Nc=2 Isc=0 Ikr=0 Rc=1
+               Cjc=14.57p Vjc=.75 Mjc=.3333 Fc=.5 Cje=26.08p Vje=.75
+               Mje=.3333 Tr=49.92n Tf=449.4p Itf=.1 Vtf=10 Xtf=2)
.model Q2N2222  NPN(Is=3.108f Xti=3 Eg=1.11 Vaf=131.5 Bf=217.5 Ne=1.541
+               Ise=190.7f Ikf=1.296 Xtb=1.5 Br=6.18 Nc=2 Isc=0 Ikr=0 Rc=1
+               Cjc=14.57p Vjc=.75 Mjc=.3333 Fc=.5 Cje=26.08p Vje=.75
+               Mje=.3333 Tr=51.35n Tf=451p Itf=.1 Vtf=10 Xtf=2)
.model Q2N2222A NPN(Is=3.108f Xti=3 Eg=1.11 Vaf=131.5 Bf=217.5 Ne=1.541
+               Ise=190.7f Ikf=1.296 Xtb=1.5 Br=6.18 Nc=2 Isc=0 Ikr=0 Rc=1
+               Cjc=14.57p Vjc=.75 Mjc=.3333 Fc=.5 Cje=26.08p Vje=.75
+               Mje=.3333 Tr=51.35n Tf=365.3p Itf=.1 Vtf=10 Xtf=2)
.model Q2N5581  NPN(Is=3.108f Xti=3 Eg=1.11 Vaf=303.2 Bf=90.37 Ne=1.541
+               Ise=459f Ikf=1.296 Xtb=1.5 Br=8.001 Nc=2 Isc=0 Ikr=0 Rc=1
+               Cjc=14.57p Vjc=.75 Mjc=.3333 Fc=.5 Cje=26.08p Vje=.75
+               Mje=.3333 Tr=49.92n Tf=449.4p Itf=.1 Vtf=10 Xtf=2)
.model Q2N2282  NPN(Is=3.108f Xti=3 Eg=1.11 Vaf=131.5 Bf=217.5 Ne=1.541
+               Ise=190.7f Ikf=1.296 Xtb=1.5 Br=6.18 Nc=2 Isc=0 Ikr=0 Rc=1
+               Cjc=14.57p Vjc=.75 Mjc=.3333 Fc=.5 Cje=26.08p Vje=.75
+               Mje=.3333 Tr=51.35n Tf=365.3p Itf=.1 Vtf=10 Xtf=2)

* 2N2905 family -------------------------------------------------------------

.model Q2N2904  PNP(Is=9.913f Xti=3 Eg=1.11 Vaf=90.7 Bf=81.67 Ne=1.975
+               Ise=4.066p Ikf=.5161 Xtb=1.5 Br=3.798 Nc=2 Isc=0 Ikr=0 Rc=1
+               Cjc=14.57p Vjc=.75 Mjc=.3333 Fc=.5 Cje=20.16p Vje=.75
+               Mje=.3333 Tr=31.94n Tf=404.3p Itf=.4 Vtf=10 Xtf=2)
.model Q2N2904A PNP(Is=9.913f Xti=3 Eg=1.11 Vaf=90.7 Bf=103.3 Ne=2.321
+               Ise=14.63p Ikf=.7853 Xtb=1.5 Br=3.614 Nc=2 Isc=0 Ikr=0 Rc=1
+               Cjc=14.57p Vjc=.75 Mjc=.3333 Fc=.5 Cje=20.16p Vje=.75
+               Mje=.3333 Tr=30.54n Tf=405.8p Itf=.4 Vtf=10 Xtf=2)
.model Q2N2905  PNP(Is=9.913f Xti=3 Eg=1.11 Vaf=90.7 Bf=197.8 Ne=2.264
+               Ise=6.191p Ikf=.7322 Xtb=1.5 Br=3.369 Nc=2 Isc=0 Ikr=0 Rc=1
+               Cjc=14.57p Vjc=.75 Mjc=.3333 Fc=.5 Cje=20.16p Vje=.75
+               Mje=.3333 Tr=29.17n Tf=405.6p Itf=.4 Vtf=10 Xtf=2)
.model Q2N2905A PNP(Is=9.913f Xti=3 Eg=1.11 Vaf=90.7 Bf=263.5 Ne=1.932
+               Ise=886.5f Ikf=.5498 Xtb=1.5 Br=3.313 Nc=2 Isc=0 Ikr=0 Rc=1
+               Cjc=14.57p Vjc=.75 Mjc=.3333 Fc=.5 Cje=20.16p Vje=.75
+               Mje=.3333 Tr=28.89n Tf=404.6p Itf=.4 Vtf=10 Xtf=2)
.model Q2N2906  PNP(Is=9.913f Xti=3 Eg=1.11 Vaf=90.7 Bf=81.67 Ne=1.975
+               Ise=4.066p Ikf=.5161 Xtb=1.5 Br=3.798 Nc=2 Isc=0 Ikr=0 Rc=1
+               Cjc=14.57p Vjc=.75 Mjc=.3333 Fc=.5 Cje=20.16p Vje=.75
+               Mje=.3333 Tr=31.94n Tf=404.3p Itf=.4 Vtf=10 Xtf=2)
.model Q2N2906A PNP(Is=9.913f Xti=3 Eg=1.11 Vaf=90.7 Bf=103.3 Ne=2.321
+               Ise=14.63p Ikf=.7853 Xtb=1.5 Br=3.614 Nc=2 Isc=0 Ikr=0 Rc=1
+               Cjc=14.57p Vjc=.75 Mjc=.3333 Fc=.5 Cje=20.16p Vje=.75
+               Mje=.3333 Tr=30.54n Tf=405.8p Itf=.4 Vtf=10 Xtf=2)
.model Q2N2907  PNP(Is=9.913f Xti=3 Eg=1.11 Vaf=90.7 Bf=197.8 Ne=2.264
+               Ise=6.191p Ikf=.7322 Xtb=1.5 Br=3.369 Nc=2 Isc=0 Ikr=0 Rc=1
+               Cjc=14.57p Vjc=.75 Mjc=.3333 Fc=.5 Cje=20.16p Vje=.75
+               Mje=.3333 Tr=29.17n Tf=405.6p Itf=.4 Vtf=10 Xtf=2)
.model Q2N2907A PNP(Is=9.913f Xti=3 Eg=1.11 Vaf=90.7 Bf=263.5 Ne=1.932
+               Ise=886.5f Ikf=.5498 Xtb=1.5 Br=3.313 Nc=2 Isc=0 Ikr=0 Rc=1
+               Cjc=14.57p Vjc=.75 Mjc=.3333 Fc=.5 Cje=20.16p Vje=.75
+               Mje=.3333 Tr=28.89n Tf=404.6p Itf=.4 Vtf=10 Xtf=2)
.model Q2N3485  PNP(Is=9.913f Xti=3 Eg=1.11 Vaf=90.7 Bf=81.67 Ne=1.975
+               Ise=4.066p Ikf=.5161 Xtb=1.5 Br=3.798 Nc=2 Isc=0 Ikr=0 Rc=1
+               Cjc=14.57p Vjc=.75 Mjc=.3333 Fc=.5 Cje=20.16p Vje=.75
+               Mje=.3333 Tr=31.94n Tf=404.3p Itf=.4 Vtf=10 Xtf=2)
.model Q2N3485A PNP(Is=9.913f Xti=3 Eg=1.11 Vaf=90.7 Bf=103.3 Ne=2.321
```

```
+               Ise=14.63p Ikf=.7853 Xtb=1.5 Br=3.614 Nc=2 Isc=0 Ikr=0 Rc=1
+               Cjc=14.57p Vjc=.75 Mjc=.3333 Fc=.5 Cje=20.16p Vje=.75
+               Mje=.3333 Tr=30.54n Tf=405.8p Itf=.4 Vtf=10 Xtf=2)
.model Q2N3486  PNP(Is=9.913f Xti=3 Eg=1.11 Vaf=90.7 Bf=197.8 Ne=2.264
+               Ise=6.191p Ikf=.7322 Xtb=1.5 Br=3.369 Nc=2 Isc=0 Ikr=0 Rc=1
+               Cjc=14.57p Vjc=.75 Mjc=.3333 Fc=.5 Cje=20.16p Vje=.75
+               Mje=.3333 Tr=29.17n Tf=405.6p Itf=.4 Vtf=10 Xtf=2)
.model Q2N3486A PNP(Is=9.913f Xti=3 Eg=1.11 Vaf=90.7 Bf=263.5 Ne=1.932
+               Ise=886.5f Ikf=.5498 Xtb=1.5 Br=3.313 Nc=2 Isc=0 Ikr=0 Rc=1
+               Cjc=14.57p Vjc=.75 Mjc=.3333 Fc=.5 Cje=20.16p Vje=.75
+               Mje=.3333 Tr=28.89n Tf=404.6p Itf=.4 Vtf=10 Xtf=2)

* 2N4402 family ------------------------------------------------------------

.model Q2N4402  PNP(Is=4.497f Xti=3 Eg=1.11 Vaf=90.7 Bf=119.6 Ne=1.963
+               Ise=5.478p Ikf=.8382 Xtb=1.5 Br=2.686 Nc=2 Isc=0 Ikr=0 Rc=.5
+               Cjc=12.82p Vjc=.75 Mjc=.3333 Fc=.5 Cje=27.27p Vje=.75
+               Mje=.3333 Tr=71.57n Tf=335p Itf=10 Vtf=10 Xtf=2)
.model Q2N4403  PNP(Is=4.497f Xti=3 Eg=1.11 Vaf=19.27 Bf=236.4 Ne=1.963
+               Ise=2.773p Ikf=.8382 Xtb=1.5 Br=2.525 Nc=2 Isc=0 Ikr=0 Rc=.5
+               Cjc=12.82p Vjc=.75 Mjc=.3333 Fc=.5 Cje=27.27p Vje=.75
+               Mje=.3333 Tr=69.18n Tf=325p Itf=10 Vtf=10 Xtf=2)

* 2N5109 -------------------------------------------------------------------

.MODEL Q2N5109  NPN(IS=5E-15 ISE=10NA NE=4 ISC=10NA NC=4 BF=90 IKF=.2A VAF=240
+               CJC=5PF CJE=10PF RB=0.25 RE=0.25 RC=1.5 TF=0.1NS TR=20NS
+               KF=1E-15)

*--------------------------------------------------------------------------

* connections:   non-inverting input
*                | inverting input
*                | | positive power supply
*                | | | negative power supply
*                | | | | open collector output
*                | | | | | output ground
*                | | | | | |
.subckt LM119    1 2 3 4 5 6
*
  f1    3  9 v1 1
  iee   7  4 dc 100.0E-6
  q1    9  2  7 qin
  q2    8  1  7 qin
  q3    9  8  3 qmo
  q4    8  8  3 qmi
.model qin NPN(Is=800.0E-18 Bf=333.3)
.model qmi PNP(Is=800.0E-18 Bf=1002)
.model qmo PNP(Is=800.0E-18 Bf=1000 Cjc=1E-15 Tr=59.42E-9)
  e1   10  6  3  9  1
  v1   10 11 dc 0
  q5    5 11  6 qoc
.model qoc NPN(Is=800.0E-18 Bf=41.38E3 Cjc=1E-15 Tf=23.91E-12 Tr=24.01E-9)
  dp    4  3 dx
  rp    3  4 5.556E3
.model dx  D(Is=800.0E-18)
*
.ends

*--------------------------------------------------------------------------
```

```
* connections:   non-inverting input
*                | inverting input
*                | | positive power supply
*                | | | negative power supply
*                | | | | open collector output
*                | | | | |
.subckt LM139    1 2 3 4 5
*
  f1     9   3 v1 1
  iee    3   7 dc 100.0E-6
  vi1   21   1 dc .75
  vi2   22   2 dc .75
  q1     9 21  7 qin
  q2     8 22  7 qin
  q3     9  8  4 qmo
  q4     8  8  4 qmi
.model qin PNP(Is=800.0E-18 Bf=2.000E3)
.model qmi NPN(Is=800.0E-18 Bf=1002)
.model qmo NPN(Is=800.0E-18 Bf=1000 Cjc=1E-15 Tr=475.4E-9)
  e1    10   4  9  4  1
  v1    10 11 dc 0
  q5     5 11  4 qoc
.model qoc NPN(Is=800.0E-18 Bf=20.69E3 Cjc=1E-15 Tf=3.540E-9 Tr=472.8E-9)
  dp     4   3 dx
  rp     3   4 1.020E3
.model dx  D(Is=800.0E-18)
*
.ends

*----------------------------------------------------------------------------
* connections:   non-inverting input
*                | inverting input
*                | | positive power supply
*                | | | negative power supply
*                | | | | open collector output
*                | | | | |
.subckt LM339    1 2 3 4 5
*
  x_lm339 1 2 3 4 5 LM139
*
* the LM339 is identical to the LM139, but has a more limited temp. range
*
.ends

*----------------------------------------------------------------------------
* connections:   non-inverting input
*                | inverting input
*                | | positive power supply
*                | | | negative power supply
*                | | | | output
*                | | | | |
.subckt uA741    1 2 3 4 5
*
  c1    11 12 8.661E-12
  c2     6  7 30.00E-12
  dc     5 53 dx
  de    54  5 dx
  dlp   90 91 dx
  dln   92 90 dx
  dp     4  3 dx
  egnd 99   0 poly(2) (3,0) (4,0) 0 .5 .5
  fb     7 99 poly(5) vb vc ve vlp vln 0 10.61E6 -10E6 10E6 10E6 -10E6
```

```
  ga    6  0 11 12 188.5E-6
  gcm   0  6 10 99 5.961E-9
  iee  10  4 dc 15.16E-6
  hlim 90  0 vlim 1K
  q1   11  2 13 qx
  q2   12  1 14 qx
  r2    6  9 100.0E3
  rc1   3 11 5.305E3
  rc2   3 12 5.305E3
  re1  13 10 1.836E3
  re2  14 10 1.836E3
  ree  10 99 13.19E6
  ro1   8  5 50
  ro2   7 99 100
  rp    3  4 18.16E3
  vb    9  0 dc 0
  vc    3 53 dc 1
  ve   54  4 dc 1
  vlim  7  8 dc 0
  vlp  91  0 dc 40
  vln   0 92 dc 40
.model dx D(Is=800.0E-18)
.model qx NPN(Is=800.0E-18 Bf=93.75)
.ends

* end of library
```

Appendix C SPICE Semiconductor Parameters

Default Parameters for Semiconductor Devices

Junction Diodes

	Name	Parameter	Units	Default
1	IS	saturation current	A	1.0 E-14
2	RS	ohmic resistance	ohms	0
3	N	emission coefficient	–	1
4	TT	transit-time	sec	0
5	CJO	zero-bias junction capacitance	F	0
6	VJ	junction potential	V	1
7	M	grading coefficient	–	0.5
8	EG	activation energy	eV	1.11
9	XTI	saturation-current temp. exp.	–	3.0
10	KF	flicker noise coefficient	–	0
11	AF	flicker noise exponent	–	1
12	FC	coefficient for forward-bias depletion capacitance formula	–	0.5
13	BV	reverse breakdown voltage	V	infinite
14	IBV	current at breakdown voltage	A	1.0 E-3

Bipolar Junction Transistors (BJT)

	Name	Parameter	Units	Default
1	IS	transport saturation current	A	1.0 E-16
2	BF	ideal forward maximum beta	–	100
3	NF	forward current emission coefficient	–	1.0
4	VAF	forward Early voltage	V	infinite
5	IKF	corner for forward beta high current roll-off	A	infinite
6	ISE	B-E leakage saturation current	A	0
7	NE	B-E leakage emission coefficient	–	1.5
8	BR	ideal maximum reverse beta	–	1
9	NR	reverse current emission coefficient	–	1
10	VAR	reverse Early voltage	V	infinite
11	IKR	corner for reverse beta high current roll-off	A	infinite
12	ISC	B-C leakage saturation current	A	0
13	NC	B-C leakage emission coefficient	–	2
14	RB	zero bias base resistance	ohms	0
15	IRB	current where base resistance falls halfway to its minimum value	A	infinite
16	RBM	minimum base resistance at high currents	ohms	RB
17	RE	emitter resistance	ohms	0
18	RC	collector resistance	ohms	0
19	CJE	B-E zero-bias depletion capacitance	F	0
20	VJE	B-E built-in potential	V	0.75
21	MJE	B-E junction exponential factor	–	0.33
22	TF	ideal forward transit time	sec	0
23	XTF	coefficient for bias dependence of TF	–	0
24	VTF	voltage describing VBC dependence of TF	V	infinite
25	ITF	high-current parameter for effect on TF	A	0
26	PTF	excess phase at freq = 1.0/(TF*2π) Hz	deg	0
27	CJC	B-C zero-bias depletion capacitance	F	0
28	VJC	B-C built-in potential	V	0.75
29	MJC	B-C junction exponential factor	–	0.33
30	XCJC	fraction of B-C depletion capacitance connected to internal base node	–	1
31	TR	ideal reverse transit time	sec	0
32	CJS	zero-bias collector-substrate capacitance	F	0
33	VJS	substrate junction built-in potential	V	0.75
34	MJS	sustrate junction exponential factor	–	0
35	XTB	forward and reverse beta temp. exp.	–	0
36	EG	energy gap for temp. effect on IS	eV	1.11
37	XTI	temperature exponent for effect on IS	–	3
38	KF	flicker noise coefficient	–	0
39	AF	flicker noise exponent	–	1
40	FC	coefficient for forward-bias depletion capacitance formula	–	0.5

Junction Field Effect Transistors (JFET)

	Name	Parameter	Units	Default
1	VTO	threshold voltage	V	−2.0
2	BETA	transconductance parameter	A/V^2	1.0 E-4
3	LAMBDA	channel length modulation parameter	1/V	0
4	RD	drain ohmic resistance	ohm	0
5	RS	source ohmic resistance	ohm	0
6	CGS	zero-bias G-S junction capacitance	F	0
7	CGD	zero-bias G-D junction capacitance	F	0
8	PB	gate junction potential	V	1
9	IS	gate junction saturation current	A	1.0 E-14
10	KF	flicker noise coefficient	–	0
11	AF	flicker noise exponent	–	1
12	FC	coefficient for forward bias depletion capacitance formula	–	0.5

Metal Oxide Silicone Field Effect Transistors (MOSFET)

	Name	Parameter	Units	Default
1	LEVEL	model index number	–	1
2	VTO	zero-bias threshold voltage	V	0
3	KP	transconductance parameter	A/V^2	2
4	GAMMA	bulk threshold parameter	$V^{0.5}$	0
5	PHI	surface potential	V	0.6
6	LAMBDA	channel length modulation (model index 1 and 2 only)	1/V	0
7	RD	drain ohmic resistance	ohm	0
8	RS	source ohmic resistance	ohm	0
9	CBD	zero-bias B-D junction capacitance	F	0
10	CBS	zero-bias B-S junction capacitance	F	0
11	IS	bulk junction saturation current	A	1
12	PB	bulk junction potential	V	0.8
13	CGSO	gate-source overlap capacitance per meter channel width	F/m	0
14	CGDO	gate-drain overlap capacitance per meter channel width	F/m	0
15	CGBO	gate-bulk overlap capacitance per meter channel width	F/m	0
16	RSH	drain and source diffusion sheet resistance	ohm/sq.	0
17	CJ	zero-bias bulk junction bottom capacitance per meter of junction parameter	F/m^2	0
18	MJ	bulk junction bottom grading coefficient	–	0.5
19	CJSW	zero-bias bulk junction sidewall capacitance per meter of junction perimeter	F/m	0
20	MJSW	bulk junction sidewall grading coefficient	–	0.3

(continued)

Metal Oxide Silicone Field Effect Transistors (MOSFET) (*cont.*)

	Name	Parameter	Units	Default
21	JS	bulk junction saturation current per square meter of junction area	A/m^2	
22	TOX	oxide thickness	meter	1
23	NSUB	substrate doping	$1/cm^3$	0
24	NSS	surface state density	$1/cm^2$	0
25	NFS	fast surface state density	$1/cm^2$	0
26	TPG	type of gate material: +1 opposite to substrate −1 same as substrate 0 Aluminum gate	–	1
27	XJ	metallurgical junction depth	meter	0
28	LD	lateral diffusion	meter	0
29	UO	surface mobility	$cm^2/V\text{-}s$	600
30	UCRIT	critical field for mobility degradation (model index 2 only)	V/cm	1
31	UEXP	critical field exponent in mobility degradation (model index 2 only)	–	0
32	UTRA	transverse field coefficient (mobility) (deleted for model index 2)	–	0
33	VMAX	maximum drift velocity of carriers	m/s	0
34	NEFF	total channel charge (fixed and mobile) coefficient (model index 2 only)	–	1
35	XQC	thin-oxide capacitance model flag and coefficient of channel charge share attributed to drain (0–0.5)	–	1
36	KF	flicker noise coefficient	–	0
37	AF	flicker noise exponent	–	1
38	FC	coefficient for forward-bias depletion capacitance formula	–	0.5
39	DELTA	width effect on threshold voltage (model index 2 and 3 only)	–	0
40	THETA	mobility modulation (model index 3 only)	1/V	0
41	ETA	static feedback (model index 3 only)	–	0

Answers To Odd-Numbered Problems

Chapter 2:

1. $P = 0.08$ watts
3. $P = 2.12 \times 10^{-4}$
5. 780.39 meters
7. Angle = 17.50°
9. Angle = 35°
11. $d = 8.48$ miles
13. $L_s = 94.49$ dB
15. $f = 70.7$ MHz

Chapter 3:

1. $Z = 47.27\ \Omega$
3. $\alpha = 232.14$ dB
5. $C_1 = 20.15$ pf/ft, $L_1 = 0.107$ μH/ft
   $C_2 = 34.12$ pf/ft, $L_2 = 0.063$ μH/ft
7. $C = 19.69$ nf, $L = 9.47$ μH
9. $\Gamma = 0.444$, Return Loss = −7.04 dB
11. $\Gamma = 0.6$, Return Loss = −4.43 dB, VSWR = 4:1
13. Length = 0.245 cm, distance = 0.152 cm

Chapter 4:

1. $A_p = 8.5$
3. EIRP = 1500 watts, Density = 0.265 watts/km
5. $A_t = 31.7$ dB
7. $P_{diss} = 61.76$ to 116.6 watts
9. Ratios of the length, L, and spacings, R, are increased from 1.428 to 2.0
11. $L = 4.26$ cm

Chapter 5:

1. DC = 0 volts,

N	Volts
1	6.366
2	0
3	2.122
4	0
5	1.273

3. DC = 0.6 volts,

N	Freq (Hz)	Volts
0	0	0.6
1	1000	1.123
2	2000	0.908
3	3000	0.606
4	4000	0.282
5	5000	0
6	6000	−0.189
7	7000	−0.261
8	8000	−0.228

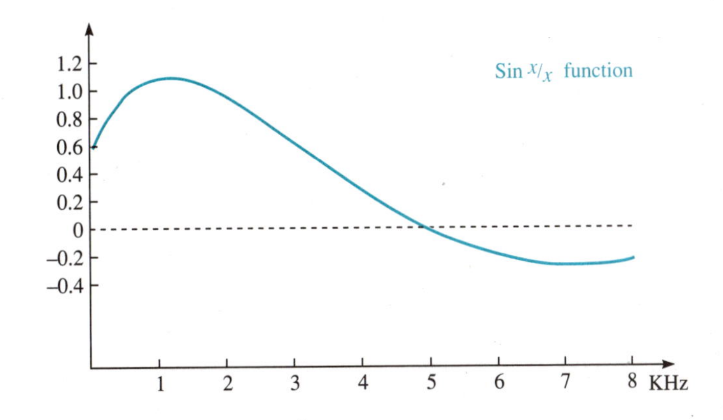

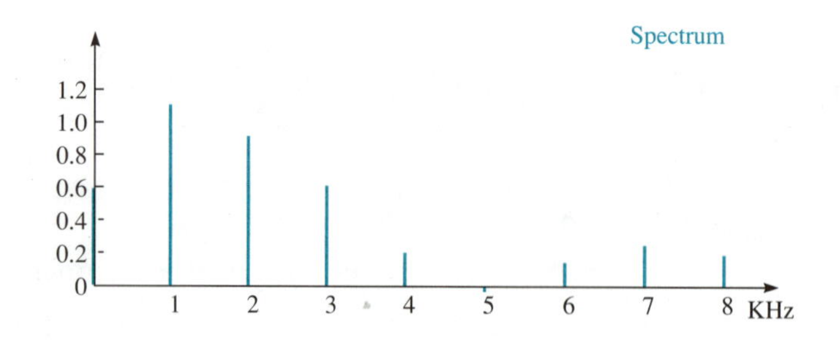

5.

N	Volts
0	0
1	3.18
2	1.59
3	1.06
4	0.795
5	0.636
6	0.530

7. 0.7 volts
9. 1.0 volt

Chapter 6:

1. $F_1 = 5$ MHz, $F_2 = 7$ MHz, Output consists of: 2 MHz, 3 MHz, 5 MHz, 7 MHz, 8 MHZ, 12 MHz, 17 MHz, 19 MHz, and 22 MHz.
3. 4.22×10^{-21} watts/Hz
5. BW = 20.83 kHz
7. BW = 5 MHz, $I_n = 75$ nA, $E_n = 0.54$ μV.
9. 33.3%
11. Factor = 1.068, $NF = 0.288$ dB
13. $F = 1.125$, $NF = 0.571$, Temp = 36 K

Chapter 7:

1. $B = 0.03$
3. $R = 43.15$ Ω
5. Freq = 13.787 kHz
7. Freq = 6.800421 MHz
9. Open bias resistor
11. Filter is open

Chapter 8:

1. Freq = 80 kHz, Frequency deviation = 90–145 kHz
3. Change in frequency = 1.193
5. F = 200 kHz
7. Range = 150 kHz

Chapter 9:

1. m = 73.3%
3. m = 62.6%
5. BW = 5 kHz
7. P_c = 18.76 W, $P_{lower} = P_{upper}$ = 3.907 W
9. m = 0.245
11. Two of the output transistors are bad
13. Gain = 26.69
15. A capacitor has shorted
17. Image = 1.450 MHz; yes, because it is exactly the image.
19. −10.5 dBm
21. 29.7 dB

Chapter 10:

1. Deviation = 20 kHz, m = 6.66 rad, m = 6.66 rad
3. 66.6%
5. 0.9972, 0.9957, 1.6357
7. 101.499 or 101.501 MHz
9. It would not modulate
11. S/N = 21.31 (13.29 dB)
13. S/N = 1.14 (0.56 dB)
15. c = 91 nf

Chapter 11:

1. m = 25%
3. 97%, not practical.
5. 2.496 to 2.504 MHz
7. 1.190 MHz and 1.210 MHz
9. Use the trig identities to form only the lower sideband

11. 97 kHz

13. 30.005 MHz

15. PEP = 0.44 Watts; Average Power = 0.22 Watts

Chapter 12:

1. 280 lines

3. 285.7 lines

5. 3.6 MHz

7. 13.1 msec

9. 1.6 msec

11. 13.33 to 33.33 MHz

13. 29.2°

15. 62°

17. $\theta_c = 70.25°$
NA = 0.574

Glossary

Absorption The transfer of energy from an electromagnetic wave to the molecules of the atmosphere. This produces a loss in the signal strength.

Acquisition The process of pulling in a signal to a phase lock loop to lock the signal to the loop.

AGC Automatic Gain Control—the process which automatically corrects for any changes in input signal in a communications receiver. The process maintains a constant gain throughout the system.

Amplifier A device which raises the amplitude of a signal.

Amplitude Modulation The process of varying the amplitude of a carrier wave by means of a modulating signal.

Analog Communications Form of electronic communications where the modulation used is a continuously varying signal that appears as such in the final signal output.

Angle Modulation The process of varying the phase or frequency of a carrier by means of a modulating signal.

Antenna Efficiency Ratio of power radiated by an antenna to the total input power to the antenna.

Antenna A device which allows power to be radiated into the air by matching the impedance of a power amplifier to that of air (377 Ω); and allows signals to be received and transferred to a receiver by once again matching the impedance of air (377 Ω) to that of the receiver.

Array An antenna which contains two (2) or more antenna elements.

Attenuation The lessening of the value of a signal by a component or device.

Balanced Line A transmission line where both conductors carry current. The currents are 180° out of phase.

Balanced Modulator Also called a product modulator. A device which multiplies signals together for modulation purposes.

BALUN BALanced to UNbalanced interface. Matches a balanced and an unbalanced transmission line.

Bandpass Filter A device which passes only a specified band of frequencies with a minimum amount of attenuation. All other frequencies are attenuated.

Bandwidth The band of frequencies passed by a filter or the band that an antenna will transmit or receive efficiently.

Barkhausen Criteria The criteria for sustaining oscillations in a circuit. It says that the gain around a feedback loop must be ≥ 1, and the phase shift must be $2n\pi$ around the loop.

Beam Width Where the antenna radiation pattern drops by 3 dB on either side of the maximum radiation point of the antenna.

Bessel Function Mathematical functions used in calculating FM systems.

Burst Noise Low frequency noise generated by bias current levels.

Capacitance The parameter set up between the center conductor and outer conductor of a transmission line.

Capture Range The range within which the phase lock loop will lock on to an incoming signal.

Carrier Rf signal that is used to transport the intelligence from one point to another.

Carrier Shift A form of AM distortion where the positive and negative alternations of the modulated signal are not equal.

Characteristic Impedance The impedance of a transmission line. In free space it is equal to $\sqrt{(\mu/\epsilon)} = 377\ \Omega$.

Coaxial The condition where one conductor is surrounded by another conductor.

Conductance The amount of leakage through the dielectric material of a cable.

Conversion Loss The amount of loss encountered in a mixer when comparing the rf input to the IF output.

Correlated Noise The noise that is generated by an input signal.

Crystal A mineral in hexagonal form in nature having piezoelectric properties that are highly useful in communications.

CW Continuous Wave—a continuous signal.

***D/d* Ratio** Ratio of the inside diameter of the outer conductor to the outside diameter of the inner conductor of a coaxial cable.

Deemphasis Process at the receiver of an FM system which restores the original signal by deemphasizing the high frequency components.

Demodulation The process of removing the intelligence from a carrier.

Deviation Sensitivity The radian/volt or Hertz/volt characteristic of an angle modulated system.

Diagonal Clipping A distortion which occurs when the RC time constant in an AM system is too long.

Dielectric A material which obstructs the movement of electromagnetic energy.

Dielectric Constant The amount of obstruction of a signal by a material relative to that of air.

Diffraction The modulation of energy within a wave when it passes near an edge.

Dipole An antenna with two poles which is usually either a quarter-wave or half-wave long.

Direct Synthesis A system where the final output signal is from multiple crystal oscillators or a single oscillator and multiple dividers.

Directive Gain The ratio of the power radiated in a specific direction to the power radiated to the same point by a reference oscillator.

Directivity Maximum directional gain of an antenna.

Director The element in an antenna array that focuses the energy from the driven element into a beam that is in the direction of interest.

Discriminator A circuit where the intelligence in an angle modulated system is removed.

Distortion A changing of the normal and usual shape of a signal.

DSB-FC Double sideband, full carrier system.

DSB-SC Double sideband, suppressed carrier system.

EIRP Effective Isotropic Radiated Power—equivalent transmit power that an isotropic antenna would radiate to result in the same power density in a certain direction and at a given point as a practical antenna.

Electromagnetic Wave A wave in which both the electric and magnetic fields are across the direction of propagation.

Emission The transmission of electromagnetic radiation.

External Noise Noise consisting of atmospheric, extraterrestrial, and man-made. Generated outside the circuit.

Far Field Area beyond the near field, also called the radiation field.

FCC Federal Communications Commission.

FDM Frequency division of multiplexing. (The transmission of a broad range of signals in individual bands.)

Feedback A process where a portion of an output signal is sent back to the input to provide control of the overall circuit.

Field Intensity The intensity of both the electric and magnetic fields of a wave.

Filter A device which allows certain frequencies to pass while attenuating all others.

Flicker Noise Noise associated with defects in the surface structure of semiconductors.

FM Stereo An FM system which transmits and receives signals which have two channels.

Fourier Series Mathematical expression used to change a time domain signal to the frequency domain.

Frequency Deviation The amount that the carrier in an FM system is changed when modulated.

Frequency Domain Amplitude of a signal is displayed as a function of frequency. (Spectrum analyzer is used for frequency domain displays.)

Frequency Modulation The process where a carrier frequency is varied by a modulating signal.

Ground Wave Electromagnetic wave that travels along the surface of the earth.

Harmonics Signals that are generated along with the fundamental. They are multiples of the fundamental.

Helix A spiral used to make communications antennas.

Hertz Antenna A dipole antenna whose entire length is one-half wavelength.

High-Pass Filter A device which passes frequencies above a cut off point and attenuates those below this point.

Hold-In Range Range of frequencies that the input to a PLL may vary and still maintain a locked condition.

Huygens' Principle A concept that states: every point on a given spherical wavefront can be considered to be a point source of electromagnetic waves from which other secondary waves are radiated.

Incident Angle The angle at which an electromagnetic wave strikes an object or surface.

Independent Sideband A single carrier modulated by two independent modulating systems.

Indirect Synthesis System that employs a PLL for synthesizer operation.

Inductance The parameter present on a transmission line due to current flowing through a metallic conductor.

Information Capacity A measure of how much information can be carried through a system in a given time period.

Informational Bandwidth The total band in which the intelligence of the modulated signal can be found.

Instantaneous Frequency Exact frequency of the carrier at a given instant of time.

Instantaneous Frequency Deviation Instantaneous change in frequency of a carrier.

Instantaneous Phase Exact phase of a carrier at a given instant of time.

Instantaneous Phase Deviation Instantaneous change in phase of a carrier at a given instant of time.

Intelligence The modulating signal in a modulation system.

Intelligence Bandwidth Bandwidth of a system that is required to propagate information through the system.

Interelement Capacitance The capacitance formed between the elements of an active device.

Interference When rays interact with one another and cause the original signal efficiency to decrease.

Intermediate Frequency The resulting frequency of mixing an rf and LO signal in a mixer circuit.

Intermodulation Distortion Distortion occurring when two input signals are applied at the same level to the input of an amplifier. These signals are within the bandwidth of the amplifier and will overload the circuit.

Internal Noise Noise generated within a device or component.

Isolation A high value of attenuation between two points.

Isotropic Source A source in which all of the rays propagate equally in all directions.

Lead-Lag Network A combination of series and parallel resonant RC circuits used in RC oscillator circuits.

Local Oscillator The high power signal in a mixing circuit.

Log-Periodic Antenna An antenna with a log tapered to its construction and a wide bandwidth.

Loop Part of a feedback system that sends a signal back to the input.

Low-Pass Filter A device which passes frequencies below a curve off point and attenuates those above this point.

Marconi Antenna A dipole antenna whose total length is one-quarter wavelength.

Medium An agent that is used to move an electromagnetic wave from one point to another.

Mixer A component which combines two different frequencies to produce a third frequency which may be either the sum or difference of the two original signals.

Modulate To cause a carrier signal to vary.

Modulating Signal The portion of the signal that contains the modulation or intelligence to be transmitted.

Modulation The process of placing the modulating signal on to the carrier.

Modulation Index Figure of merit of a modulation system that tells the percentage of modulation being used.

Modulator A circuit which combines the carrier and modulating signal to provide the appropriate modulation scheme.

Near Field The area which is close to an antenna. Also called the induction field.

Noise Any unwanted electrical signal within a communications system that interferes with the sound or image being communicated.

Noise Bandwidth The frequency range over which the noise is being characterized.

Noise Current Current which depends on an electron charge, DC bias current, and the bandwidth of the system.

Noise Factor The ratio of the input SNR to the output SNR.

Noise Figure The logarithmic representation of the noise factor, F. (10 log F)

Noise Power Density Noise in a 1 Hz bandwidth. Equal to the Boltzmann's constant (K) times the temperature in Kelvin.

Noise Temperature Temperature, in Kelvin, which represents the noise characteristics of a device or system.

Nonlinear When an input signal does not result in a corresponding proportional output signal.

Oscillator Device employing positive feedback and frequency determining devices that produces an output that fluctuates between two states, positive and negative.

Overmodulation When the modulating signal causes the carrier to have gaps in the final modulated signal.

Parabolic Antenna An antenna with a parabolic reflector attached to it.

Peak Detector AM demodulator which responds to the peak of the modulated signal.

Permeability Value of inductance per unit length of a transmission line.

Permittivity Value of capacitance per unit length of a transmission line. Sometimes called the dielectric constant.

Phase Comparator A circuit that compares the phase of two signals applied to its input.

Phase-Lock Loop A circuit with a feedback loop that is designed to take a reference frequency and lock an incoming signal to that frequency.

Phase Modulation Process where the phase of a carrier is changed by a modulating signal.

Piezoelectric The process where a voltage causes a mechanical vibration and a mechanical vibration produces a corresponding voltage.

Polarization How the electric field that is radiated is oriented.

Power Density The amount of energy radiated into space, expressed in watts/cm^2.

Power Gain The same as directive gain of an antenna except total power fed to the antenna is used.

Preemphasis Process in an FM transmitter which increases the high frequencies of the signal.

Preselector A filter arrangement which selects a certain band of frequencies.

Product Modulator Also called a balanced modulator, multiplies signals together for modulation purposes.

Propagation The process of sending electromagnetic waves through a medium.

Pull-In Range Equal to one-half the capture range of a PLL.

Radiation Pattern Pattern of an antenna which describes where the device is radiating and where the signals are attenuated.

Radiation Resistance An ac antenna resistance.

Rays Electromagnetic waves.

Rectangular Pulse A short burst of energy with fast rise and fall times.

Rectifier Distortion A form of distortion which occurs when the RC time in an AM system is too short.

Reflection When the second medium is so dense that the energy which strikes it is sent back to the source.

Reflection Coefficient Percentage of power reflected back from a mismatched load impedance.

Refraction Change in direction of rays as they pass obliquely from one medium to another.

Regenerative Feedback Positive feedback where the feedback reinforces the original signal.

Resistance Parameter on a transmission line which results in a loss in signal through the line.

Resonance Frequency where the inductive reactance and capacitive reactance in a circuit are equal.

Resonator A device which will resonate at a single frequency and attenuate others.

Return Loss A measure of the amount of signal that is lost due to a mismatch.

Rf Source A generator of an rf signal.

Shield A protective cover on a transmission line which keeps external signals out and internal signals within the transmission line.

Shot Noise Noise generated when electrons strike a metallic device.

Sideband Signals which develop when the modulation process is performed.

Signal-to-Noise Ratio Ratio of the signal level applied to a device compared to the noise level.

Sky Wave Waves that are radiated into the sky and are reflected or refracted back to earth.

Smith Chart A transmission line calculator that is used to characterize transmission lines.

Snell's Law Optical law that states that the index of refraction times the sine of the incident angle is equal to the index of refraction of a second medium times the sine of the angle of reflection.

Space Wave Electromagnetic wave that travels through the air in a direct line of sight (LOS) or from the transmitter to the ground and back to a receiver.

Spectra A group of signals that make up a complex signal such as a square wave, triangular wave, pulse, etc.

SSB-RC Single sideband, reduced carrier.

SSB-SC Single sideband, suppressed carrier.

Stray Capacitance Capacitance set up by components at high frequencies if they are raised above the circuit board or have excessively long leads.

Symmetry A proportion in size and form of the parts of a signal to each other.

Synthesizer A device that produces a large number of frequencies at its output and usually is built around a PLL.

TEM Mode Mode of an electromagnetic wave in which the electric and magnetic fields are transverse (across) to the direction of propagation.

Temperature Coefficient Property of a device which accounts for its parameter change with temperature.

Thermal Noise Noise generated when a current flows through a resistive device.

Time Domain The representation of a signal amplitude as a function of time. The oscilloscope is an instrument that displays signals in a time domain.

Total Harmonic Distortion The total contribution by all the harmonics to a signal output.

Transducer A component that converts electrical energy to mechanical energy or mechanical energy to electrical energy in a mechanical filter.

Transmission Line A component that moves energy from one point to another efficiently.

Transverse Describing a component that lies across or is crossing from side to side.

Trapezoidal Pattern Pattern used to determine the percent of modulation in a system.

Ultimate Attenuation Point on a filter curve which indicates a minimum value of attenuation needed to some specified frequency.

Unbalanced Line Transmission line where one line is at ground potential while the other one carries current.

Uncorrelated Noise Noise that is present even when there is no signal at the input.

Varactor Diode A two element device which varies its capacitance as a function of dc bias voltage.

VCO Voltage Controlled Oscillator—the frequency can be varied by changing a dc voltage.

Velocity of Propagation The velocity of the electromagnetic wave as it propagates through a medium.

Vestigial Sideband A form of AM where the carrier is transmitted at full power along with one complete sideband and part of another.

VSWR Voltage Standing Wave Ratio.

Wavefront The surface of constant phase of a wave.

Wavelength The length of one cycle of a signal.

Wien-Bridge A circuit which has two alternative arms balanced so that there is no voltage across the center during a balanced condition. When an unbalance occurs, there is a voltage across the center.

Bibliography

Adamson, Thomas, *Electronic Communications,* Delmar Publishing, Albany, New York, 1988.

Banzhof, Walter, *Computer-Aided Circuit Analysis Using SPICE,* Prentice-Hall, Englewood Cliffs, N.J., 1989.

Berlin, Howard M., *The Illustrated Electronics Dictionary,* Merrill, Columbus, Ohio, 1986.

Couch, II, Leon W., *Digital and Analog Communications,* 3rd ed., Macmillan, New York, 1987.

Graf, Rudolf F., *Modern Dictionary of Electronic Terms,* 5th ed., Howard Sams, Indianapolis, 1977.

Killan, Harold B., *Modern Electronic Communication Techniques,* Macmillan, New York, 1985.

Laverghetta, Thomas S., *Modern Microwave Measurements and Techniques,* Artech House, Norwood, Ma., 1988.

Miller, Gary M., *Modern Electronic Communications,* 3rd ed., Prentice-Hall, Englewood Cliffs, N.J., 1988.

Schoenbeck, Robert J., *Electronic Communications, Modulation and Transmission,* Merrill, Columbus, Ohio, 1988.

Sklar, Bernard, *Digital Communications, Fundamentals and Applications,* Prentice-Hall, Englewood Cliffs, N.J., 1988.

Tena, Lloyd, *Electronic Communications,* Schaum Outline Series, McGraw-Hill, New York, 1979.

Tomasi, Wayne, *Fundamentals of Electronic Communications Systems,* Prentice-Hall, Englewood Cliffs, N.J., 1988.

Vladimirescu, A., Zhang, Kaihe, Newton, A.R., Pederson, D.O., and Sangiovanni-Vincentelli, A., *SPICE, Version 2G, User's Guide,* University of California, Berkeley, Ca., 1981.

Young, Paul H., *Electronic Communications Techniques,* Merrill, Columbus, Ohio, 1985.

Photo Credits

Series page, p. ii/Preface, p. v/Contents, p. xi/Chapter 1 opener, p. xvi: Vernon Valley, N.J., earth stations beaming messages to communications satellite in orbit over a fixed spot on Earth. (Robert Perro, Photo Researchers)

Chapter 2 opener, p. 10: HP 5400 Series oscilloscope, depicting an acquired AM signal. (Courtesy Hewlett-Packard)

Chapter 3 opener, p. 32: Transmission lines. (Jean-Claude Lejeune, Stock, Boston)

Chapter 4 opener, p. 68: Television antennas. (Hugh Rogers, Monkmeyer Press Photo Service)

Chapter 6 opener, p. 118: Newark, N.J. airport. (Michael Dwyer, Stock, Boston)

Chapter 7 opener, p. 144: Printed circuit board. (Chips & Technology, Inc.)

Chapter 8 opener, p. 182: Electronic circuitry. (Comstock)

Chapter 9 opener, p. 208: Electronic signal instrumentation. (Radio Shack, a division of Tandy Corporation)

Chapter 10 opener, p. 260: WCRB radio announcer, Waltham, Mass. (Eric Neurath, Stock, Boston)

Chapter 12 opener, p. 324: Television news broadcast, WTNH. (Gale Zucker, Stock, Boston)

Appendices A, B, and C openers, pp. 375, 400, 406: Earth stations. (Pete Turner, The Image Bank)

Index